Learning Flash CS4 Professional

Rich Shupe

O'REILLY®
Beijing • Cambridge • Farnham • Köln • Sebastopol • Taipei • Tokyo

Learning Flash CS4 Professional

by Rich Shupe

Printed in Canada.

Published by O'Reilly Media, Inc., 1005 Gravenstein Highway North, Sebastopol, CA 95472.

O'Reilly Media books may be purchased for educational, business, or sales promotional use. Online editions are also available for most titles (*safari.oreilly.com*). For more information, contact our corporate/institutional sales department: 800-998-9938 or *corporate@oreilly.com*.

Editor: Steve Weiss

Production Editors: Chris Meredith and Rachel Monaghan

Developmental Editor: Linda Laflamme

Copy Editor: Amy Thomson

Technical Reviewers: Thomas Yeh and Anselm Bradford

Proofreader: Rachel Monaghan

Interior Designer: Ron Bilodeau

Cover Designer: Karen Montgomery

Indexer: Julie Hawks

Print History:

April 2009: First Edition.

ISBN: 978-0-596-15976-4
[F]

Adobe Developer Library, a copublishing partnership between O'Reilly Media Inc. and Adobe Systems, Inc., is the authoritative resource for developers using Adobe technologies. These comprehensive resources offer learning solutions to help developers create cutting-edge interactive web applications that can reach virtually anyone on any platform.

With top-quality books and innovative online resources covering the latest tools for rich-Internet application development, the *Adobe Developer Library* delivers expert training, straight from the source. Topics include ActionScript, Adobe Flex®, Adobe Flash®, and Adobe Acrobat® software.

Get the latest news about books, online resources, and more at *adobedeveloper-library.com*.

CONTENTS

FOREWORD

At the time of this writing, Flash CS4 has been out in the world for approximately six months. In that time, one of my responsibilities at Adobe has been to help the Flash community get up to speed with all of the new changes. Like any software release that includes a significant number of new features, it has been a bumpy road for some people as they adapt to the new ways of doing things. Flash CS4 has undergone some major changes to existing features, in addition to introducing some brand new capabilities.

Just for some background, the central focus for each release of the Flash authoring tool has tended to be cyclical in nature. Flash MX 2004 was primarily focused on providing support in the tool for the then-new ActionScript 2.0 language. Following that, Flash 8 unveiled a slew of new designer options, like filter effects, bitmap manipulation, and alpha channel video support. Keeping the cycle going, Flash CS3's main thrust was to provide support for the great new ActionScript 3.0 language. Noticing the pattern here? Well, designers will be happy to know that the cycle remains intact. Flash CS4 is packed with features dedicated to the creative professional.

Probably the biggest change, and the thing you will notice right off the bat, is the entirely new animation and timeline model. For years the Flash community has been asking for an animation system similar to that found in tools like After Effects. Well, Adobe has listened and has completely revamped the way in which you create timeline animations. The repetitive creation and modification of keyframes is a thing of the past, as Flash will now automatically create the necessary keyframes as you manipulate your objects on the timeline. A full-featured motion editor has also been added, giving you fine control over your animations.

Another request that we have gotten for years is for 3D support within Flash. With Flash Player 10, we now have the capability to manipulate any display object in 3D space. There are some new tools in Flash CS4 to support these features, and they make creating 3D effects no more difficult than creating any other type of animation.

Character animators will be very happy to find support for inverse kinematics (IK) in Flash CS4. IK allows designers to create bones between movie clips, and within shapes, to create animations that adhere to the constraints of a skeletal structure. This system is not just for characters, either, and can be used to create a wide array of animation effects.

I could go on and on talking about the features in this exciting new version of Flash. I could talk about the new user interface, the new Spray and Deco drawing tools, the new Project panel, and the enhanced integration with Adobe Flex. I could do that, but I'm going to let a much better writer than myself, Rich Shupe, take over from here.

I remember seeing Rich speak at my very first Flash Forward conference in New York and was extremely impressed with his knowledge and methods of teaching. Through my own years of teaching Flash, I know how hard it is to explain some of the seemingly ***unexplainable*** parts of our amazing platform. Rich definitely has a knack for finding ways to make these areas understandable.

In addition to the book that you're holding right now, Rich is also the writer of *Learning ActionScript 3.0: A Beginner's Guide* (O'Reilly), which I consider to be one of the best ActionScript books ever written. In fact, you will find a quote from me on the cover of newer versions of the book stating just that. Rich's writing style and true mastery of Flash makes learning the new versions of Flash and ActionScript easier than I ever imagined. That book also makes an excellent companion to this one, giving you a well-rounded library of knowledge on Flash.

I would also like to take a moment to thank O'Reilly for being committed to producing this book in full color. I can't tell you how much more engaging a book like this is when the code samples and screenshots look exactly as they do on your own computer screen. Beyond just being pretty to look at, the full-color treatment greatly enhances the learning experience as well.

Rich has structured this book in a project format, which is no easy task. I find it to be really refreshing that readers will actually be building something useful at the same time that they are learning the new software. While reading a prerelease version of this book, I actually learned many new techniques that I didn't know about—and I work for Adobe—so I'm confident that you will find it an extremely rewarding experience.

—Lee Brimelow

Flash Platform Evangelist, Adobe

San Francisco, 2009

PREFACE

A Dynamic Duo

The book you hold in your hand is one of a pair. It's not a left sock or a single cufflink, because it's unique and it's still of use without its mate. It's more like a fraternal twin or, perhaps more appropriately, the A–M volume of two-tome encyclopedia. The volumes are linked, and they share common ancestry and a common vision, but they exist perfectly well on their own.

The book you're reading now is your first ace in the proverbial hole, and focuses on the Flash CS4 Professional application. This book will teach you how to draw and import graphics; use text, sound, and video; and even position objects in 3D space and create a moveable skeletal arm—all without any programming at all.

The second bullet is *Learning ActionScript 3.0* (O'Reilly) and covers ActionScript, Flash's internal scripting language responsible for making Flash projects interactive. *Learning ActionScript 3.0* picks up ActionScript where this book leaves off. It takes an introductory chapter nestled within 13 other chapters of application goodness and expands it into coverage that includes the bulk of ActionScript 3.0's core features.

Why were these books conceived as a pair? Simply put, each book was designed to better deliver its share of the knowledge lode. Each book works better independently than if both volumes were encumbered by tying everything together into one 700-page disc-slipper. This book introduces the Flash application to users who may never have seen it before. Its companion volume assumes familiarity with the Flash interface and is, therefore, able to concentrate solely on the ActionScript language.

Can I See Your Portfolio?

Don't be deceived by these carefully laid plans. In addition to having an independent streak, this book also teaches its fair share of ActionScript. Why bother when its companion volume is waiting to be read? For one reason, ActionScript is as much a part of Flash as its Timeline panel. Writing a book about Flash without talking about ActionScript is like writing about Sherlock Holmes and neglecting to mention Watson.

For another reason, this book is project-based and you can't get very far with a project of any significant scale without using ActionScript. Its project is a leading player in this book, to be sure, but it's not the only act in town. In each chapter, topics are first introduced to you in short, digestible exercises that convey an idea, demonstrate a tool, or explain a script's syntax. Only at the end of each chapter, in a reinforcing, real-world scenario, are these skills applied to the ongoing portfolio you will build by book's end.

The portfolio itself is unique, too. Designed not to hinge on a conventional project that trudges through every chapter, this book takes a different path on its way to your studio. The project herein was conceived to be more than a standard top- or left-frame navigation website. The portfolio was created to highlight all the major features that Flash CS4 Professional has to offer and to push the limits of the average Flash authoring experience.

Assets were intentionally designed to create problems to solve, such as how best to add expressive movement to a complex animation and how to optimize large files for quicker download. Design ideas were chosen because they offered opportunities to solve these and other problems in creative ways.

Yes, the book was planned so that you don't have to invest a lot of time in a project you have no interest in building. Yet if you do choose to practice what you've learned in each chapter by applying your new skills to on ongoing project, you'll hone your chops that much quicker.

Who This Book Is For

This book is aimed at Flash designers and developers who are picking up Flash CS4 Professional for the first time, as well as users upgrading from prior versions who are looking to acquaint themselves with the version's new features.

No prior experience with Flash is necessary to enjoy this book, as you learn the Flash interface from the ground up, but there are plenty of new features of which veteran Flash designers and developers can take advantage.

Reading through this preface and looking over a few sample chapters will increase the chances that you'll be happy with the content and straightforward approach adopted herein.

What Is—and Isn't—in This Book

More than anything else, a book's preface should be able to give you a few hints as to whether or not the book is right for you. Here are two sections for quick perusal that will, ideally, help you evaluate your interest in this volume.

What's In

It's always helpful to pore over the table of contents of a book to see how much of the material catches your eye. Here, detail is replaced by a descriptive sentence or two that may give you some additional insight on the book's content.

Chapter 1, *Interface Essentials*

This chapter covers the Flash CS4 Professional interface, introducing you to menus, panels, and windows. It also covers how to customize the interface to your liking.

Chapter 2, *Creating Graphics*

Chapter 2 concentrates on drawing assets within Flash, including both Flash's natural drawing tools and more traditional object-based drawing techniques.

Chapter 3, *Using Symbols*

Symbols, such as movie clips and buttons, are at the center of this chapter, including those created by Flash CS4's new Spray Brush and Deco tools.

Chapter 4, *Importing Graphics*

Chapter 4 covers importing graphics into Flash, including Flash's integration with other Adobe Creative Suite tools, and the import of Photoshop PSD files and Illustrator AI files.

Chapter 5, *Animation*

This chapter describes how to animate assets using Flash's Timeline, including using Flash CS4's new Motion Editor.

Chapter 6, *ActionScript Basics*

Chapter 6 introduces you to the ActionScript language and lays the groundwork for topic-specific scripting in the chapters to come.

Chapter 7, *Filters and Blend Modes*

Filters and blend modes, like Photoshop's drop shadow and overlay, respectively, are the focus of Chapter 7. Also included in this chapter is a technique for creating soft-edge masks.

Chapter 8, *3D*

Flash CS4's new integrated 3D features are discussed in Chapter 8. Simple use of x-, y-, and z-axes for both rotation and movement are the center of the discussion, but a parallax scrolling environment is also added to the portfolio project.

Chapter 9, *Components*

Completing the portfolio gallery, Chapter 9 demonstrates the loading of external assets with components using little to no ActionScript.

Chapter 10, *Inverse Kinematics*

Adding a skeletal infrastructure to movie clip and shape animations is at the heart of Chapter 10. Inverse kinematics is put to use to constrain character movement to a range of motion that mimics that of human joints.

Chapter 11, *Text*

Chapter 11 details how to create text fields and format text, including loading external HTML and CSS files. The chapter content concludes with a demonstration of the XFL file format for opening InDesign assets in Flash.

Chapter 12, *Audio*

Playback of digital audio from external files isn't the only thing covered in Chapter 12. Also in the spotlight is real-time sound visualization of stereo amplitudes.

Chapter 13, *Video*

Chapter 13 examines adding video to the portfolio gallery, but not before you'll learn how to compress video assets using Adobe Media Encoder.

Chapter 14, *Publishing and Deploying*

Finally, the book ends with a chapter explaining how to deploy your project to the Web, as well as how to create a desktop application with AIR.

What's Not

Unfortunately, it's inevitable that material must be left out of a book about such a large topic, and this volume is no exception. Scope and space limitations simply do not allow a complete and exhaustive account of Flash and ActionScript features. Here are a handful of things these pages don't deliver:

Flash for mobile devices

A big topic of study unto itself, there just wasn't room to include coverage of Flash Lite, Adobe's Flash playback engine for mobile devices. This book focuses on creating assets for online (Flash Player) and desktop (AIR, or Adobe Integrated Runtime) delivery, but not for handhelds or consumer electronics.

Other Flash platform technologies

Because this book focuses on Flash CS4 Professional, other Flash Platform technologies such as Flex and Flash Media Server were necessarily omitted. AIR makes an appearance in Chapter 14 because an AIR authoring workflow is integrated into Flash CS4 Professional. However, the focus of that appearance is on deploying your project to the desktop, and no significant discussion of AIR-specific features is included.

Script Assist

No coverage of the GUI (graphical user interface) script editor, Script Assist, appears in this book.

Alternative ActionScript editors

Again, as this book is Flash-centric, it doesn't cover the use of FlexBuilder, FlashDevelop, FDT, or other external ActionScript editors.

ActionScript 2.0

Although you can still develop in Flash CS4 Professional using ActionScript 2.0, this book and its companion volume concentrate on the newer, faster, more powerful version of the language.

The other 90% of ActionScript 3.0

Since this book's thrust is on the Flash CS4 Professional authoring environment, only a limited amount of ActionScript made it into these pages. If you're versed in Flash already and are considering reading this book to brush up your knowledge, know that it will provide help with the Flash interface but may be insufficient on the scripting front. This book can get you started, but if you want a resource more dedicated to AS3, try the companion volume.

Object-oriented programming

Object-oriented programming (OOP) is outside the scope of this book. This book introduces AS3 in a procedural context using scripts written in the Timeline. If you want to learn OOP using AS3, take a look at the companion volume to see whether it includes enough OOP for your liking.

Companion Website

All the exercises included in this book are available for download from the book's companion website, *http://www.LearningFlashCS4.com*. Supplemental materials are also available and will be introduced over time through the site blog. Finally, although I never quite know how quickly I can reply, I can be reached directly through the companion website.

Related Resources

Here is a small list of additional resources to help you as you learn Flash.

Adobe

Important starting points from the makers of Flash:

- Flash product page: *http://www.adobe.com/products/flash/*
- Flash downloads: *http://www.adobe.com/support/flash/downloads.html*
- Flash support center: *http://www.adobe.com/support/flash/*

AS2 Migration

If you were attracted to this book because you're upgrading to Flash CS4 Professional from a prior version of Flash and you have experience with ActionScript 2.0, this book may help you smooth over the transition to ActionScript 3.0: *The ActionScript 3.0 Quick Reference Guide* by Jen deHaan, Darren Richardson, Rich Shupe, and David Stiller (O'Reilly).

- Book site: *http://oreilly.com/catalog/9780596517359/*
- Sample chapters: *http://www.adobe.com/devnet/actionscript/articles/as3_quick_ref.html*

Training

These destinations offer online and in-person training resources for Flash and ActionScript:

- gotoAndLearn(): *http://www.gotoAndLearn.com*
- Lynda.com: *http://www.lynda.com*
- FMA: *http://www.fmaonline.com*

Mailing Lists/Forums

For ongoing community support, try these mailing lists and forums:

- Flash Newbies: *http://chattyfig.figleaf.com/mailman/options/flashnewbie/*
- Flash Tiger: *http://groups.yahoo.com/group/Flash_tiger/*
- Kirupa: *http://www.kirupa.com*

Conferences

There's no substitute for meeting other members of the community and learning from Flash leaders from all over the world. Here are a few conferences that are really worth attending:

- Flash on the Beach: *http://www.flashonthebeach.com*
- Flashbelt: *http://www.flashbelt.com*
- Flash on Tap: *http://www.flashontap.com*

Conventions Used in This Book

The following typographical conventions are used in this book:

Italic

Indicates new terms, URLs, email addresses, filenames, file extensions, pathnames, and directories.

`Constant width`

Indicates ActionScript code, text output from executing scripts, XML tags, HTML tags, and the contents of files.

`Constant width bold`

Shows commands or other text that should be entered literally.

The scripts herein are color-coded by syntax to match the way your scripts look in Flash. Script basics (such as operators, punctuation, and so on), as well as custom names (such as for variables and functions), are black. Special words and symbols are printed in different colors: keywords and identifiers are blue, strings are green, and comments are gray. For example:

```
//Happy Birthday, Claire!
var cas:String = "Claire";
```

NOTE

A note gives additional information, such as resources or a more detailed explanation.

WARNING

This box indicates a warning or caution.

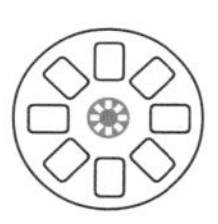

Project Progress

This icon precedes any work you'll need to do to further the cumulative portfolio project. Without this icon, discussions will sometimes describe actions you may take in the Flash interface, that don't ultimately lead to you creating an exercise file. Further, when exercises are suggested, they are always optional.

Using Code Examples

This book is here to help you get your job done. In general, you may use the code in this book in your programs and documentation. You do not need to contact us for permission unless you're reproducing a significant portion of the code. For example, writing a program that uses several chunks of code from this book does not require permission. Selling or distributing a CD-ROM of examples from O'Reilly books does require permission. Answering a question by citing this book and quoting example code does not require permission. Incorporating a significant amount of example code from this book into your product's documentation does require permission.

We appreciate, but do not require, attribution. An attribution usually includes the title, author, publisher, and ISBN. For example: *Learning Flash CS4 Professional* by Rich Shupe. Copyright 2009 Rich Shupe, 978-0-596-15976-4.

If you feel your use of code examples falls outside fair use or the permission given above, feel free to contact us at *permissions@oreilly.com*.

We'd Like to Hear from You

Please address comments and questions concerning this book to the publisher:

O'Reilly Media, Inc.
1005 Gravenstein Highway North
Sebastopol, CA 95472
(800) 998-9938 (in the United States or Canada)
(707) 829-0515 (international or local)
(707) 829-0104 (fax)

We have a web page for this book where we list errata, examples, and any additional information. You can access this page at:

http://www.oreilly.com/catalog/9780596159764

To comment or ask technical questions about this book, send email to:

bookquestions@oreilly.com

For more information about our books, conferences, Resource Centers, and the O'Reilly Network, see our website at:

http://www.oreilly.com

Safari® Books Online

When you see a Safari Books Online icon on the cover of your favorite technology book, it means the book is available online through the O'Reilly Network Safari Bookshelf.

Safari offers a solution that's better than e-books. It's a virtual library that lets you search thousands of top tech books easily, cut and paste code samples, download chapters, and find quick answers when you need the most accurate, current information. Try it for free at *http://my.safaribooksonline.com*.

Acknowledgments

Before diving in, I want to thank my favorite collaborator, Zevan Rosser, who created the art for the portfolio project. Any time I can rope Zevan into something I'm doing, I'll come out looking good.

I'd also like to thank Lee Brimelow for donating his time and effort to writing the foreword for this book. I've always felt that Lee and I shared similar goals when it came to teaching, and his video tutorials and selfless contributions to the Flash community have consistently inspired me. When I grow up, I want to be just like Lee.

On the next pedestal, I seat the O'Reilly crew that made this book possible (it's a big pedestal). Many thanks to Laurel Ackerman, Ron Bilodeau, Dan Brodnitz, Suzanne Caballero, Michele Filshie, Dennis Fitzgerald, Edie Freedman, Julie Hawks, Marsee Henon, Linda Laflamme, Mike Leonard and the O'Reilly sales team, Chris Meredith, Rachel Monaghan, Karen Montgomery, Mark Paglietti, Sara Peyton, Marlowe Shaeffer, Amy Thomson, Rachel Thurow, Betsy Waliszewski, Chris Walker, Steve Weiss, Joe Wikert, and Adam Witwer. Honestly, I can think of a few reasons why this crowd should throw their hands up and be done with me.

Included in the list of contributors to this book are my indispensible tech editors, Anselm Bradford and Thomas Yeh, and beta readers Wei-Chu Chen (Beryl), Aaron Crouch, Rajiv Ganesan, Valerie Guinn, Steven Mattson Hayhurst, and Anita Ramroop.

I would also like to thank:

- Mark Anders, Mike Chambers, Jen deHaan, Mike Downey, Richard Galvan, Mally Gardiner, Jeff Kamerer, San Khong, Sean Kranzberg, John Mayhew, Ted Patrick, Nivesh Rajbhandari, and all at Adobe.
- Bruce Wands, Joe Dellinger, Russet Lederman, Jaryd Lowder, Diane Field, The School of Visual Arts, and all of my students.
- Paul Kent, Kathy Moran, and IDG; John Davey and Flash on the Beach; Dave Schroeder and Flashbelt; Chris and Rebecca Allen and Flash on Tap; Susan Horowitz, William Morrison, Colin Macdonald, and all at University of Hawaii's Outreach program.
- Hudson Ansley, Jean-Charles Carelli, Colin Holgate, Tyler Larson, Lisa Larson-Kelley, and all at FlashCodersNY.
- Aral Balkan, Pete Barr-Watson, Brendan Dawes, Peter Elst, Brandan Hall, Mario Klingemann, Seb Lee-Delisle, Ralph Hauwert, Thibault Imbert, André Michelle, Dom Minns, Erik Natzke, Keith Peters, Darren Schall, Senocular, Sephiroth, Grant Skinner, Geoff Stearns, David Stiller, Craig Swann, Jared Tarbell, Tink, Josh Tynjala, Carlos Ulloa, and no doubt others that I'm forgetting, for support and/or inspiration.
- The Jungle, and any pals I still have after such shameful neglect (I promise, it's nothing personal).
- Most importantly, I'd like to thank my immediate family: Jodi Rotondo, Sally Shupe, and Claire Shupe, as well as ankle-biter wrangler and long-time friend Mike Wills; Steve Shupe and Cindy Shupe; and Brian Shupe and Abigail Janssens; for allowing me to spend so much time writing. I also owe a big dose of gratitude to my extended family for supporting me even when I'm thankless.

Len, Cass, Ry, and Annabel.... All the best!

About the Author

Rich Shupe is the founder and president of FMA—a full-service multimedia development company and training facility in New York City. Rich teaches a variety of digital technologies in academic and commercial environments, and has frequently lectured on these topics at Flash on the Beach, Flashbelt, Flash on Tap, FlashForward, Macworld, and other national and international events. He is a faculty member of New York's School of Visual Arts' MFA Computer Art Department. Rich is also the author or coauthor of multiple books, including *Learning ActionScript 3.0* (O'Reilly), *The ActionScript 3.0 Quick Reference Guide* (O'Reilly), *Flash CS3 Professional Video Training Book* (Peachpit Press), *CS3 Web and Design Workflow Guides* (Adobe). He also presents video training on Flash and other topics for Lynda.com. Visit Rich's website at *http://www.fmaonline.com*.

Colophon

Our look is the result of reader comments, our own experimentation, and feedback from distribution channels. Distinctive covers complement our distinctive approach to technical topics, breathing personality and life into potentially dry subjects. The text font is Linotype Birka; the heading font is Adobe Myriad Pro; and the code font is LucasFont'sTheSansMonoCondensed.

CHAPTER 1

INTERFACE ESSENTIALS

Introduction

IN THIS CHAPTER

Like many creative technologies, the Flash platform (not just the software, but also the community) has exploded over the last several years, fueled by faster, less expensive computers, the arrival of economical broadband access, and the growing role the Internet plays in our personal and professional lives. If you're reading this book—whether you're a developer or designer (or perhaps a member of the recently dubbed *devigner* camp: a programming creative that defies easy description)—it's because Flash does, or will, play a part in your productivity. No matter what your experience level, you'll find that Flash is a big application with a lot to offer.

The latest version, Flash CS4 Professional, has added even more tools to your potential toolbox. Flash's internal scripting language, ActionScript 3.0, has been enhanced in key areas, and many new controls have been added to the application interface. Such features as simple 3D asset manipulation, inverse kinematics for animation, and a brand-new motion editor are among major changes available to all Flash users, with or without programming skills.

With all of this power, however, comes complexity. There's no denying that Flash has a significant amount of breadth and depth, and learning your way around can be a challenge if you try too much, too fast. As discussed in the preface, this book will introduce you to the highlights of Flash's essential features and apply examples of their use to an ongoing portfolio project. You'll focus primarily on the Flash interface, with a measure of ActionScript thrown in to get the job done.

NOTE

When you are ready to take your projects to the next level and learn more ActionScript, this book's companion volume, Learning ActionScript 3.0: A Beginner's Guide *(O'Reilly), will help. For more information, see the Preface.*

Getting to Know the Flash CS4 Interface

The first step in learning Flash is exploring the application interface—sometimes referred to as an *integrated development environment* (IDE). Since Adobe's acquisition of Flash, the interface has undergone changes to improve consistency and interoperability with other Adobe Creative Suite applications. Flash CS4 Professional has come closer to this goal of common user interface elements than any previous version of Flash. Before you can investigate further, however, you'll need to create a new document.

Creating a New Document

If you're new to Flash, you may be surprised when you are immediately prompted with a Welcome screen upon launch. Shown in Figure 1-1, this screen is divided into five main areas.

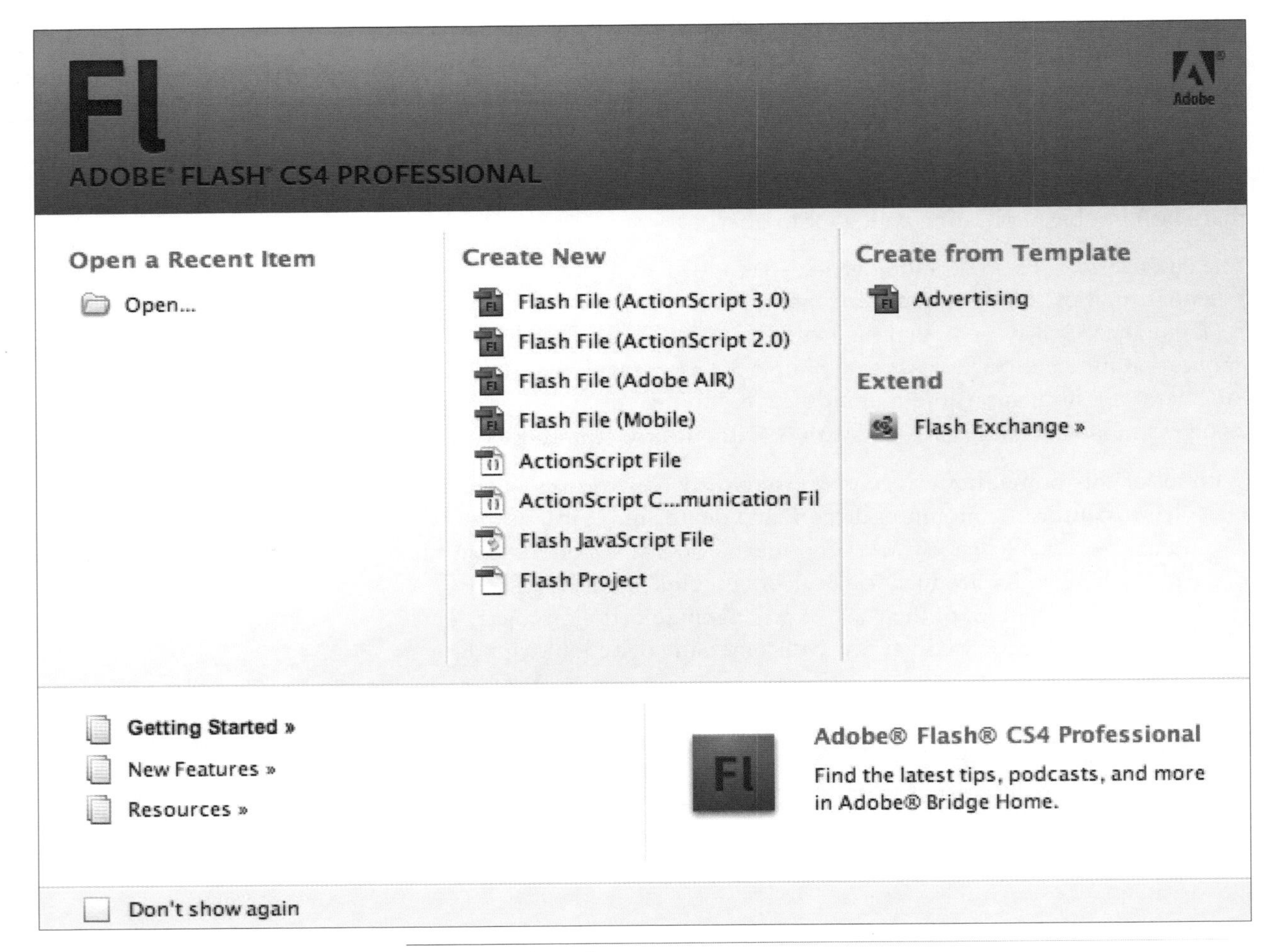

Figure 1-1. The Welcome screen

Open a Recent Item

A short list of previously opened documents at your fingertips. Upon first launch, this list will be empty, but it will maintain a list of your eight most recently used files, as well as an option to open any existing file. To open a listed recent item, click its name. To open an existing document that is not listed, click the Open button, and a standard file browser will appear, allowing you to find and open the document.

Create New

You will likely select from this list most often and, when working with this book, you will usually choose the Flash File (ActionScript 3.0) option. Clicking this button will create a native Flash file (often abbreviated as FLA and pronounced "flah" because of its *.fla* filename extension) that is preconfigured to use ActionScript 3.0 as its scripting language. This choice tells Flash to use the appropriate syntax when checking and compiling your scripts.

Create from Template

Clicking this button opens a dialog displaying ready-made templates optimized for various advertising dimensions. Clicking any template will open a file that is preconfigured with the template assets. You will create and use a template later in this chapter.

Extend

The link in this section connects to the online Adobe Exchange, which allows you to download extensions that add functionality to Flash.

Links area

The ribbon across the bottom of the Welcome screen provides links to additional online material covering introductory material, new features, further resources, tips, podcasts, and more.

If you prefer not to use the Welcome screen, you can disable it by checking the "Don't show again" feature in the lower-left corner of the screen. All of the features on the Welcome screen are available in the File and Help menus, and you can restore the Welcome screen later in the application preferences, if you wish.

NOTE

Although the limited ActionScript covered in this book will focus exclusively on version 3.0, it is still possible to create ActionScript 1.0- or 2.0-based files by choosing Flash File (ActionScript 2.0) during the file creation process.

Selecting which version of ActionScript is used in each file is very important. Flash Player 9 and later, for example, can play Flash files created in any of three versions of ActionScript: 1.0, 2.0, and 3.0. Versions 1.0 and 2.0 can coexist and are represented collectively by the ActionScript 2.0 option. However, it is not possible for a single FLA to include both ActionScript 3.0 and either previous version.

The good news is, you're not locked in to this decision when creating a new file. It's possible to change which version of ActionScript a file uses after you have created the file. This won't change any of the code you've written, but it will tell Flash how to correctly check and compile your scripts.

Using the New Application Window

Depending on how extensively you use Flash and how much you accomplish within the application itself (as opposed to creating assets in other applications, for example), the interface can quickly become cluttered with windows and panels. Flash CS4 Professional helps address this issue by originating most of its features within one main application window. You can then adjust this configuration in a number of ways, making it easier to organize your work environment while still customizing it to your liking.

If you've used Flash before, you may be surprised at the setup of the work area the first time you launch the application. While some areas of the interface remain familiar, others have changed location, shape, and even functionality. Take a moment to look at the default layout of the interface in Figure 1-2.

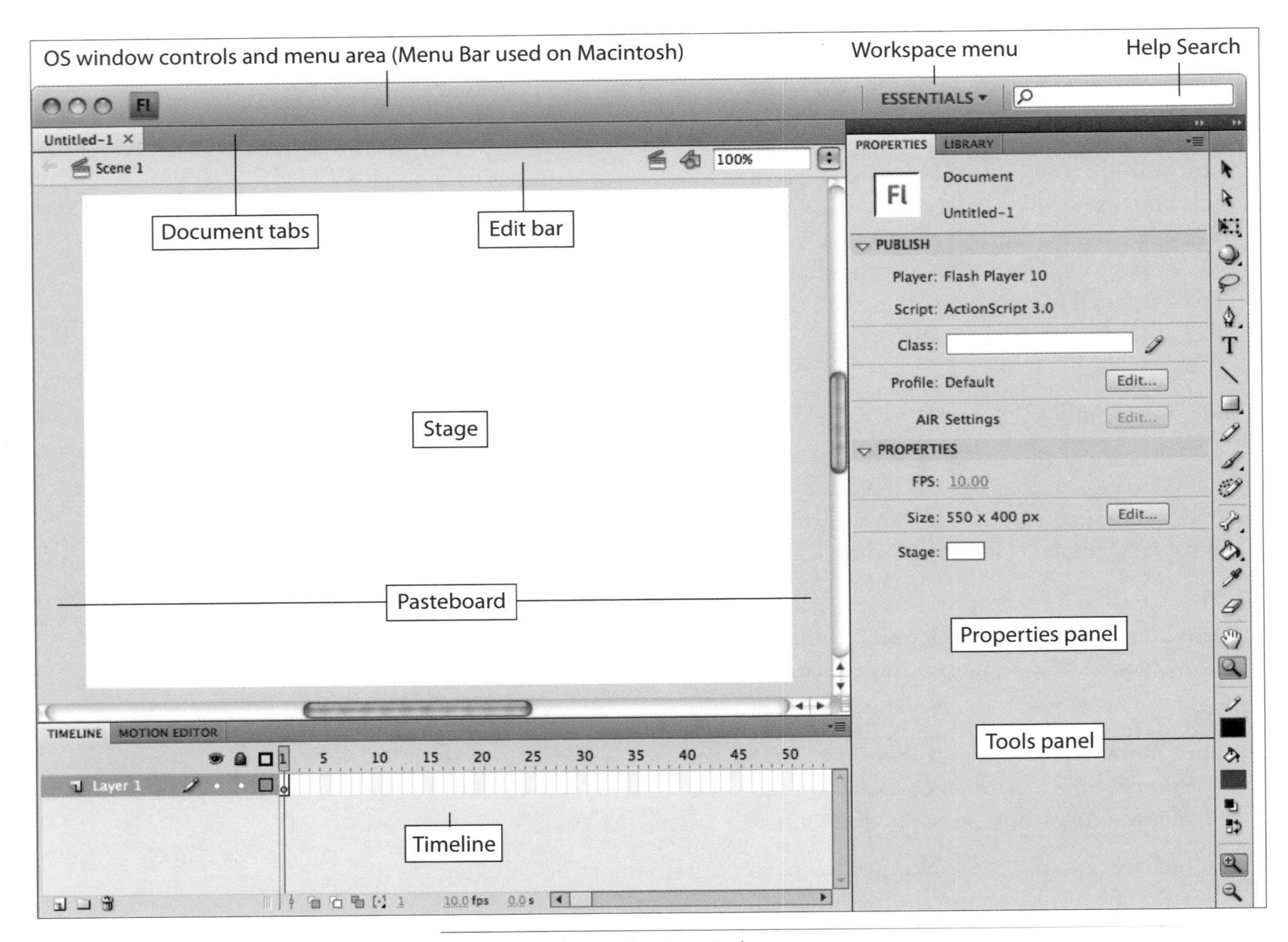

Figure 1-2. The application window

NOTE

Figure 1-2 depicts the Essential workspace. You can change the workspace to a number of preset layouts, as well as customize it to your own liking. See "Selecting and Editing a Workspace," later in this chapter.

The focus of the default layout is the coupled *Stage* and *Pasteboard*. The Stage is the area of your final file that will be visible during playback. The Pasteboard is the gray area surrounding the Stage that will not be displayed at runtime. This is handy for moving assets in and out of view. For example, an asset that sits on the Pasteboard will not initially be visible, and can be animated onto the Stage. Similarly, an asset can be moved to the Pasteboard to be hidden from view.

As you read this book, it will help you to liken the Flash world to the real-life world of theater or film. Think of the Stage as the stage or set on which a play or film is performed. Think of the Pasteboard as the wings of the theater or the soundstage, used by performers and crew for props, costume changes, and entrances and exits.

By default, each time a Flash document is opened, it is displayed as a tab just above the Stage inside the main application window. All Flash documents are collected as tabs in a layout just above the Stage inside the main application window. You can drag these tabs to reorder them, drag them out of the tabbed layout to create new document windows, or drag them back into the tabbed layout to consolidate windows.

For example, if you open two documents, they will appear as tabs in the order in which they were opened—*MyDocument1.fla* and *MyDocument2.fla*, for example. If you drag *MyDocument1.fla* to the right of the tab for *MyDocument2.fla*, the tabs will reorder. If you drag a document tab down away from its original location, the document will transfer to its own window. Later, you could drag the window bar back to the series of tabs, and the document would reappear as a tab, discarding the unused window.

Each document window contains a Stage and Pasteboard, as well as a small group of navigation and zoom tools collectively known as the *Edit Bar*. The Edit Bar helps you navigate through nested assets the same way your operating system lets you navigate through nested folders—maintaining a breadcrumb path of your progress as you descend through the levels of nested content. It also allows you to select *scenes*, which are analogous to the scenes of your play or film, and *symbols*, which are analogous to actors in a play. You'll learn more about these in later chapters. Finally, you can use the Zoom menu in the right corner of the Edit Bar to view your document at various sizes. The menu offers sizes between 25 and 800%, and you can enter values manually ranging from 8% to 2000% zoom.

NOTE

Scenes are used to break up very long timelines into manageable chunks. You can use the Scenes panel (Window→Other Panels→Scenes) to add, rename, navigate to, and delete scenes.

Above the Stage is the main window bar, containing operating system controls—window controls for both platforms, and the full application menu on the Windows platform (the standard operating system Menu Bar is used on the Macintosh platform.) Also included in this window bar is an interface configuration menu and a Help search field.

Beneath the Stage is the *Timeline*. As its name implies, the Timeline uses a time-based metaphor for playing through your file and is analogous to a reel of film. By default, the Flash *playhead*, or current frame marker, scrolls through the Timeline during playback, displaying each linear frame of your project the way light projects each frame of film sequentially onto a screen.

Each time you place an asset on the stage during authoring, it will appear in the timeline much the way an event is represented in a historical timeline. A short event, for instance, might be represented as a dot, while a line spanning several years might reference a longer event. The Flash Timeline works in a similar way.

You can control when assets are displayed in your file by placing them in specific frames of the Timeline. For example, if you want a background image to appear throughout your presentation, that asset should span every frame used. A menu, on the other hand, might only appear in one frame. Further, if you want something to be visible for a short time, you can remove it from the timeline after a short span of frames.

Unlike film, however, the time is not fixed in Timeline playback. Not only can you change the frame rate of your file—a bit like slow motion or fast-forward—but you can also use ActionScript to jump the Flash playhead, moving it from any frame to any other frame, and even stopping it on a given frame for any length of time. In these ways, playing a Flash file can be thought of more like watching a movie on DVD.

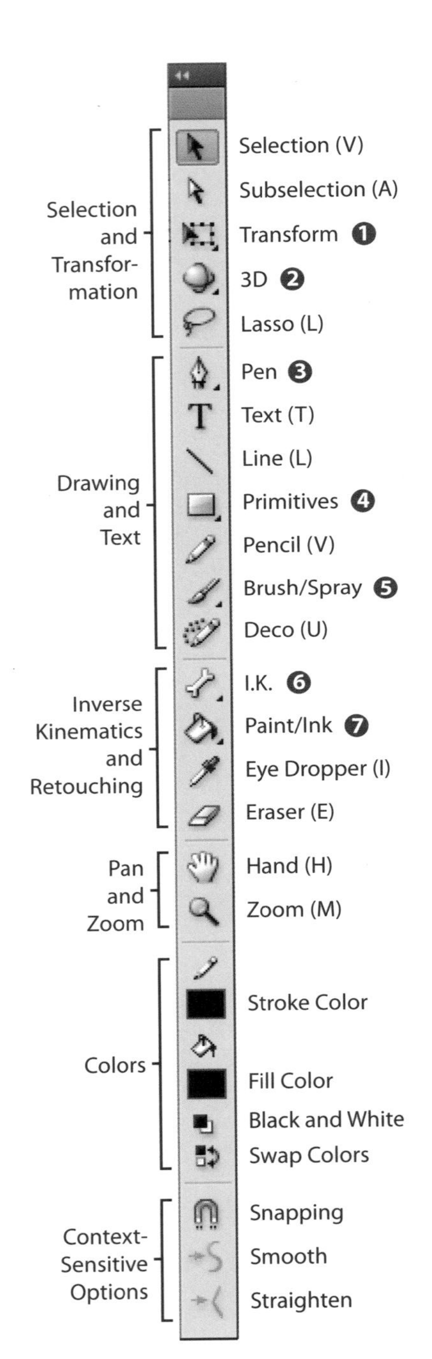

Figure 1-3. The Tools panel; keyboard shortcuts are listed in parentheses

WARNING

Another way that the Flash Timeline model differs from film is that Flash animations on older computers can play back slower than the requested frame rate. So, it's important to test your work on a variety of computers whenever possible.

Docked to the right of the Stage and Timeline (as seen previously in Figure 1-2) are the *Properties*, *Library*, and *Tools* panels. Panels display groups of related parameters and controls for easy access. You'll use panels frequently and you can rearrange them to suit your workflow. You can show or hide them based on need, arrange them in logical groups, and rearrange them at will.

The Tools panel contains a variety of tools for creating and manipulating assets. The Library panel is used to organize Flash symbols and similar assets and is, therefore, somewhat akin to a theater dressing room holding actors for a performance. Finally, the Properties panel is where you will most often adjust the many document and asset properties, such as size and position. These three panels, as well as a few more, are discussed in greater detail in the following section.

Understanding Panels

Panels are the real workhorses of the Flash interface, and bear the brunt of the poking and prodding you'll do while working on a file. Throughout the book, you'll work with these panels extensively, so this discussion is meant to give you a brief functional overview of some of the most commonly used panels.

Tools

Divided into six main sections, Flash's Tools panel is the go-to panel that contains the tools you'll use throughout your work (Figure 1-3). The Tools panel also contains menus of related tools (Figure 1-4), only one of which can be used at a given time. The following list describes the Tools panel's six sections.

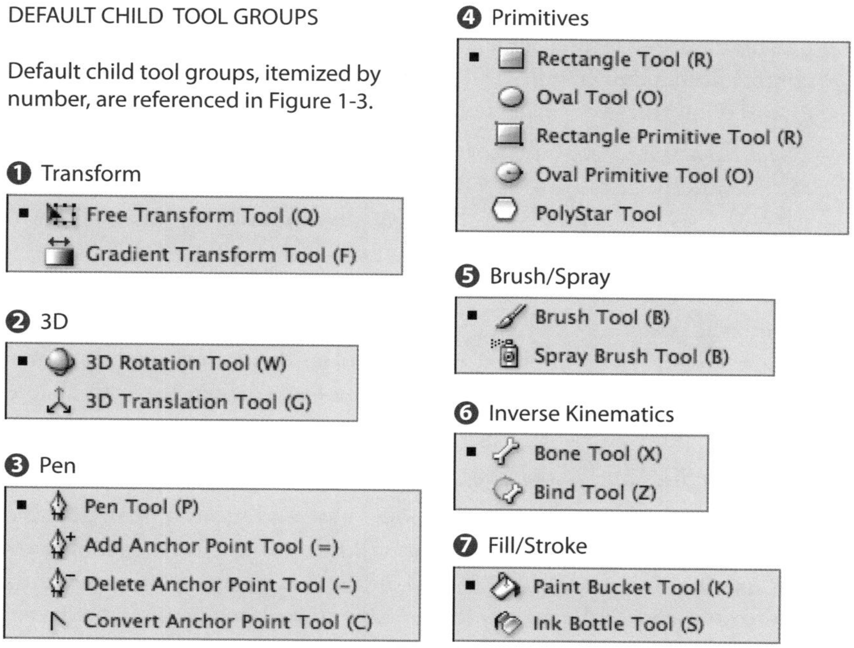

DEFAULT CHILD TOOL GROUPS

Default child tool groups, itemized by number, are referenced in Figure 1-3.

❶ Transform

- ▪ Free Transform Tool (Q)
- Gradient Transform Tool (F)

❷ 3D

- ▪ 3D Rotation Tool (W)
- 3D Translation Tool (G)

❸ Pen

- ▪ Pen Tool (P)
- Add Anchor Point Tool (=)
- Delete Anchor Point Tool (-)
- Convert Anchor Point Tool (C)

❹ Primitives

- ▪ Rectangle Tool (R)
- Oval Tool (O)
- Rectangle Primitive Tool (R)
- Oval Primitive Tool (O)
- PolyStar Tool

❺ Brush/Spray

- ▪ Brush Tool (B)
- Spray Brush Tool (B)

❻ Inverse Kinematics

- ▪ Bone Tool (X)
- Bind Tool (Z)

❼ Fill/Stroke

- ▪ Paint Bucket Tool (K)
- Ink Bottle Tool (S)

Figure 1-4. Default tool menus accessed from the Tools panel; keyboard shortcuts are listed in parentheses

Selection and transformation

The first section of the Tools panel is dedicated to selection and transformation tools. Here you will find three ways to select elements for manipulation. The *Selection* and *Lasso* tools function the same way they do in most applications, allowing you to select objects by clicking or dragging over them, respectively. The *Subselection* tool allows you to select individual points and control handles of vectors rather than the entire object.

Also in this section are two transformation tool menus. The *Transform* menu contains the 2D tools *Free Transform* (used for scaling, rotating, and skewing objects) and *Gradient Transform* (used for performing similar manipulations on gradient fills). The *3D* menu, new to Flash CS4, contains the *3D Rotation* tool (used to rotate objects around the x-, y-, and z-axes in 3D space) and the *3D Translation* tool (used to move the positions of objects along the same three axes in 3D space).

Drawing and text

This section of the Tools panel contains creation tools, such as *Text* and *Line*. The Text tool creates text elements, and the Line tool creates lines (also commonly called *strokes* in Flash). Here you'll also find a menu of primitive shapes, including rectangles, ellipses, polygons, and stars. The equally common *Pencil*, *Brush*, and *Pen* tools function much the same way they do in other applications, except that in Flash, all drawing tools create vectors rather than pixels.

Pencil and Brush are primarily freehand tools, while Pen is used to draw *Bézier* curves—computer-interpolated curves that are shaped by dragging control points and handles. The Pen menu contains additional tools for granular control over corners and smoothing of these curves.

The Brush menu contains a tool that is new to CS4 called the *Spray Brush*. Inherited from Adobe Illustrator, the Spray Brush allows you to spray shapes and symbols on the Stage, placing them automatically with possible adjustments to scale and rotation.

In a similar vein, the *Deco* tool distributes shapes or symbols along a grid, symmetric pattern, or even a growing vine algorithm. The latter provides options for leaf and flower art, as well as advanced options such as branch rotation and segment length. You'll use the Deco tool in the next chapter to create your first interactive exercise.

Inverse kinematics and retouching

New to CS4, you can now use the *Bones* tool, and its accompanying *Bind* tool, to create armatures for animating linked objects. For example, you can use the Bones tool to link and animate a robotic arm. Moving the claw at the end of the arm automatically moves each arm segment, with optional constraints such as joint rotation limits. This is known as *inverse kinematics*.

In Flash, the *Eye Dropper*, *Eraser*, and *Paint Bucket* tools work in nontraditional ways. As with the Brush and Pencil tools, they manipulate vectors exclusively, not pixels. The Paint Bucket has a cousin tool called the *Ink Well*, which applies color and other characteristics to lines the way a Paint Bucket affects fills.

Pan and zoom

The last batch of dedicated tools is tried and true, consisting of the *Hand* tool for moving the stage around within the window, and the *Zoom* tool (magnifying glass) for changing the degrees of magnification.

Colors

The colors segment of the Tools panel works together with many other tools, providing quick access to predefined colors for strokes and fills. As with other applications, you can also select a default black-and-white color set and swap the two colors with the click of a button.

Context-sensitive options

The final segment of the Tools panel is a context-sensitive area that varies depending on which selection is active. Figure 1-3 shows the options for the Selection tool, which include turning snapping on and off and smoothing or straightening selected lines.

Properties

The *Properties* panel (also called the *Property Inspector*) is the newly expanded primary location for adjusting values for such object properties as size, location, orientation, and color effects. Users of previous Flash versions will notice a big change here: the new panel is vertical and now offers convenient access to a much larger number of properties. It is still context-sensitive, hiding and revealing properties according to which object is selected. To accommodate a large number of properties, the panel provides scroll bars when needed, and is grouped into categories that can be collapsed or expanded.

When you select a tool in the Tools panel, its properties become active in the Properties panel. If a tool has no configurable attributes, the Properties panel shows document properties, such as stage size and color. You'll configure these properties when you create your first FLA, but it's helpful to become accustomed to the context-sensitive nature of the Properties panel.

Library

As you saw from the tabs at the top of the panel area in Figure 1-2, the Library is grouped with the Properties panel in the default workspace layout. To select another panel in a group, simply click its tab in the topmost portion of the panel group. The selected panel will come to the front.

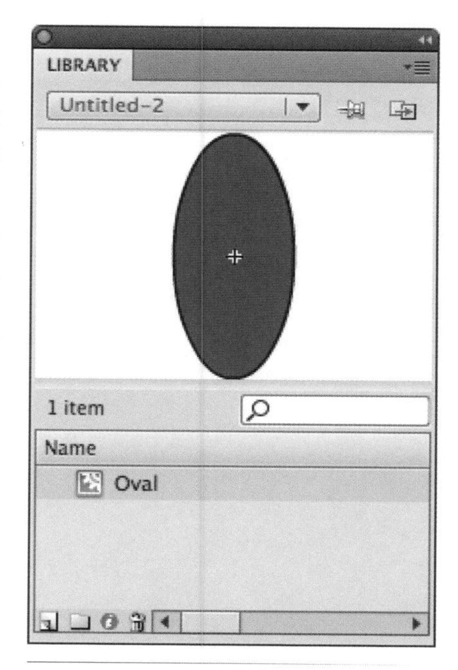

Figure 1-5. The Library panel

The Library (shown in Figure 1-5) is an essential repository of reusable assets such as bitmaps, sounds, and native Flash asset types such as *buttons* and *movie clips*. A movie clip, for example, is an animation asset that has its own dedicated Timeline. You can create an animation in a movie clip and then conveniently treat the entire animation as a single asset.

Symbols will be discussed in greater detail in Chapter 3, but the key thing to remember now is that symbols contribute mightily to the efficiency of your file. Symbols can be used many times in a file without contributing noticeably to file size beyond their first use. In addition, you can manipulate a single instance of a symbol on the Stage without affecting other instances of that symbol.

The lower-left corner of the Library features buttons for (from left to right) quickly creating new symbols, creating folders to organize your internal assets, editing symbol properties, and deleting symbols. Libraries from all open files will appear in the menu at the top of the Library, allowing you to switch between libraries easily and drag assets from any open Library to any open file.

The two buttons in the upper-right corner of the Library can help in this regard. Using the *New Library* button (shown in blue), you can create another window for the Library to allow dragging assets from one Library to another (without having to add content to the Stage). You can also *pin* a library to a Flash file (using the button that looks like a pushpin). This ensures that, when switching between files, the selected Library will always become active. This is helpful when frequently moving between many open documents.

You can now search the Library using the input search field between the asset list and thumbnail. New to Flash CS4 Professional, this feature filters the assets displayed in the Library by name. Using the column headers at the top of the asset list, you can also sort the Library by name, *linkage* name/class (used with ActionScript and discussed in Chapter 6), use count, date modified, and asset type.

Other common panels

Here are several additional, commonly used panels, all of which you can access from the Window menu:

Align

Align is an impressive tool that will align, distribute, and space assets according to desired constraints, as well as match asset sizes.

Info/Transform

These two are a duo of related transformation panels that allow you to alter specific properties of selected assets. The *Info* panel displays the position (x- and y-coordinates) and color (red, green, blue, and alpha values) under the mouse, and lets you edit the width, height, x-, and y-coordinates of a selected object. The *Transform* panel lets you alter the scale, rotation, skew, 3D rotation point, and 3D transform point of an asset.

Color/Swatches/Kuler

In this trio of related color panels, *Color* is the most powerful and is the panel in which you will most often define not only custom colors, but also gradients. The *Swatches* panel contains a collection of swatches that you can edit, save, and load, and this is where you will store custom colors and gradients for quick recall. *Kuler*, a Flash panel extension that ships with Flash, is an online system for sharing color families. The panel is found in the Windows→Extensions submenu, and an Internet connection is required for use.

Motion Editor/Motion Presets

The *Motion Editor* is a powerful companion to the Timeline and is the editing panel for the new CS4 animation model. The *Motion Presets* panel contains a list and preview animation of many predesigned motion sets, such as bounce, fly, and zoom in and/or out of frame. These panels will be discussed at length in Chapter 5.

Components/Component Inspector

Components and the *Component Inspector* provide ways to add complex functionality and user interface (UI) elements to your projects using little to no ActionScript. You will learn more about components in Chapter 9.

Actions/Output/Compiler Errors/Debug Console/Variables

These panels are used when authoring ActionScript. The *Actions* panel is Flash's internal script editor, the *Output* and *Compiler Errors* panels are used to monitor script integrity and output, and the *Debug Console* and *Variables* panels are used during debugging to trace through script execution and monitor variable values. The latter two panels are found in the Windows→Debug Panels submenu.

History

The *History* panel maintains a list of nearly everything you do in Flash and allows you to backtrack through previously completed tasks. If you are familiar with the History panels in Adobe Photoshop or Adobe Illustrator, you will feel right at home with this panel.

Although this list doesn't cover every panel available in Flash CS4 Professional, working with several of these utilities should give you the experience required to explore additional panels when needed.

Understanding the Timeline

As discussed previously, the Timeline is a time-based representation of your project file and can be likened to a roll of film. As the author of a Flash document, you are the director of a film. You will use the Timeline to navigate through your file, composing scenes, positioning actors, and controlling their entrances and exits.

Take a look at Figure 1-6, a detailed view of the Timeline. The red rectangle highlighting the number 1 and the corresponding thin red line that spans the height of the Timeline comprise the *playhead*. This marker indicates the current frame displayed to your audience. Viewers of your file will experience content—see visual assets, hear sounds, and so on—placed in this frame.

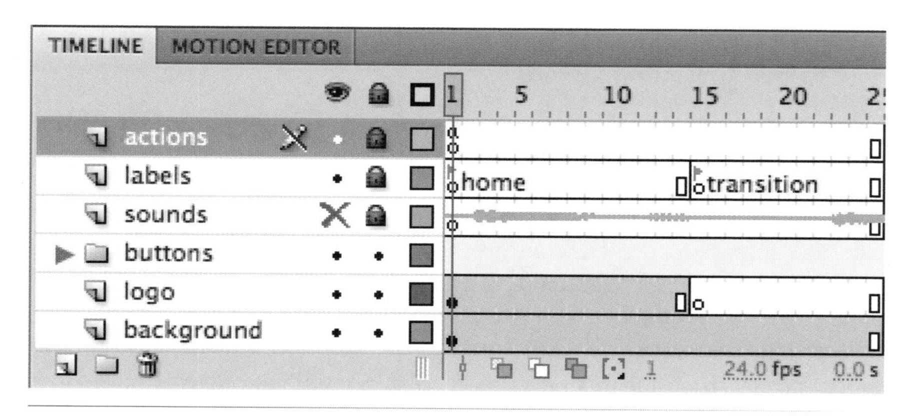

Figure 1-6. The Timeline

To work with other frames of a multi-frame Timeline, you can move the playhead by clicking or dragging along the strip of numbers at the top of the Timeline. Time increases when moving to the right, and you'll notice the frame numbers climb, to 5, 10, and so on, until the figure ends at frame 25. Positioning the playhead in any frame allows you to preview or edit the frame's content, including manipulating assets on the Stage, as well as editing the Timeline itself. For example, you will learn how to draw art on the Stage in Chapter 2, how to add sound to the timeline in Chapter 13, and how to add ActionScript to a frame in Chapter 6.

The three numbers in the lower-right corner of Figure 1-6 display information regarding the current time and rate of playback of your file. The first number shows that the playhead is currently in frame 1. The next number indicates that the *frame rate*, or speed at which this Timeline will play, is 24 frames per second. The last number shows the *elapsed time*, or the time-based equivalent of the current frame. In this case, the current frame is frame 1, which resides at 0.0 seconds. If the playhead were at frame 24, the elapsed time would read 1.0 second because the frame rate is 24 frames per second.

To the left of the time indicators, just below the playhead marker, is an icon that looks like a tiny version of the playhead. Clicking this button will auto-scroll your Timeline to center the current frame. The icons to the right of this button allow you to display and edit multiple frames at once. This will be discussed in the context of animation in Chapter 5.

Layers

So far, discussion of the Timeline has been focused on the horizontal, as the playhead moves through frames with time. There is also a vertical component of the Timeline panel. If you look again at Figure 1-6, you'll see that the Timeline is divided into rows (called *layers*). Serving a similar purpose as the layers in Adobe Photoshop or Illustrator, Timeline layers provide one way of defining a visual stacking order in your projects. The bottommost layer is at the bottom of the stack, and the topmost layer appears on top of all others. For example, in frame 1 of Figure 1-6, the contents of the *background* layer appear below the contents of the *logo* layer.

Although most layers contain some type of visual content, they serve other purposes, too, some of which are depicted in Figure 1-6. For instance, you can add sounds to the Timeline for audio playback without having to do any programming. You can name your frames in order to locate a portion of your movie easily or to perform scripting navigation using ActionScript. The scripts you write are often added to the Timeline, and can be consolidated in a dedicated layer for this purpose. It's also possible to organize layers into folders to keep things tidy.

When organizing your content, you can add or delete layers or folders using the icons in the lower-left corner of the panel, and you can rename a layer by double-clicking its text. Finally, you can hide, lock, or display the contents of layers as outlines by clicking the appropriate column in the layer (eye, lock, and box icons, respectively).

Keyframes and interpolated frames

It's important for you as a director to control many aspects of your movie. Not only must you position your actors in the foreground and your sets in the background (using layers), you must also cue the appearances of your characters. This is accomplished with *keyframes*—special frames that you add to a layer to break it into segments and control its content.

For example, keyframes appear in frames 1 and 15 of the logo layer, indicated by the circle icons found in these frames. A filled circle indicates there is content in the frame, while an empty circle shows that the frame is empty. After you define keyframes, you can let Flash do all the work of calculating the appearance of the interim frames. This will be discussed at length in Chapter 5 when you read more about animation. The small vertical rectangles in frames 14 and 25 of the logo layer simply mark the end of a frame span.

Customizing Your Interface

Now that you've seen the high points of the Flash CS4 Professional interface, it's time to make it your own. If you've used Flash before, you may be feeling a little disoriented due to the size and location changes of various panels. The good news is, the interface is highly customizable and it's very easy to adjust to your liking.

Selecting and Editing a Workspace

When customizing the interface, a good place to start is viewing the available workspace presets. In the upper-right corner of the main application window, next to the Help search field, you will find a menu of presets (shown in Figure 1-7) that are optimized for a variety of user profiles.

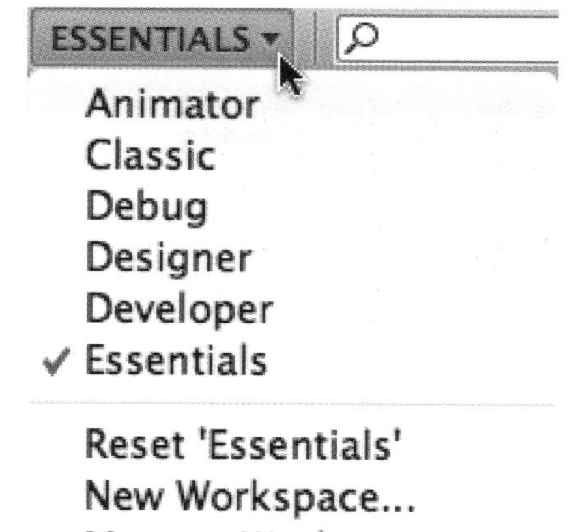

Figure 1-7. Select and manage workspaces with the Workspace menu

Start by looking at how each preset is configured and determine which is closest to your liking before editing. Users accustomed to prior versions of Flash, for example, may wish to start with the Classic workspace. This configuration places the Tools panel on the right, the Timeline at the top, and a default set of panels to the right of the stage, reproducing application layouts of prior Flash versions.

If you wish to return to the original profile at any time during the customizing process, simply choose the Reset option from the menu of presets. After you customize your interface, you can use this menu to save your layout as a new preset. You can also use the *Manage Workspaces* option if you need to rename or delete presets.

Minimizing Panels

You can view all panels at full size or in one of two minimized states: icon-and-name view or icon-only view. The default minimized state is icon-and-name view. Figure 1-8 shows all three view options.

Clicking the double-arrow icon in the upper right corner of each panel toggles between the panel's full-size view and your choice of minimized states. When minimized, you can drag a panel from one or both vertical edges (depending on where the panel is docked) to reduce its appearance to an icon-only view.

In either minimized state, clicking on the icon or name (when present) flies out the full-size panel for use, as shown in Figure 1-9. Based on your choice of preference setting, the panel can automatically minimize when you're working with any other interface element, or remain open until you choose to close it. See the upcoming section "The Preferences Dialog" for more information.

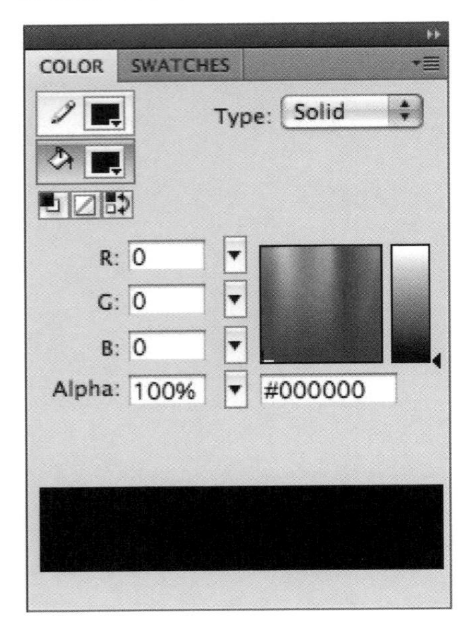

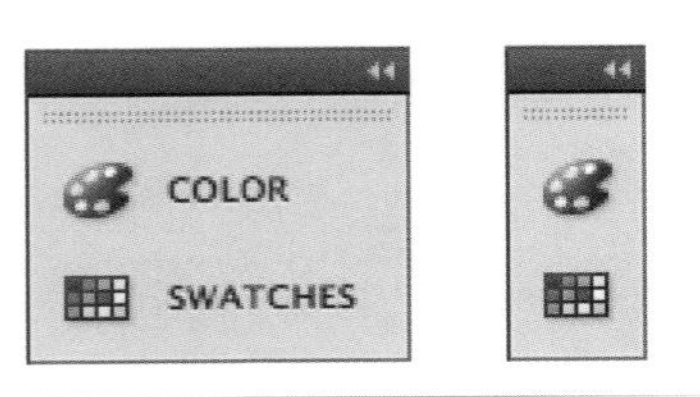

Figure 1-8. Panels shown in full-size, icon-and-name, and icon-only views

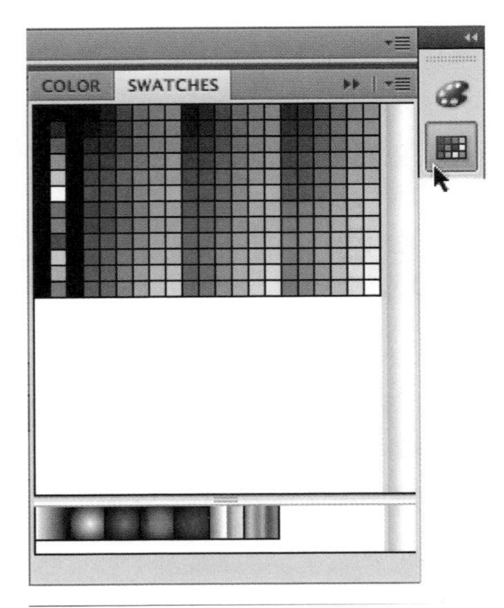

Figure 1-9. Panels can fly out from name or icon view

Grouping and Docking Panels

Some panels will already be docked to the interface in each workspace preset, and others will open in a free-floating state. Regardless of its original position, any panel can be grouped, ungrouped, and docked (snapped to a set location such as the top, bottom, or sides of a window, rather than free-floating) by simply dragging the panel by its tab to the desired destination.

To group panels, drag one panel to another. In Figure 1-10, the standalone Swatches panel is grouped into a panel set with the Color panel. The destination panel group is highlighted with a blue outline that helps show where the panel is being dragged. You can remove a panel from a group by dragging it away from the set by its tab.

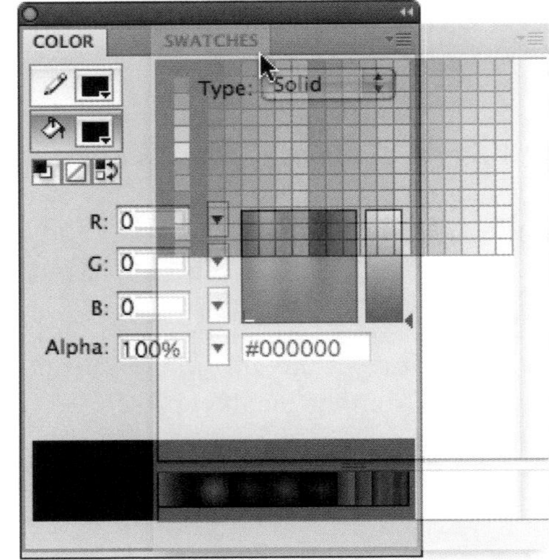

Figure 1-10. Dragging a panel to an existing panel group

To dock a panel, drag it to one of the edges of the destination panel or panel group. Figure 1-11 shows the blue outline feedback that Flash provides when you dock or group panels. The blue line indicates where the panel will end up. As the figure shows, you can dock above, below, to the left, or to the right of a panel or set. You also add a panel to a group, even when the group is minimized.

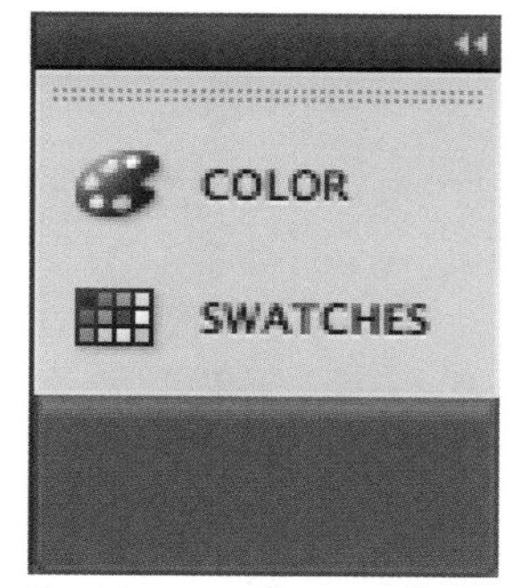

Docking panel vertically, below an existing panel group. Panels can also be docked above an existing group.

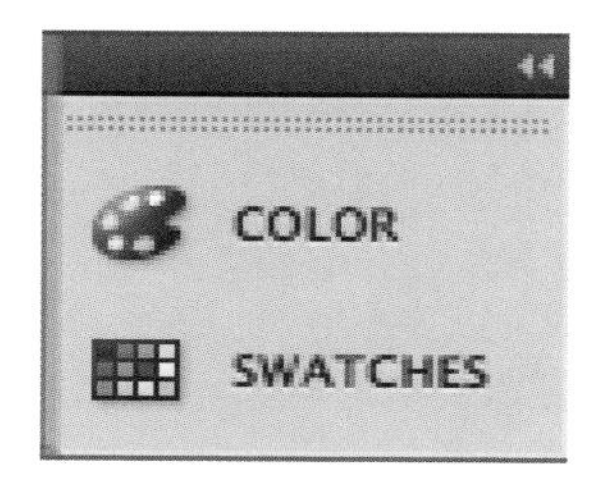

Docking panel horizontally, to the left of an existing panel group. Docking to the right of a group is also possible.

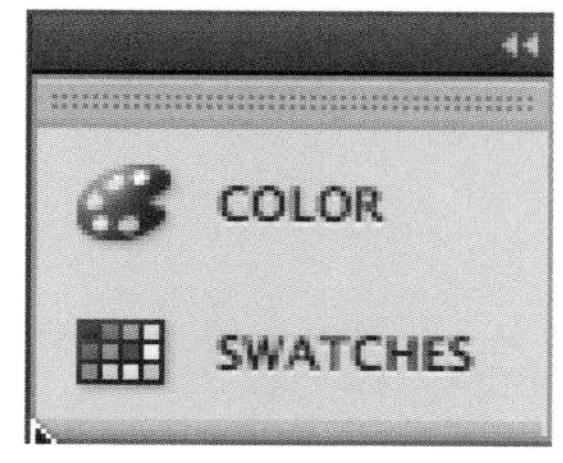

Adding panel to the bottom of an existing panel group.

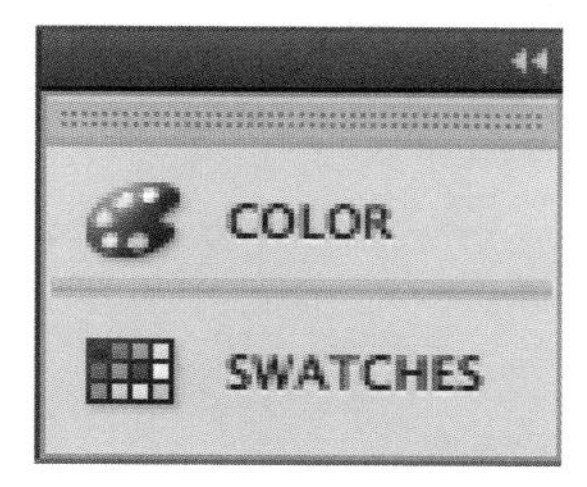

Panels can also be added to other positions within a group, such as between two panels, as shown here.

Figure 1-11. Panel dragging feedback

Dock or group?

Choosing to dock a panel on its own or to group it with other panels is purely a matter of preference. Without a very large monitor, the interface would be too cluttered to use if every panel were free-floating. Docking panels helps you wrest control over your work area because you can easily minimize panels without losing track of them.

Grouping panels together can also help you create a single place to find related panels that you will need on a regular basis. For example, it is common to find the Color and Swatches panels grouped, as well as to find the Align, Transform, and Info panels collected into one group.

NOTE

Remember that most Flash interface elements are panels, so they can be free-floating or docked in just about any configuration of rows and columns. Panels can also be resized by dragging their edges. For example, Flash CS4 Professional is the first version of the application that can organize its Tools panel in a nonvertical layout. The Tools panel is approximately square in the Designer workspace preset, but is a single horizontal row in the Developer workspace preset.

The Preferences Dialog

Because of its complexity, Flash has a large number of preference settings—too many to discuss in depth here. Instead, I'll show you some of the settings relevant to the discussions in this chapter. We'll examine additional preference settings in later chapters when applicable. To access the Preferences dialog (Figure 1-12), select the Preferences option from the Flash Professional application menu (Mac) or the Edit menu (Windows).

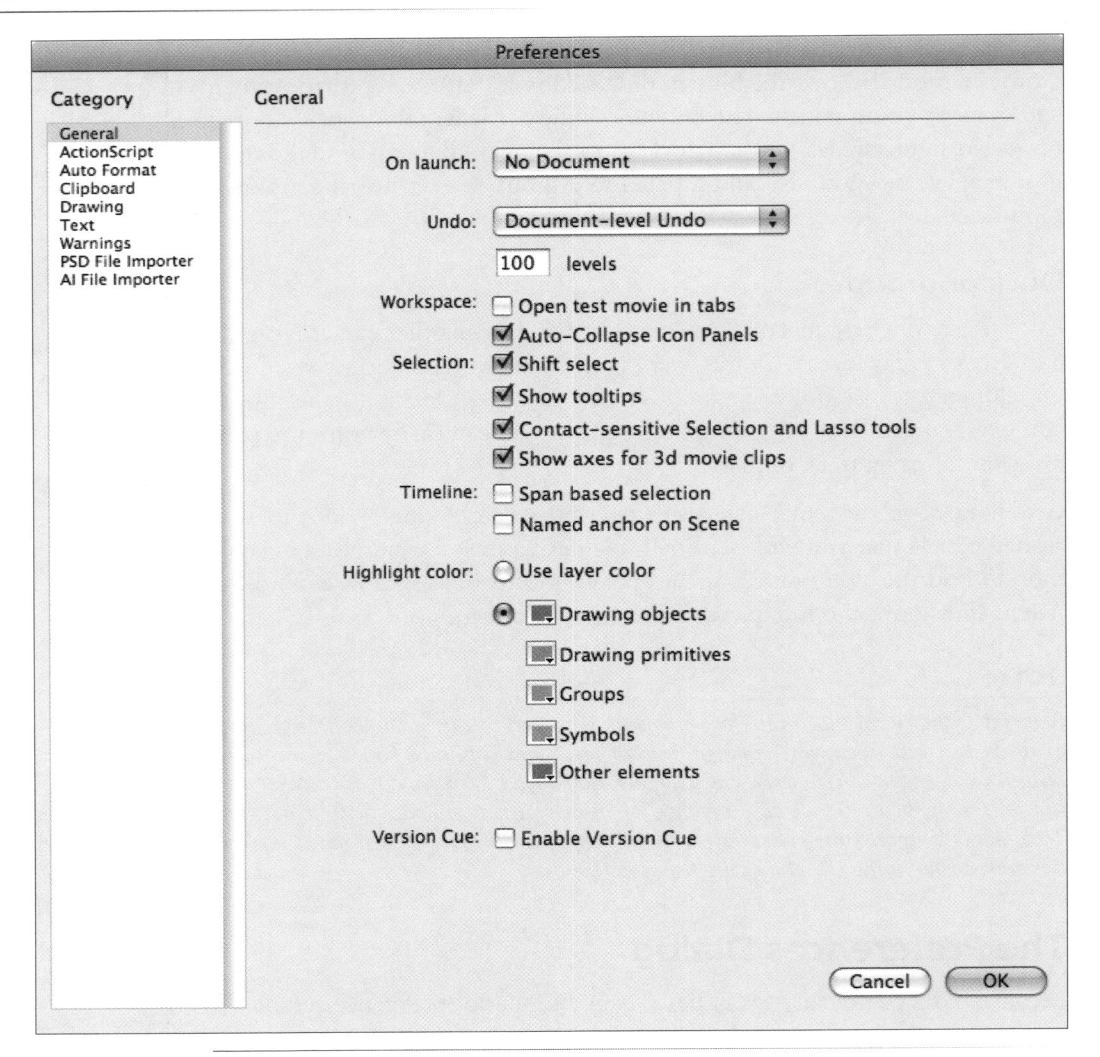

Figure 1-12. Application preferences

Within the Preferences dialog, preference settings are organized into several pages. In this chapter, we'll look at the General options:

On launch

The On launch setting allows you to dictate what occurs when the application is first opened. You may elect to show no document, create a new document, open the most recently edited document, or display the Welcome screen. If you previously elected to dismiss the Welcome screen and wish to restore it, this is the setting for you.

Undo

Flash offers two systems of storing undo operations. You can specify a *Document-level Undo* history, in which one history list is maintained for each document, or an *Object-level Undo* history, in which a list of prior activity is recorded for each major object in the application. Although the latter provides a more granular degree of undo, it will also increase the size of your authoring files substantially (it has no effect on the size of runtime files). You are probably better off staying with the default Document-level Undo option to start, and then determining later if you need to switch to storing object-level histories. In both cases, you can set the number of undo levels maintained in each history list.

Open test movie in tabs

Once you start working with FLA files, you will need to check your work by compiling test SWF files. For expediency, you will likely want to preview the test files within the Flash interface, and only occasionally test them in a browser. When viewing a SWF in Flash, the *Open test movie in tabs* option allows you to specify whether the resulting test file opens in a new tab, grouped with all other open document tabs, or in a new window.

The best choice here is a matter of taste, and relies heavily on the behavior of other elements of the Flash interface—specifically, which panels remain visible while a test file has focus and how that behavior differs when the SWF is in a tab or its own window. As such, I recommend that you try the setting both ways and determine which approach you prefer.

Auto-Collapse Icon Panels

The *Auto-Collapse Icon Panels* setting determines how active panels behave when activated from a minimized dock. When this option is enabled, the panel minimizes again the moment you interact with any other interface element. When the option is disabled, the panel remains open until you manually close it.

Customizing the Tools Panel

After you have spent some time with the Flash interface, you may find that you are itching to rearrange the tools in the Tools panel. For instance, you may want to make the Subselection tool a child of the Selection tool, or you may want to group tools of like functionality together. You may find that you want to move the Lasso tool immediately below the selection tools, or move the Text tool to the bottom of the drawing and text group. This is easily accomplished using the Customize Tools menu item in the Flash Professional menu (Mac) or Edit menu (Windows).

As shown in Figure 1-13, the current Tools panel layout appears at the left of the dialog. To change this configuration, first select a slot in the Tools panel, and then choose which of the available tools will appear in that slot. The topmost tool will appear in the panel, and all additional tools will appear in a submenu. You can also remove tools from the selected slot. Finally, if you make a terrible mess of things, you can always restore the default configuration using the Restore Default button.

NOTE

Regardless of the orientation of your Tools panel, the tools still appear in sequence, corresponding to sequential rows in this dialog. For example, compare the order of the tools in the dialog shown in Figure 1-13 with the order of the vertical tool strip shown in Figure 1-2.

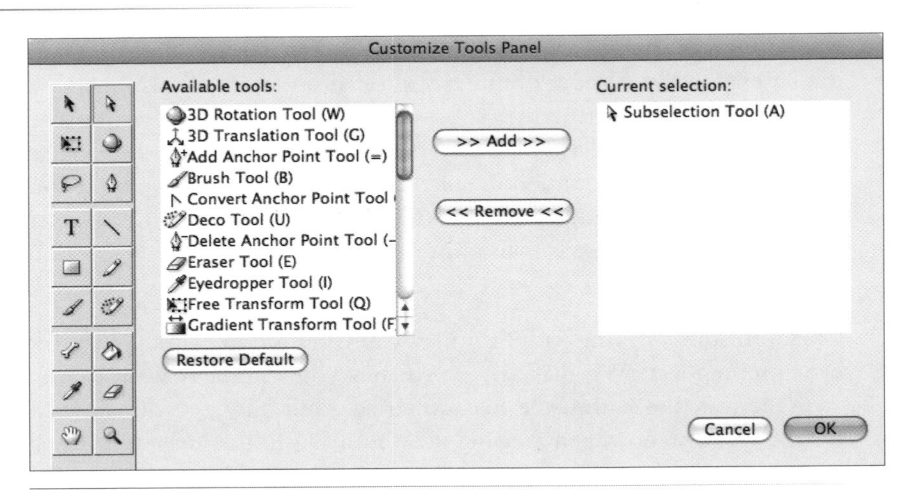

Figure 1-13. Customizing the Tools panel

Keyboard Shortcuts

The Keyboard Shortcuts feature (Figure 1-14) in the Flash Professional menu (Mac) or Edit menu (Windows) allows you to configure which keyboard shortcuts trigger which menu items.

The *Current Set* field includes preset keyboard shortcut configurations modeled after other applications. If you are a frequent user of any of the featured applications, you may wish to mimic one of those keyboard configurations. You can duplicate, rename, export, and delete any of these presets. To prevent permanent loss to shortcut configurations, you must duplicate a present, or create a new preset, before you can modify its settings.

If you wish to change the keyboard shortcut attributed to any menu, you must first select the appropriate menu category. Like many applications, the Flash interface is context-sensitive. Next, expand the menu name and select the desired menu item. Finally, select an existing shortcut to change or remove its value, assign a new shortcut, or even add a shortcut if more than one is helpful. When changing or adding selected values, type the new shortcut in the *Press key* field and click the *Change* button.

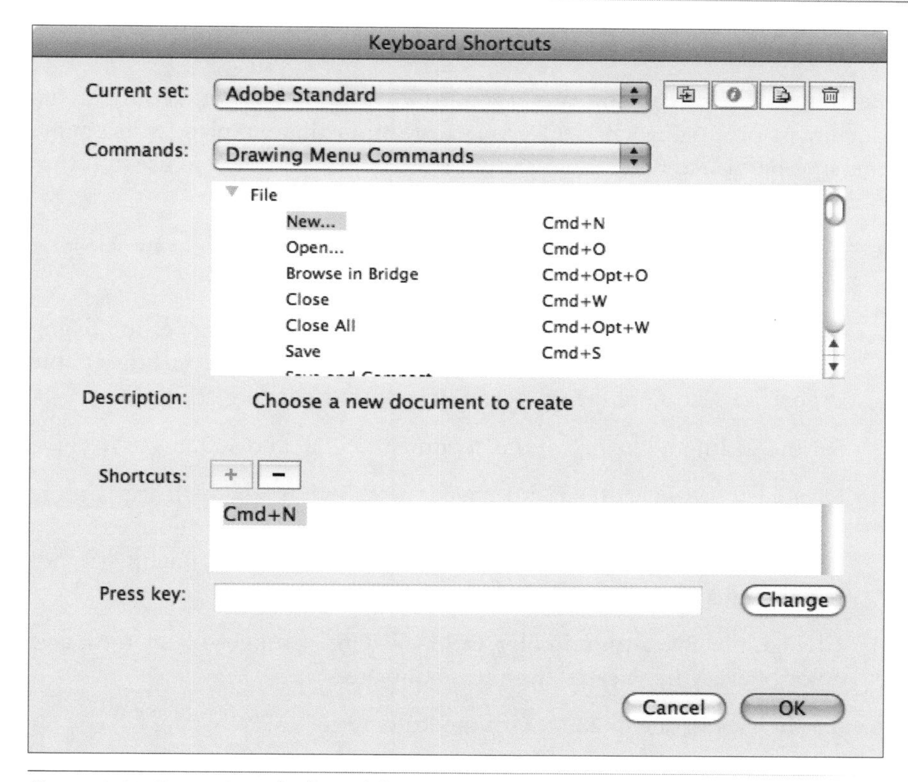

Figure 1-14. Customizing keyboard shortcuts

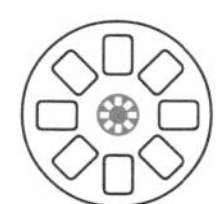

Project Progress

Throughout this book, you will often use your creativity, experiment with specific features, and try to accomplish isolated tasks. However, Flash is a very big program, and it's easy to wind up with a lot of basic skills and no way to use them. With this in mind, a book-wide project will wind its way through the chapters. This approach to learning not only allows you to review what you've read in a specific chapter, but also ties it all together to create a Flash website or application.

The project chosen for this book is a designer's portfolio. It will remain simple, helping to focus your attention on the content at hand, but the project goal is to cover as much of the material discussed as possible. In this chapter, you'll create a template that will simplify the creation of other project assets in later chapters.

Creating Your First FLA

Before you can save a template, you must create a new file and configure its document properties. Files that you create from this template will assume these properties, so the settings should match those you want to use in future files.

1. Create a new ActionScript 3.0 file on the Welcome screen, or by selecting File→New.
2. Select Modify→Document to access the Document Properties dialog (Figure 1-15). You can also access this dialog by clicking Edit in the Properties section of the Properties panel, after clicking on the Stage.
3. Set the width and height of the document to **750** and **500**, respectively.
4. Disable the Adjust 3D Perspective Angle option. You will learn more about this in Chapter 8, but, briefly, disabling this feature ensures that a 3D setting that affects perspective is based on the new document size, not the previous document size.
5. Change the Background color to black. This setting will set the Stage color of every file created from this template.
6. Set the Frame rate to **24** and set the Ruler units to Pixels.

NOTE

The project dimensions 750×500 were chosen so your project will display easily for the largest number of users. This size will fit in a browser window, without scrolling, on a monitor with a resolution as low as 800×600.

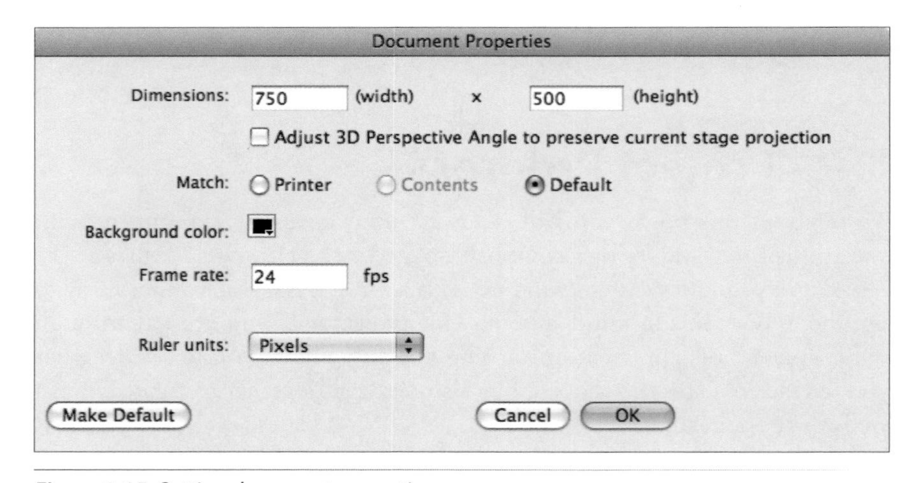

Figure 1-15. Setting document properties

Importing Your First Asset

You will learn more about creating new content in Chapter 2, and importing content in Chapter 3, but a quick venture into these areas now will improve the functionality and demonstrate a feature of your template.

1. If you haven't already done so, download the source files for this chapter from the companion website, *www.LearningFlashCS4.com*.

2. Select the first frame of the Timeline layer and import the provided *content_ui_guide.jpg* graphic by selecting File→Import→Import to Stage. Select the bitmap on Stage and, using the Properties panel, set its *x* and *y* properties to `0`.

3. Double-click the layer name and rename it `guide`. Lock the layer by clicking the dot across from the layer name in the column under the padlock.

Creating Your First Shape

Next you will create a placeholder asset that you can use to quickly add content to future files. You will learn more about drawing in Flash and creating native Flash asset types in the next few chapters. Just follow along carefully for now and be sure to test your progress through this exercise. When you're done, compare your results with the provided sample files.

1. Create a new layer in the timeline and name it `content`. You should have two layers: the locked guide and empty content layers. Click the frame in the content layer so you will be ready to create your next asset.

2. Select the Rectangle tool from the Tools panel and draw a rectangle anywhere on the stage. You will adjust this shape in just a moment.

3. Switch to the Selection tool and drag over the rectangle to select the shape. Use the Fill color chip to select a bright color.

4. Create a movie clip asset type by selecting Modify→Convert to Symbol. When the Convert to Symbol dialog appears, name the asset `content` and choose Movie Clip from the Type menu. Finally, click the tiny box in the upper-left corner of the nine-box grid next to the Registration option (you will learn much more about these options in the chapters to come). Click OK to finish the process. The rectangle you created in step 2 has been converted to a Flash movie clip symbol.

5. Using the Selection tool, click on the movie clip you just created and look at the Properties panel. If the panel is not open, access it by selecting Window→Properties. The top of the panel will show that you have selected your movie clip. Look in the Position and Size section of the Properties panel and click on the blue links next to the x and y properties. Set both of these to `0`, as shown in Figure 1-16. Your movie clip has now been moved to the upper-left corner of the Stage.

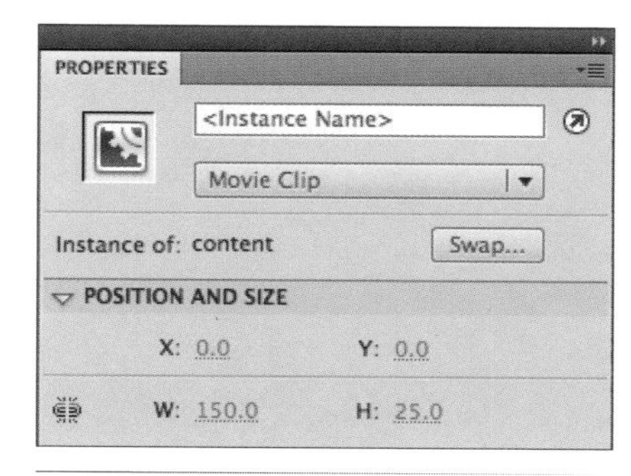

Figure 1-16. Movie clip instance properties in the Properties panel

If you can easily see your placeholder asset against the guide graphic you imported, consider this part of the exercise finished. If not, you can double-click the placeholder to edit it. After double-clicking, use the Selection tool to drag over the entire rectangle to select all of it, and then drag it with the mouse to a more visible location. You can also experiment with other drawing tools to make the placeholder more visible. The sample file, for example, added text to the movie clip. You'll learn a simple way to work with text in the next chapter.

NOTE

Text entry fields that look like web links, such as those pictured in Figure 1-16, are called hot text *fields. They work like normal text fields when clicked, but you can also drag your mouse left and right to adjust their values interactively.*

This loose approach to the placeholder asset is fine because you will delete this asset later when replacing it with project content.

Testing Your File

Now it's time to test your file. You will do this many, many times throughout the book, so it's a good idea to start now. Select Control→Test Movie, and Flash will compile your authoring-only FLA file into a final file with the extension *.swf*, suitable for distribution. Because of the extension, these documents are usually called *swiff* files.

NOTE

Although not required every time you test, it helps to save your file at least once before testing. By default, SWF files are compiled to the same directory, and given the same name, as your FLA. Knowing where the FLA is saved will help you find the SWF later, if needed. Saving your work also increases the likelihood of recovering from a potential crash.

You should see everything you added to your file, including the guide graphic that you imported and the movie clip that you created. Close the window or tab with your SWF in it, and return to your FLA. If you're happy with your results, continue with the project. If not, open the *template_01.fla* file from the companion source files and continue on with that file. This template document is provided for your convenience.

Creating a Guide Layer

When creating a template, you will want to design it so that a minimum amount of fuss is required to create usable new files. The placeholder movie clip is useful because it has already been positioned and you can easily edit it to add new content. The guide graphic, however, should not appear in newly published files.

To solve this problem you need to convert the normal layer in which the guide graphic resides into a special layer type called (appropriately enough) a *guide layer*. Double-click the icon to the left of the guide layer name and, in the Layer Properties dialog, choose the Guide type. Click OK to close the dialog and test your movie again.

You should now see only the placeholder movie clip. The guide layer is visible in authoring mode to help you position assets and guide your design and development, but it is not included in the final published SWF file. Check your work against *template_02.fla*. Continue with the project if you're satisfied, or use the provided source file from this point forward.

Adding Utility Layers

The last things to add to your file are three empty layers that you'll make use of later in newly created files.

1. Select the content layer to make sure the new layers appear at the top of the layer stack.
2. Click the New layer button in the lower-left corner of the Timeline panel, and create three new layers.
3. Name the layers, from the top down, `actions`, `labels`, and `sounds`.
4. Lock the layers to prevent unwanted editing and compare your work against the source file, *template_03.fla*.

Saving Your File As a Template

All that remains now is saving your file as a template. Instead of using the standard Save option, use File→Save as Template. The Save as Template dialog will appear, and you can name the file, assign it to a category, and write a simple description. Follow the settings in Figure 1-17, naming the template `Content` and creating a category called `Learning Flash CS4`.

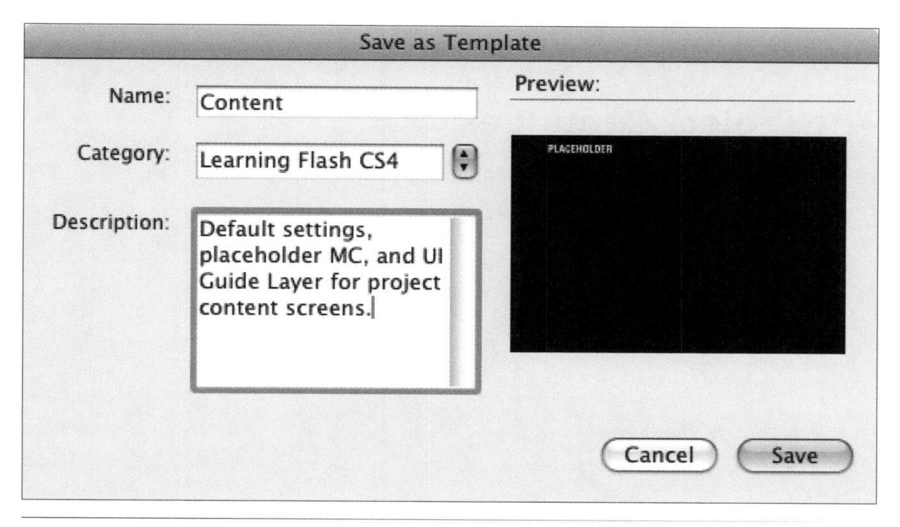

Figure 1-17. Saving a FLA as a template

From this point forward, you can create a new file from this template, instead of an empty new file, as shown in Figure 1-18. Try this now, and you'll find that the new file is correctly sized and contains the correct layers, the placeholder movie clip, and the guide layer in place.

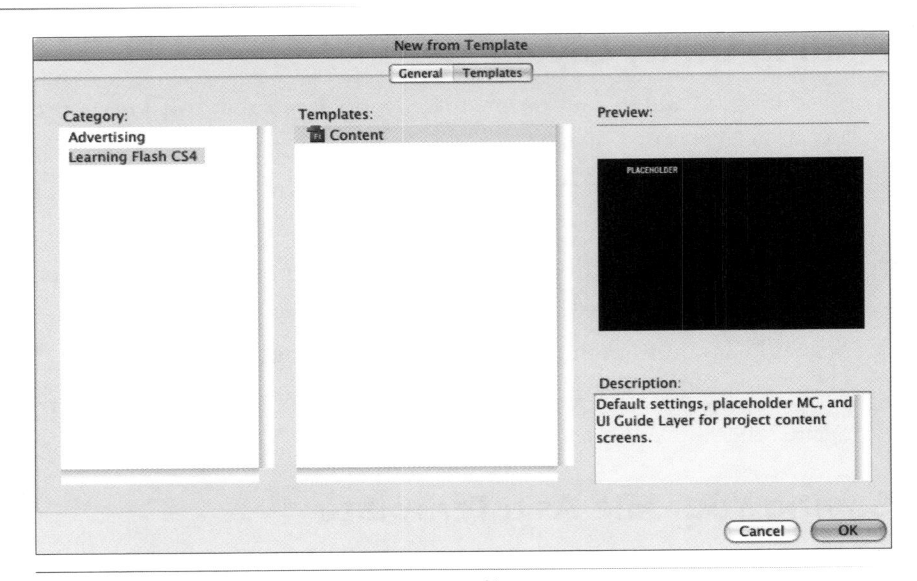

Figure 1-18. Using a template to create a new file

If you weren't entirely successful, don't worry about it. It's still early in the development of the project, and the upcoming chapters will explain the steps you took here in much greater detail. If you prefer, open the provided source file *template_03.fla* and save that as a new template for your ongoing efforts.

The Project Continues...

In the next chapter, you'll start creating content for your Portfolio.

CHAPTER 2

CREATING GRAPHICS

Introduction

One of the things about Flash that makes it a bit hard to define to new users is that it wears many hats. Although it is known for its ability to compile and load assets created in other applications, it is not just an authoring tool. That is, one of Flash's strengths is that it can be used to create assets all on its own. In fact, one of the distinctions between Flash and Flex (another Flash Platform technology) has always been that Flex authoring tools (like Flex Builder) don't have Flash's graphics drawing environment and don't allow you to draw custom assets while authoring.

Believe it or not, some designers use Flash as a primary illustration tool in specific cases, creating art in Flash and saving it to several image formats and even to video files. Part of Flash's advantage as an illustration tool is its unique drawing modes—editing configurations that allow object-based drawing, similar to Adobe Illustrator, and that contain drawing tools that treat vectors with the casual familiarity of pixels.

Flash's primary graphic building blocks, vectors, are points joined by lines used to describe shapes. You may have worked with vector assets if you've used EPS (Encapsulated PostScript) or PDF (Portable Document Format) files. The following pages give you an overview of how to approach drawing with Flash's tool set.

At the end of the chapter, you'll use a lot of what you've learned to work on the ongoing portfolio project. In the meantime, you may get the most from this material by experimenting often. As you read the descriptions in this chapter, keep Flash open and tinker as you go. You will occasionally find descriptive references about drawing, or even outright suggestions to try tasks as you read along. Unless specified otherwise, however, it is not necessary to save your work. You will not begin working on the cumulative project until the end of the chapter.

IN THIS CHAPTER

Drawing Modes

Flash has two distinct drawing modes: Merge Drawing mode and Object Drawing mode. *Merge Drawing mode* is the original drawing mode and allows shapes in the same layer to overwrite and join with each other. Introduced in Flash 8, *Object Drawing mode* prevents interaction between shapes. You can toggle between the two modes by clicking the Object Drawing button on or off. This button is in the context-sensitive tool options at the bottom of the Tools panel whenever a drawing tool is active (Figure 2-1). Examples of tools with access to this option include Pen, Line, all basic shapes (Rectangle, Oval, and so on), Pencil, and Brush.

Figure 2-1. Drawing mode toggle, showing Object Drawing mode on

Merge Drawing Mode

In Merge Drawing mode, overlapping fills or strokes of like color will be joined and differing colors can be used to destroy vector segments. When used properly, merge drawing is a powerful Flash feature. If you're careless, however, it can ruin your artwork.

NOTE

Unlike most object-based drawing applications, such as Adobe Illustrator, Flash treats fills (the inside of a square, for instance) and strokes (the lines surrounding a square) as easily separable, discrete objects. It's very easy, for example, to select a stroke and drag it to a new location, ending up with a strokeless fill and a fill-less stroke. It's also possible to use strokes to subdivide fills, to paint fills but not strokes, and to convert strokes to fills, among other things. It may take some getting used to, but, as you'll learn in this chapter, this unique treatment of strokes and fills can be a valuable illustration tool.

To understand Merge Drawing mode, it helps to understand how strokes and fills behave in Flash. You will learn more about this throughout the chapter, but for now you can see the difference between the two drawing modes with a few simple shapes. If you want to try things along with the text, start by selecting the Rectangle tool and turning Object Drawing mode off.

When you draw shapes in Merge Drawing mode, they are what you might call "unprotected" because there is nothing to prevent them from interacting with other shapes in the same layer. If you draw a rectangle on the stage using the Rectangle tool and then switch to the Selection tool, you will see that you can select each segment of the shape's stroke and fill with a separate single click. You'll learn more about selection techniques later, but for now just notice that the shape is vulnerable to your selecting and even editing specific areas without manipulating the underlining vector path directly.

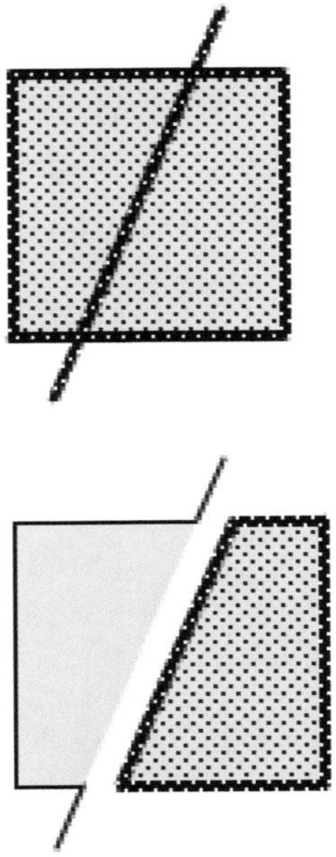

If you use the Line tool to draw a line all the way through the middle of the shape, the shapes interact and the line divides the rectangle in two. At first, this is not immediately apparent. As you would expect, you can drag the Selection tool over all the artwork to select both the rectangle and the line, as shown at the top of Figure 2-2.

Figure 2-2. Selecting a square and line created with Merge Drawing mode (top) and selecting and moving a portion of the merged shapes (bottom).

When you click on either side of the shape fill, however, only that segment of the fill is selected. Furthermore, when you double-click the fill, the fill and the surrounding strokes, *including the new dividing line*, are selected. You can even drag with the mouse or use the arrow keys on your keyboard to move the selected area away from the original shape, as shown in the bottom of Figure 2-2.

If you consider the combined shape before pulling it apart, the strokes and fill are divided at every intersection. The fill can be isolated into two parts, and, perhaps less obviously, the stroke can be isolated into nine parts. The four sides of the rectangle result in six parts (because two sides are divided by the line passing through them), but the dividing line is also subject to interaction from another shape. Accordingly, it is divided into three parts: above, inside, and below the rectangle.

Union, intersection, and deselecting

Within the same Timeline layer, strokes and fills of like color can join to form a union, and strokes and fills of different colors can intersect and eliminate the underlying artwork—but only when you deselect. You can always change your mind before a shape change takes place, as long as you don't deselect.

For example, if you overlap a circle and a rectangle shape of the *same* color, both without strokes, the two shapes would become one after deselecting. If you moved the topmost shape away from the bottom shape before deselecting, no change would occur.

Alternately, if you centered a small circle on top of a larger circle of a *different* color (both circles without strokes) and deselected, you could delete the inner circle and end up with a donut. Again, if you aborted the process prior to deselecting, no change would occur.

The same is true of strokes. Strokes of the same color in the same layer can merge into connected strokes if they are deselected. Strokes of different color, or strokes intersecting fills, can divide the underlying shape.

Object Drawing Mode

When Object Drawing mode is enabled, you are protecting shapes from interacting. This behavior is similar to drawing two discrete objects in Adobe Illustrator. Without intervention from you, those two objects won't typically join or intersect each other.

When you draw a shape with Object Drawing mode on, Flash encapsulates the shape in a wrapper of sorts—hermetically sealed for your protection. You can still edit the shape, but the object wrapper prevents shapes from interacting the way they would if Merge Drawing mode were enabled.

Remember the Merge Drawing mode example of the line dividing a rectangle? This time, turn Object Drawing mode on, create a circle with the Oval

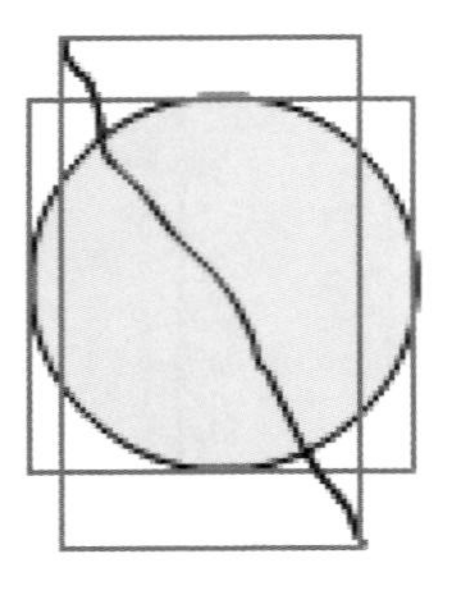

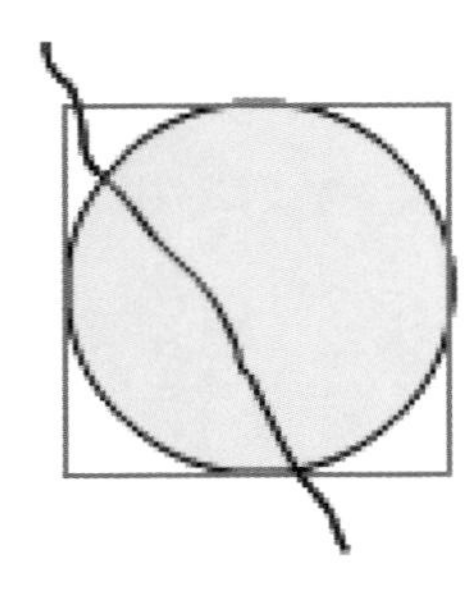

Figure 2-3. Selecting a circle and line created with Object Drawing mode (top) and selecting and moving only the circle (bottom); shapes remain discrete and can be moved independently

tool and use the Pencil tool to draw a line over it. Because the object wrapper protects the two shapes from interacting, clicking on one half of the circle will select the entire circle, rather than a portion of it. Moving the selected shape to the right (as in the Merge Drawing mode example), both shapes remain intact and the line's position is unchanged. This contrasting behavior can be seen in Figure 2-3.

NOTE

Both the Rectangle and Oval tools have companion tools called Rectangle Primitive and Oval Primitive, respectively, which have special powers. The former can control the radius of each of its four corners independently (allowing you to create tab-shaped buttons with rounded corners, for example), and the latter can control its starting and ending angle, as well as inner radius (allowing you to create pie pieces, arcs, and donut shapes). Check the companion website for examples.

Grouping and Breaking Apart

The Drawing Mode toggle button in the Tools panel doesn't change existing assets from Merge Drawing mode to Object Drawing mode, or the other way around. Changing the setting affects newly drawn assets only. Using the Modify menu's Group and Break Apart options, however, you can achieve an effect similar to changing an asset's drawing mode.

When you have an unprotected shape that was created using Merge Drawing mode, you can select and *group* the asset (Modify→Group) to protect it. Although it makes more sense to think of grouping as collecting more than one object into a single unit, you can also group a single shape to prevent it from interacting with other shapes.

When you have a Drawing Object (a shape created while Object Drawing mode was enabled), you can convert it to an unprotected shape by breaking it apart (Modify→Break Apart). This is handy when you want to join or cookie-cut two shapes that were previously created as Drawing Objects.

NOTE

If you break a Drawing Object into a shape, you can't later return it to a Drawing Object (short of using Edit→Undo). However, you can group the shape to protect it from further edits.

You will see this process appear more than once throughout this book, since you can use it to degrade a complex object into a less complex object. For example, you can break a text field into individual editable letters, and then break those letters again into vector shapes.

Alternatively, you can edit the contents of a group or Drawing Object without breaking it apart. To do this, double-click the object. The application will enter an editing mode without radically changing the interface. This allows you to edit groups and Drawing Objects in the context of any surrounding art. While in editing mode, double-clicking any unoccupied area of the Stage or Pasteboard will return you to the Stage.

The Edit Bar just above the Stage shows a "breadcrumb trail" of buttons leading up to your current location in a chain of possible nested objects you are editing. For example, if you edit a Drawing Object placed on the Stage, the Edit Bar shows a button called *Scene 1* (the default name for the main Timeline), followed by the text *Drawing Object* to mark your current location. Clicking the *Scene 1* button will exit editing mode, and you can again access the Stage.

If you group two Drawing Objects together and want to edit one, you must first double-click the group to edit its contents and then double-click the Drawing Object to edit the shape therein. The Edit Bar ultimately includes buttons that say *Scene 1* and *Group*, followed by the text *Drawing Object*—again showing where you are. In this case, you can click the Edit Bar's *Group* button to back out one level in editing mode, or the *Scene 1* button to again access the Stage.

Drawing au Naturel

Some users who are new to Flash go through a short period of adjustment when it comes to the drawing tools. In Flash, the traditional object-based model is not the primary drawing technique. Unlike Adobe Illustrator and others, Flash allows you to work with vectors in a very fluid, natural way. Instead of manipulating curves with vertices and control handles (common to object-based graphics), you can just grab hold of a line and drag it (Figure 2-4).

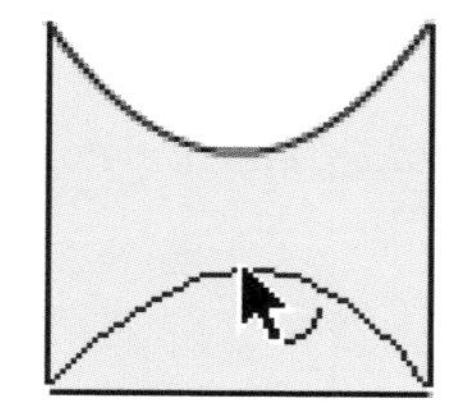

Figure 2-4. Manipulating strokes by dragging

Figure 2-4 shows just such a manipulation in progress. To accomplish this, start by placing your cursor over the shape you wish to manipulate. To help identify which object you might be clicking with the Selection tool, a small icon, often called a *cursor badge*, appears in the lower-right corner of the cursor. A curve appears when you roll over a line segment, and a corner appears when you roll over a corner point. In Figure 2-4, you can see a curve to the lower right of the cursor, reaffirming that the cursor is over a line, not a corner. If you click and immediately drag the mouse, you can manipulate the selected stroke or fill. By contrast, if you click and let go, you select the stroke or fill, and subsequent drags will move the selected element to a new location.

Further in the "natural drawing" vein, you can create vectors with a paintbrush or remove them with an eraser. Each time you paint with a brush, it may feel like you're painting with pixels, but the resulting shape will be comprised of vectors.

Drawing with Bézier Curves

Grabbing and dragging lines may eventually become your preferred method of finessing a curve. However, it is also possible to manipulate curves with the same precision afforded in other object-based drawing applications. Flash offers the same Pen and Subselection tools as Adobe Illustrator and allows you to create, add, subtract, and transform vertices.

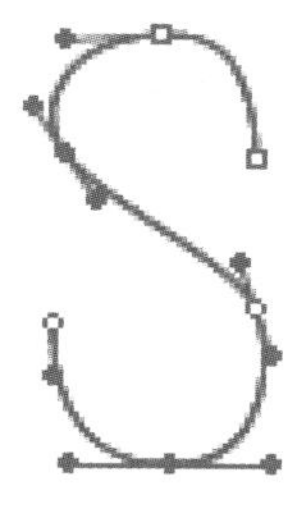

Figure 2-5. Manipulating strokes by dragging

To create the s-shaped Bézier curve shown in Figure 2-5, for example, you use the Pen tool. Starting at the upper-right corner of the S, you can drag a bit

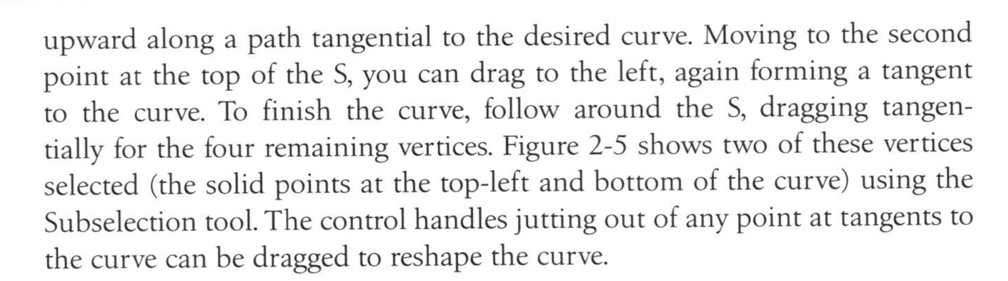

upward along a path tangential to the desired curve. Moving to the second point at the top of the S, you can drag to the left, again forming a tangent to the curve. To finish the curve, follow around the S, dragging tangentially for the four remaining vertices. Figure 2-5 shows two of these vertices selected (the solid points at the top-left and bottom of the curve) using the Subselection tool. The control handles jutting out of any point at tangents to the curve can be dragged to reshape the curve.

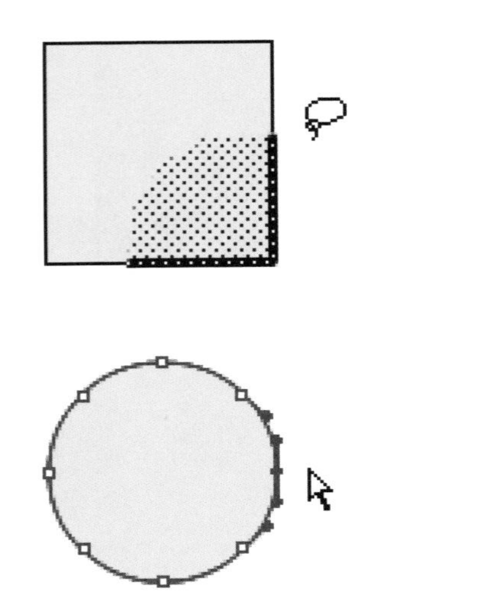

Figure 2-6. Selecting using the Lasso (top) and Subselection tools

Selecting

As with drawing, selecting objects in Flash can be approached from both the "natural" and "precise" viewpoints. You'll learn specific selection subtleties, as they apply to strokes and fills, in a moment. However, it's worth mentioning in the context of drawing styles that you can select individual vertices if required or simply drag the Lasso tool over a shape, as shown in Figure 2-6. Despite the fact that you are dragging your Lasso tool over a vector shape, you can select a subset of that shape with the ease of surrounding pixels in a bitmap.

Using Fills and Strokes

As previously discussed, Flash treats the fill and stroke of a single shape as discrete elements almost unto themselves. Bear this in mind when you select them with the Selection tool, and you'll be less likely to end up with unexpected results.

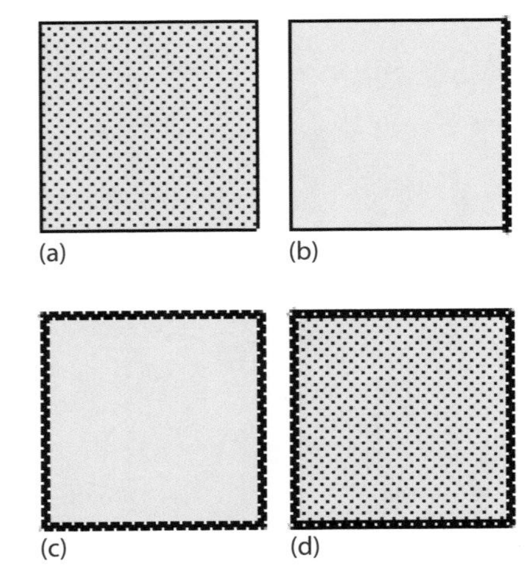

Figure 2-7. Selecting fill only (a), stroke segment only (b), complete stroke only (c), and fill and stroke (d)

Selecting Separately

You will quickly realize that some attention is required when selecting shapes if you don't intend to leave fragments of your object behind. This is because you can select objects in Flash with a great deal of granularity. For example, it is not only possible, but even simple, to select a shape's fill and stroke separately. It is even possible to select a fragment of a stroke very easily, even accidentally.

Figure 2-7 shows the visual feedback Flash provides upon various selection actions. Figure 2-7(a) is the result of a single click in the interior of the shape. Look carefully, and you'll see that only the fill is selected. If you were to try to move this shape, you would leave behind, and possibly overwrite, its stroke.

Figure 2-7(b) shows the result of a single click on the shape's stroke—the right side in this case. A single click on a stroke selects only the clicked-upon fragment between two corner or intersection points. Only for ellipses or circles will the entire stroke be selected, because these contain no corners.

Figure 2-7(c) shows the result of double-clicking the stroke. A double-click selects all contiguous parts of the stroke. In the example, all four sides of the square are selected, despite the intervening corner points.

Finally, Figure 2-7(d) shows the result of double-clicking the fill of the shape. This has the effect of selecting not only the entire contiguous fill, but also the entire surrounding stroke. If you want to be sure to grab all parts of the shape, use this method. Alternately, you can drag over the object with the Selection or Lasso tools to achieve the same result.

Stroke Properties

The Properties panel (Window→Properties) is among the most often used Flash panels. It is context-sensitive and provides access to most properties that are editable within the Flash interface. Fill and stroke properties (Figure 2-8) can be found in the Fill and Stroke section of the panel when a shape is selected and are itemized in the following list:

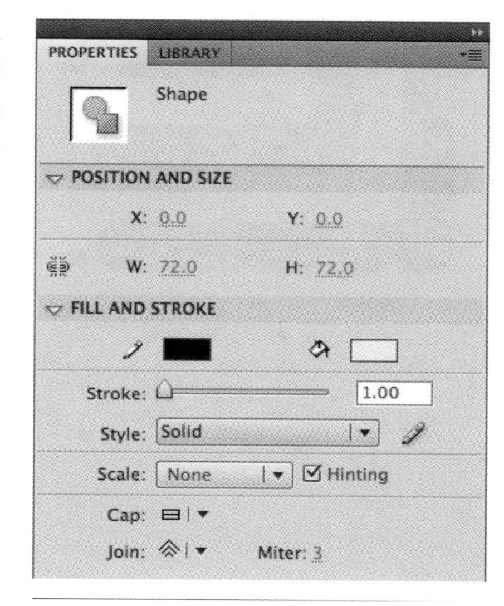

Figure 2-8. Fill and stroke properties

Stroke

Stroke is the weight or thickness of the stroke in pixels.

Style

The menu for the *Style* option offers seven preset line styles: Hairline, Solid, Dashed, Dotted, Ragged, Stippled, and Hatched. Clicking the pencil to the right of the menu opens an editing panel that allows you to configure advanced properties of these line styles, such as the space between dashes or dots.

Scale

When a movie clip is scaled, strokes scale accordingly by default. This can distort the appearance of artwork because the stroke can thin or thicken as a result of stroke scaling. A new stroke feature, *Scale*, lets you dictate how strokes are scaled when an object is resized. This setting affects the thinning and thickening of the stroke only when the object as a whole is enlarged or reduced. You can specify None (no change to the stroke thickness), Horizontal or Vertical (adjusting strokes only when the object is scaled in the specified direction), or Normal (the default behavior, in which strokes are always scaled anytime the object as a whole is scaled).

For example, say you wanted to create a custom vertical scroll bar. If the overall asset is scaled in both directions, you may want the strokes in the scroll bar to scale because the entire asset is being reduced or enlarged. However, you may not want the scroll bar strokes to thicken if the scroll bar is only scaled vertically. This would allow you to elongate a scrolling text field without thickening the scroll bar strokes. To accomplish this, you would set the Scale property to Horizontal so the stroke is only scaled when enlarging the asset horizontally or both horizontally and vertically.

NOTE

The book's companion website, http://www.LearningFlashCS4.com, has examples of line scaling in action.

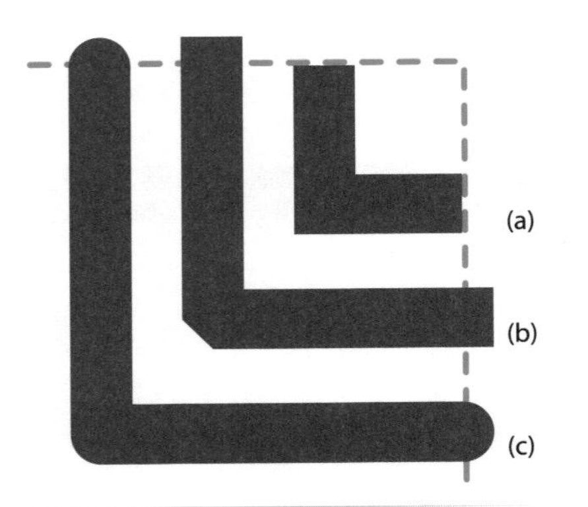

Figure 2-9. Stroke caps and joins; (a) miter join and no caps, (b) bevel join and square caps, and (c) round join and round caps.

Hinting

When enabled, *Hinting* attempts to nudge stroke positions to whole pixels, as opposed to subpixel positioning (decimal values for x and y properties). This prevents lines from thinning or thickening due to antialiasing side effects.

Cap

Cap affects the shape of line end caps. This property can be set to Round (default), Square, or None. Round and Square both add a cap to the end of the line, while None adds nothing, matching the exact length of the line. The three corner types are shown in Figure 2-9.

Join

Join affects the corners of line joints. This property can be set to Miter, Round, or Bevel. Examples of these joins are shown in Figure 2-9.

Miter

The *Miter* property affects the sharpness of corner joins when the Join property is set to Miter.

Mining Properties with Tools

Although the Properties panel is typically the go-to way of assigning most properties in the Flash interface, three tools are available to make this assignment process quick and easy for strokes and fills. The Eyedropper tool in Flash is unique among similar tools in other applications. It not only retrieves color values, but also retrieves all stroke or fill properties and automatically switches to a tool useful for assigning these properties to another stroke or fill.

After using the Eyedropper, Flash automatically switches to either the Paint Bucket or Ink Well tool, depending on whether you clicked on a fill or stroke, respectively. Both tools work in a similar fashion, applying properties to the object on which you click, but the Paint Bucket affects only fills, and the Ink Well affects only strokes.

When hovering over an object, the Eyedropper shows a cursor badge icon in the lower-right corner of the cursor. When over a line, a small line appears; when over a fill, a small paintbrush appears. Although the Eyedropper prepares only properties relevant to the element on which you clicked (prepping stroke color only when clicking on a stroke, for example), it is possible to query color from a stroke and apply it to a fill, and vice versa. If you hold down the Shift key when using the Eyedropper, Flash will populate both stroke and fill colors for your subsequent use of the Paint Bucket or Ink Well tools. For instance, this allows you to pull a color from a stroke and then apply it to a fill without an interim trip to the color tools.

Working with Color

There are several ways to create and apply colors in Flash. You can use existing color libraries, create your own custom colors (either before they are needed or on-the-fly), save color libraries for later use, and even retrieve color families from other users online.

Pop-Up Palette

Perhaps the quickest way to access or create a color is by using the pop-up color palette. This palette is available in the Tools, Properties, and Color panels, and anywhere a color chip is available. Figure 2-10 shows the palette accessed from the Fill color chip in the Tools panel.

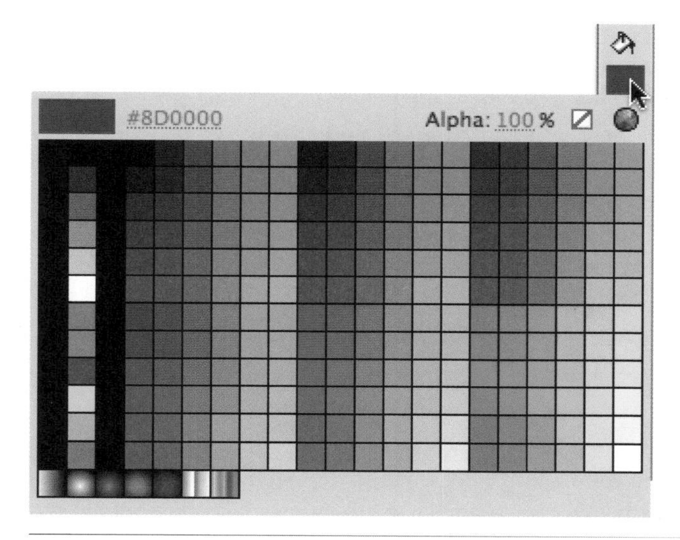

Figure 2-10. Pop-up color palette, accessed from the Tools panel

Using the panel is quite simple. You can select from any precreated solid colors (found in the middle of the palette) or gradients (found at the bottom of the palette). You'll learn how to add swatches in the next section.

You can also create custom colors on the fly using the tools along the top of the palette. The first item is a preview of the color you're selecting. The second item serves two functions. It is a preview of the hexadecimal value of the color you're selecting and allows you to change the value on the fly.

The next item in the row is the alpha value, or percent of transparency in any color. A maximum value of 100 is opaque, while a minimum value of 0 is transparent. The red line button indicates no value, allowing you to remove color from an object. Finally, the color wheel opens the operating system color picker as an alternate color-selection tool.

NOTE

Hexadecimal numbers are used to specify color values in HTML and as additional color systems in applications. They consist of three character pairs that represent the red, green, and blue values of a color. Each character can be 0–9 or A–F, 0 being the lowest, and F being the highest. Together, each pair represents a number from 0 to 255. So, #FF0000 is all red, no green, and no blue. Further, #FFFFFF is the maximum of all colors, or white, and #000000 is no color, or black.

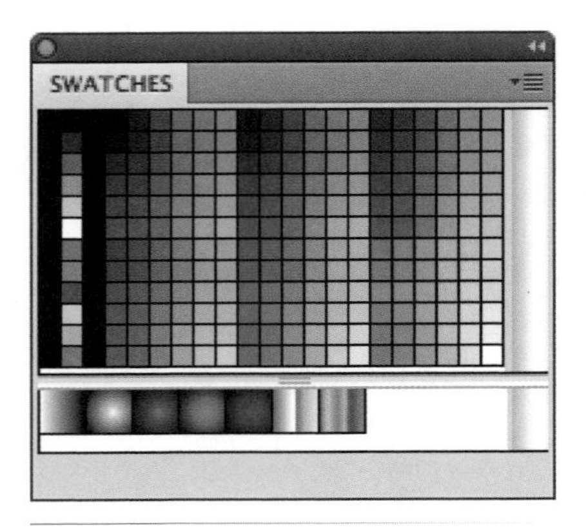

Figure 2-11. Swatches panel

Swatches Panel

The Swatches panel is the repository of color swatches that appear in the pop-up palette. You can leave the Swatches panel open and select from it if you prefer not to use the pop-up palette, and anytime you create a custom color (solid or gradient), you can add it to the swatches in this panel. With the color active, the cursor will change to a Paint Bucket when rolling over an empty area of the appropriate section (solids or gradients) of this panel. Clicking will add the color to the end of that area of the palette. Thereafter, the color will become available to any interface element that uses the pop-up palette. Using the panel-specific menu (the button for which can be seen in the upper-right corner of the panel in Figure 2-11), you can also load color swatches from, or save to, an external source.

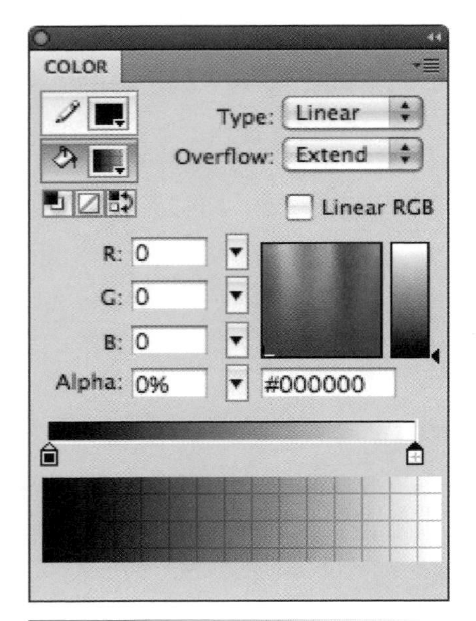

Figure 2-12. Color panel

Color Panel

The Color panel is where you are most likely to create or edit custom colors. As in the Tools panel, here you can choose to edit stroke or fill colors, but the feature set of the Color panel is more complete. In addition to offering input for hexadecimal color values and percentage alpha values, you can also edit standard 0–255 range RGB color values—both by pop-up slider and text input. You can also use a built-in color picker and lighten or darken any color. While editing, the color appears in a large horizontal stripe at the very bottom of the panel for easy preview. In Figure 2-12, for example, the panel is configured to edit a linear gradient.

Creating gradients

To create a gradient in the Color panel, change the color's *Type* setting from Solid to Linear or Radial. Doing so adds three new interface elements to the panel.

The primary new interface element is the gradient definition bar that appears between the color picker and large preview stripe at the bottom of the panel. Figure 2-12 shows a gradient featuring two colors (evidenced by the color chips immediately beneath the bar), but this bar allows you to manipulate up to 15 colors in any gradient. To add a color, click the definition bar, and an editing chip will appear.

Selecting any chip allows you to edit that color portion of the gradient using the color picker and numeric input fields in the panel. Double-clicking a chip displays the pop-up palette for quick selection of preexisting colors. Any changes are reflected in the preview stripe at the bottom of the panel.

Figure 2-12 shows the editing of the rightmost color (note the darker triangle at the top of the chip pointing to the color being edited). Because this color has a low alpha value (0, in this case) you can see a grid through the color. This is a helpful indicator to show you how much transparency is applied to the color.

The next interface element the gradient color type adds to the panel is the Overflow menu near the panel's upper-right corner. This menu has three options: Extend, Reflect, and Repeat. *Extend* continues the first and last colors of the gradient infinitely. *Reflect* continues to flip the gradient end to end, creating a smooth transition as though seen in a mirror. A gradient from black to white, for example, would read black, white, white, black, and so on. *Repeat* cycles the gradient without reflecting it each time, creating a hard edge. The same example gradient would read black, white, black, white, and so on. These options come into play when the scale of your gradient is smaller than the fill it occupies. You'll learn more about scaling your gradient with the Gradient Transform tool later on.

The final gradient-specific addition to the panel is the Linear RGB option. This creates linear or radial gradients that are compatible with the SVG (Scalable Vector Graphics) standard.

Kuler Panel

Shown in Figure 2-13, the Kuler panel can be found in the Windows→Extensions submenu and offers access to an online repository of user-defined color families. This panel requires an Internet connection and connects to Adobe's Kuler service. You can browse, search, and submit families of up to five colors for use in your work. Unlike the Color panel, this extension allows you to use several color theories when creating families in which colors affect one another in a variety of ways. When you've found a family you like, you can automatically add it to your Swatches panel. For more information, visit the Kuler site, *http://kuler.adobe.com/*.

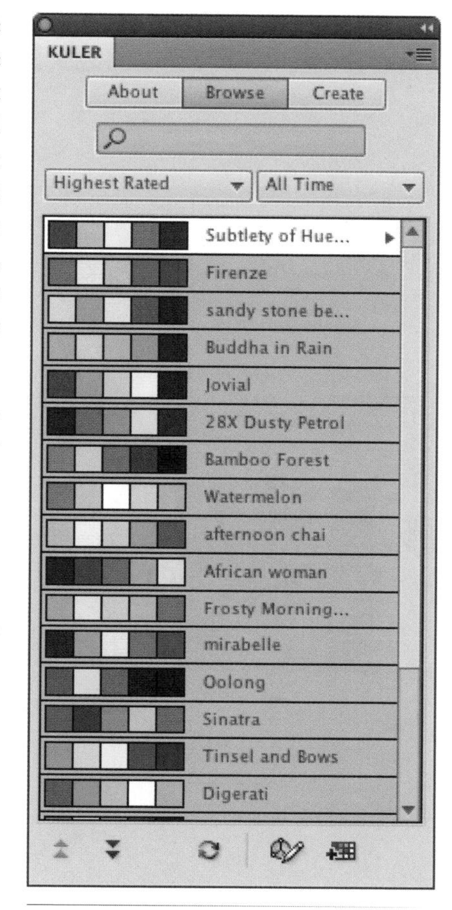

Figure 2-13. Kuler panel

Using Context-Sensitive Tool Options

Once you've had some time to play with the Flash Tools panel, you'll see that many tools have context-sensitive options that appear at the bottom of the panel. Most are self-explanatory, but some are unique to Flash's natural drawing techniques.

Paint Modes

When working with a brush, you can choose a paint mode that will affect what is colored by that brush. You can access these modes by the first context-sensitive option, shown in Figure 2-14.

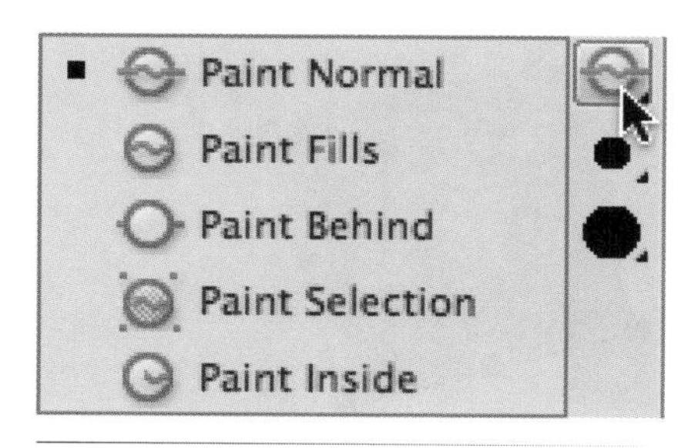

Figure 2-14. Brush paint modes

The modes are fairly self-explanatory and apply primarily to unprotected shapes (created in Merge Drawing mode or broken apart). Some work with Drawing Objects and some have a few subtleties worth mentioning. Paint Normal offers no special effect and allows you to return to this setting after using another option. Paint Fills paints only fills, leaving strokes untouched.

Paint Behind affects only areas around the shape, leaving fill and strokes intact. Paint Selection affects only selected fills and strokes. Paint Inside paints only within a shape, as long as you originate your painting within the same shape.

Erase Modes

Erase modes (available as an option of the Eraser tool) control how the eraser affects shape strokes and fills, and are similar to Paint modes. Erase Normal, Erase Fills, and Erase Inside behave the same way as their painting counterparts. Erase Lines affects only strokes, and Erase Selected Fills erases only selected fills, but not selected strokes (differing slightly from Paint Selection). A related context-sensitive tool is also available—the Faucet tool automatically erases any contiguous stroke or fill.

Transforming Assets

Although there are many meanings for the word *transform*, when referring to the Flash user interface, transforming typically means scaling, rotating, skewing, or moving (sometimes referred to as *translating*) an object. There are three essential ways to transform an asset.

Free Transform Tool

The Free Transform tool applies a series of handles and a transformation point on any selected object. Depending on which handle you grab with your mouse, you can alter the appearance of the object in different ways. To help you with this process, the cursor changes to a shape that is specific to each available task.

Figure 2-15 shows all the possible cursors when using the Free Transform tool. Moving clockwise from the lower left, the cursors represent move, skew vertically, skew horizontally, rotate, scale horizontally, scale vertically and horizontally, and scale vertically. Finally, the cursor in the center of the shape is for moving the point around which transformations occur.

Figure 2-16 shows the effect of moving the transformation point. The top graphic shows the process of moving the transformation point from the default center of the object to the upper-left corner. The bottom graphic shows that, when rotating, the object is rotated around the new transformation point rather than the center of the object as before.

Figure 2-15. Free Transform tool cursor feedback

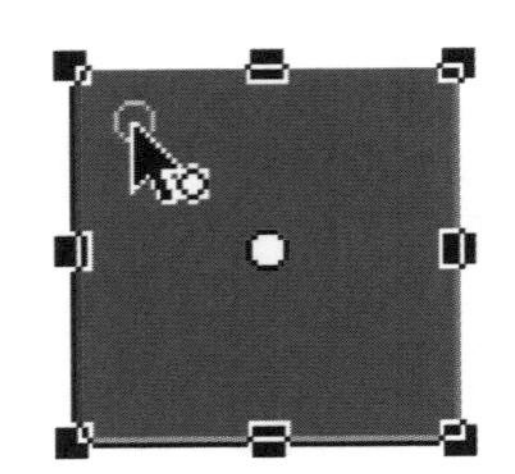

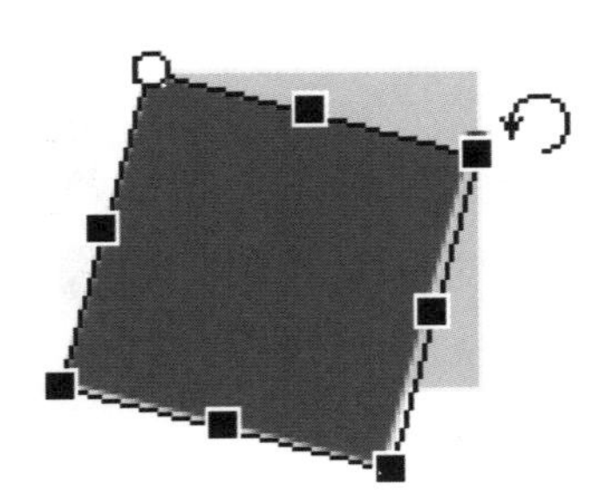

Figure 2-16. Changing the transformation point

Transform Panel

The Transform panel (Figure 2-17) offers options to scale horizontally and vertically (independently or proportionally), rotate, and skew an object. If the object is an instance of a symbol, such as a movie clip, you can also rotate on any 3D axis and move the 3D center point of the instance along any 3D axis, both of which you'll learn more about in Chapter 8.

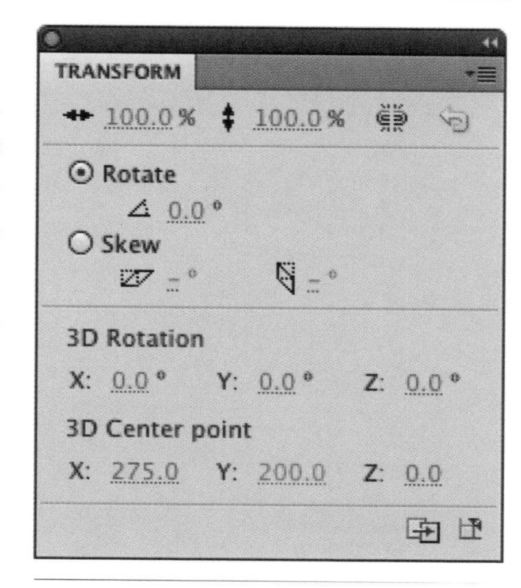

Figure 2-17. Transform panel

There are three compelling reasons to use the Transform panel. First, numeric entry for property values allows for greater precision. Reasons two and three are the unsung heroes in the lower-right corner of the panel. The first of these two buttons duplicates your most recently completed selection and transformation, akin to step and repeat in other applications. For example, you could create a series of copies of a movie clip that are progressively larger in size. The second button removes all transformations and returns the instance to its original state.

Gradient Transform Tool

The final transform tool, called the *Gradient Transform* tool, is used to edit gradients. It can be found in the same transform tool menu as the Free Transform tool. To more easily understand how this tool works, create a rectangle shape with a radial gradient fill. Figure 2-18 shows a rectangle with a red stroke and red-to-white radial gradient so that you can easily differentiate it from the black interface elements of the Gradient Transform tool.

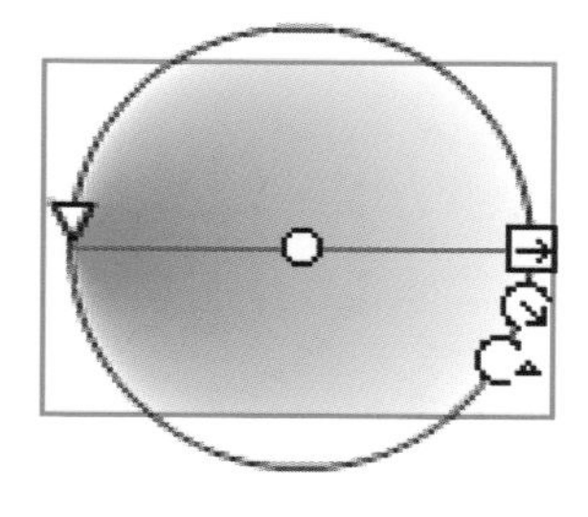

Figure 2-18. Transforming a gradient

With the Gradient Transform tool active, click the radial gradient fill you created. A circle with a line across its diameter and a handful of icons will appear, as shown in Figure 2-18.

If you can't see the tool's interface, it probably means that the scale of the gradient is too big to fit in your view. Look for this symptom when you see what looks like a solid color instead of your expected gradient (if you zoom in closely to a gradient, it will look like a single color). Use the Zoom tool (or Zoom menu in the Edit Bar) to zoom out until you see the Gradient Transform tool interface and scale the gradient down. In this case, use the middle of the three icons on the circumference of the circle. This scales the gradient proportionally.

Once you have the interface in view, look at the three grouped icons at the lower right. The first icon, an arrow inside a square, scales the gradient in only one direction, always along the diameter line. Try this first so your radial gradient is no longer a perfect circle.

You can use the diameter scaling option in conjunction with the bottom of the three icons (a rotation arrow), which rotates the gradient. Using these two tools, you can scale a gradient along any angle. As mentioned previously, you can scale the gradient proportionally by using the middle of the three icons, an arrow inside a circle.

Finally, the center of the interface contains two icons. Dragging the circle icon drags the location of the gradient within the fill. Dragging the triangle above the circle skews the gradient in either direction along the diameter line. Figure 2-18 shows the arrow all the way to the left of the fill, skewing the gradient to the left. This can create an illusion of movement. As with scaling, rotating the gradient first allows you to skew the gradient in any direction.

Creating Static Text

You'll learn a lot about text in Chapter 11, particularly when it comes to ActionScript control and user input. However, it's a good idea to get used to working with text early, and a good way to start is to think about text as a graphics object. That may sound a bit odd, but you don't have to give up the ability to edit the text during authoring.

To put this into perspective, think about the common problem of custom typefaces (also called *fonts*). Text that uses a custom font usually looks fine on the machine that created it because the font is installed on that machine. However, when that text is viewed on machines without the required font, or even on the author's machine if the font is disabled, text using the missing font will not appear.

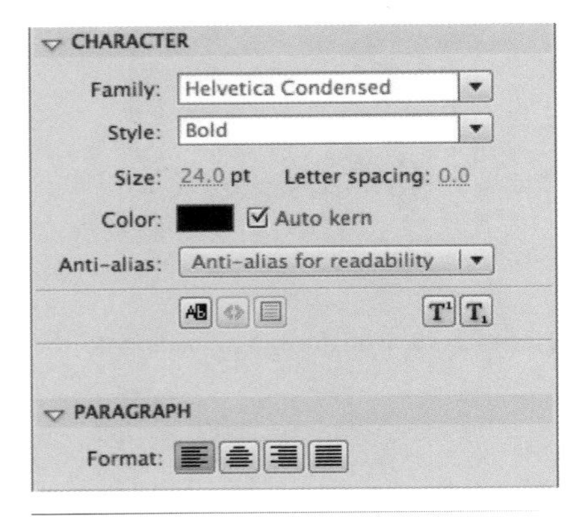

Figure 2-19. Detail of Properties panel showing select Static text properties

Flash provides a couple of ways around this, some of which you'll learn in Chapter 11. The simplest solution applies when the text is *static*, or does not have to change at runtime. In this case, Flash provides a text type, appropriately called Static Text, which is editable at the time of authoring but treated as a graphic at runtime.

To create a Static text field, choose the Text tool, click on the Stage, and type something. If you don't type something to start with, the text field will disappear (because Static text fields are somewhat akin to graphics, this is the equivalent of clicking on the Stage with the Rectangle tool, but not drawing anything). With the text field still active, make sure Static Text is selected in the menu at the top of the Properties panel, as shown in Figure 2-19.

The Properties panel contains several properties that you can adjust for Static Text fields. You'll focus on a select group of these properties for now, and learn the remainder in Chapter 11. In the Character properties group, you can change these properties:

Family

The *Family* property specifies the font you want to use. Because you are working with static text, any font that will display in Flash is appropriate.

Style

The *Style* menu includes all font styles (bold, italic, and so on) available for the typeface you chose.

Size

Size is the type size, in points.

Letter spacing

The *Letter Spacing* setting controls the space between *all* the letters of the text field. This feature is sometimes called *tracking* in other applications. This is in contrast to *kerning*, which is the space between *two* letters.

Color

Color is the color of the text, and provides access to the pop-up color palette.

Selectable

The final row of settings in the Character section of the Properties panel is a series of small buttons. The leftmost button controls whether the text is selectable at runtime. This is handy in some cases, such as for copy and paste support. However, if this is not a feature you want to include, it can be annoying because text-selection highlighting and/or the standard I-beam text cursor may appear and distract your users from design or animation. Typically, you should disable this setting.

For paragraphs you can alter:

Format

The *Format* property controls the alignment of the text within the field, and can be left, center, right, or full justified.

NOTE

You can use the Properties panel to apply individual settings to subselections of a text field. For example, different words or paragraphs within the same text field can have different settings.

Using the Spray Brush

New to Flash CS4 Professional is the *Spray Brush*, previously available to Adobe Illustrator users. Found in the Brush submenu of the Tools panel, the Spray Brush automatically adds shapes or symbols to the Stage as long as the mouse button is pressed. It's like an advanced can of spray paint, spraying graphics instead of plain color. The results of each spray are conveniently collected in a group for easy management.

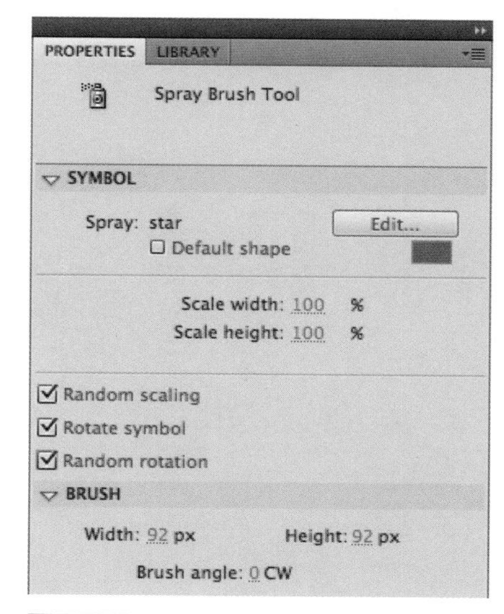

Figure 2-20. Spray Brush tool properties

Figure 2-20 shows a detail of the Properties panel and an example configuration of the Spray Brush. The first setting of the brush dictates whether the brush will use a default shape (a small square) of your color choice, or a symbol from the Library. More on that in a moment.

The next section controls the horizontal and vertical scale of the shape or symbol to spray. The final three settings of the panel's Symbol section control the orientation and randomness applied to the symbol instance as you spray them. *Random scaling*, as the name implies, scales each symbol instance randomly. *Rotate symbol* rotates the symbol as it leaves the virtual spray nozzle, orienting the instance based on the movement of the mouse. *Rotate randomly* rotates each instance randomly, regardless of mouse movement.

Finally, you can use the Brush section of the Properties panel to change the width, height, and angle of the Spray Brush.

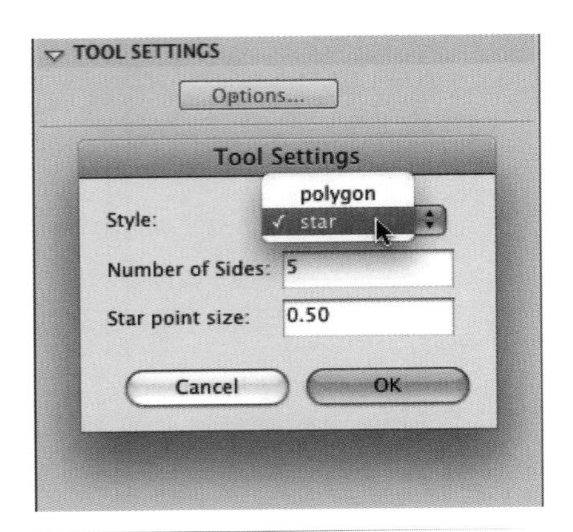

Figure 2-21. The PolyStar tool

Creating a Symbol

Now consider the idea that you can spray a symbol of your choosing instead of the default small square. You learned a bit about symbols in Chapter 1 and you will learn much more about them in the next chapter. Take a quick moment now, however, to review how to create a movie clip symbol. For this exercise, you'll use a star.

In the Primitives menu of the Tools panel, where the tried-and-true Rectangle and Oval tools reside, you'll also see the PolyStar tool. This tool can create multisided shapes (polygons) and multipointed shapes (stars, starbursts, and so on). To specify the shape you want to use, deselect any active selection by switching to the Selection tool and clicking on any unoccupied location of the Stage or Pasteboard. With nothing selected, choose the PolyStar tool. A Tool Settings section with an Options button appears in the Properties panel.

Clicking this button brings up a straightforward dialog, shown in Figure 2-21. Using this dialog, you can switch between polygon and star, and dictate the number of sides or size of the star points accordingly.

Figure 2-22. Spray Brush results

After you've configured the PolyStar tool, drag your mouse on the stage to draw a small star, approximately 20 pixels × 20 pixels. Remember, this is going to be sprayed many times from a virtual spray can, so it needs to be small enough to look good in numbers, but big enough to see.

Next, switch to the Selection tool and drag over the entire star to select everything. As you did in Chapter 1, use the Modify→Convert to Symbol menu command. When the resulting dialog appears, choose Movie Clip for the Type property, a center registration point, and name the symbol `Star`.

Now that you have a symbol in your Library, you can use it instead of the Spray Brush's default shape. Do so by clicking the Edit button at the top of the Symbol section of the Properties panel while the Spray Brush Tool is active. Figure 2-22 shows the result of spraying with the setting shown in Figure 2-21.

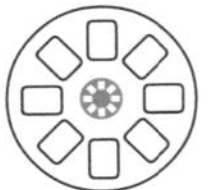

Project Progress

In Chapter 1 you created a template that you will use throughout the asset creation phases, described in this and later chapters. In the following step-by-step instructions, you will focus on creating the assets for the home page, as well as the user interface widget used to control sound playback. If you haven't already, download the source files from the companion website discussed in the Preface.

As discussed briefly in Chapter 1, it is sometimes useful—especially within teams of multiple Flash designers and developers—to create assets in a dedicated FLA first and then later move them to the final project file. This provides a convenient way for many people to contribute simultaneously to a large job without having more than one person share a single file. In addition, this approach will reinforce skills you'll learn in future chapters, such as working with multiple libraries and asset positioning.

Figure 2-23 shows the finished version of the file you're creating. Keep in mind that the interface visible in this figure, including the controls, logo, frame, and paper background, are all part of the guide layer you created in Chapter 1. The guide image has been muted here to remind you that you'll be focusing on the content within the frame.

NOTE

Throughout this project, numerical values are provided for entry into the Properties panel. In all cases, feel free to adjust these values to suit your preferences or better match a provided image. Many small factors can contribute to discrepancies in these numbers. For example, your choice of typeface may alter the size of text fields. The exact numbers are not always important. If adjustments don't disrupt your progress, focus on what looks good and works well for you.

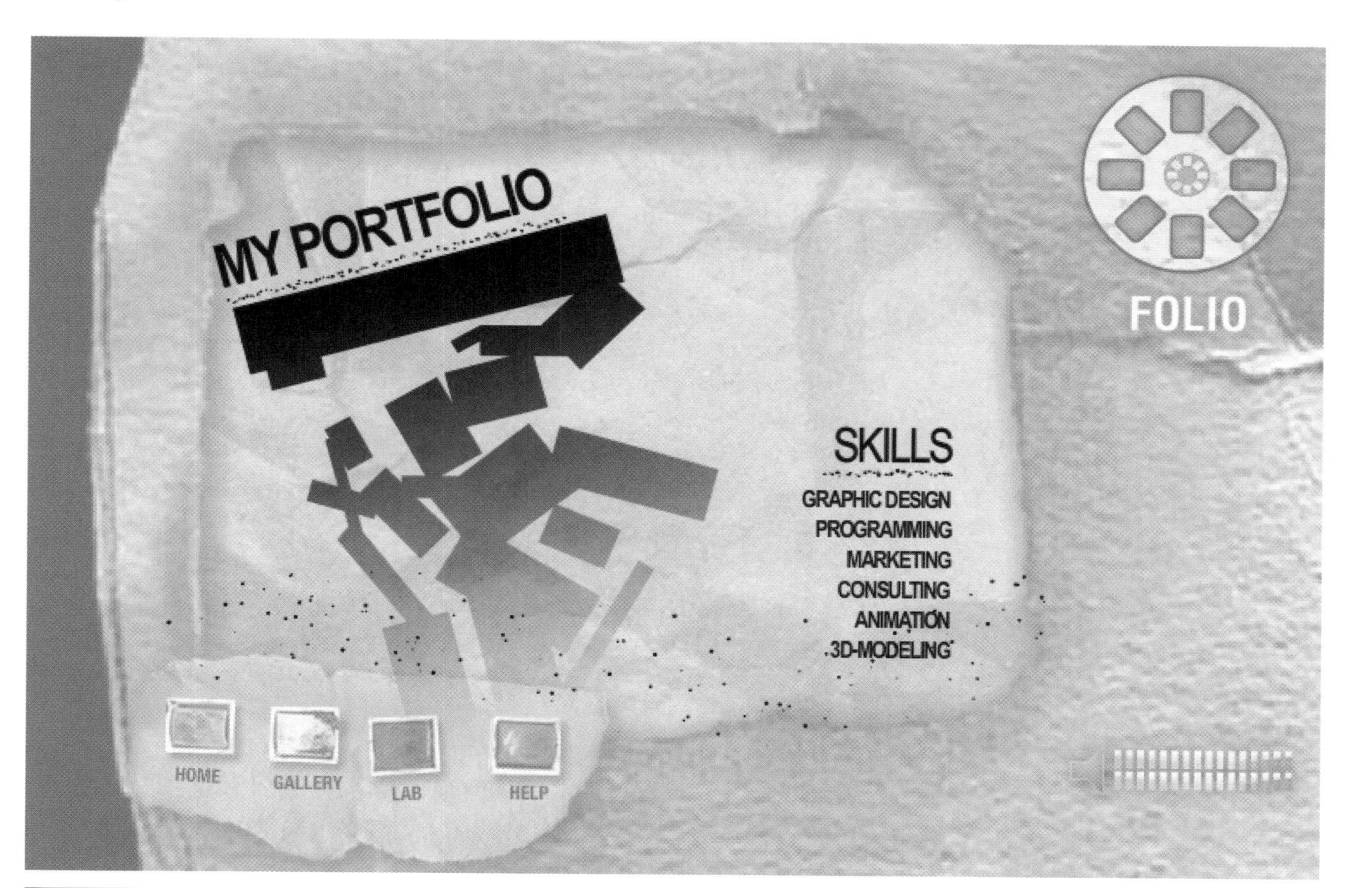

***Figure 2-23.** The home page you will create in this segment of Project Progress*

Creating a New File and Container Movie Clip

Knowing that the final project dimensions are 750×500 (as described in Chapter 1), it's helpful to work from your template file. This not only allows you to preview your work in an appropriate setting, but it also contains a correctly positioned movie clip within which you can add your content. You will soon see that this will make it much easier to transfer your work from one file to another.

Create a new file (File→New) using the Content template you created in Chapter 1.

You will be creating the content for the *Home* page and adding this content later to an ongoing project file. To keep your assets organized, rename the container movie clip by opening the Library and double-clicking the name field of the *content* movie clip in the vertical list to edit its name. Rename it `HomePage`.

Moving your attention to the Stage, double-click the placeholder content symbol to open the movie clip for editing (you will delete the placeholder artwork in the next section, after you have added some content). By looking at the breadcrumb trail in the Edit Bar, you should see that you're inside the newly renamed *HomePage* movie clip (for more information on the breadcrumb trail, see Chapter 1). If you see that the movie clip is still named *content*, consider trying step 2 again. The name of the symbol will not affect anything in this FLA, but if you neglect to give each symbol a unique name, it will be harder to manage your assets when you combine them into a larger master file.

Adding Title Text and Underline

Most prominent on the home page is a title announcing the purpose of the project: an artist's portfolio. As such, it needs a bold treatment to draw the viewer's eye. A title will give way to falling shapes leading the user's focus directly to the project navigation.

NOTE

When adjusting the size of a text field, do not use the Free Transform tool unless you want to distort the text. Use the handles visible when the Selection or Text tools are active. These handles will adjust the size of the field but not the text therein.

For more information about creating text fields or configuring their properties, see the section "Creating Static Text," earlier in this chapter.

1. Use the Timeline New layer button or the Modify Timeline New Layer menu to add a new layer called `title` to the Timeline of the *HomePage* movie clip, and place all art created in this section into this layer.
2. Use the Text tool and Properties panel to create a Static text field on the Stage and type `MY PORTFOLIO` in the field. The field should be approximately 210 pixels wide × 100 pixels tall. However, this is only relevant because the size of the field may slightly affect numeric input values specified in these instructions. If your text field is larger or smaller, either adjust the field size or adjust the numbers as you enter them to whatever looks best.

3. Select the text field with the Selection tool. Using the Properties panel while the text field is selected, set the Character properties of the field. For the Family setting, specify a strong font worthy of a title. The sample files use Arial Narrow, not as a design choice, but because it is common on both Mac and Windows platforms. Feel free to choose another font. Use a bold Style, and a Size of 36 points. Feel free to change these settings based on your font choice. Use black for the Color and turn off the Selectable option by deselecting the button at the bottom left of the Character properties (if it is enabled). Optionally, set tighter Letter spacing (adjacent to the Size property) if your chosen font warrants this adjustment.

4. Set the Paragraph properties of the field, choosing Align left for Format.

5. With the text field selected, use the Transform panel to rotate the field to –15 degrees by clicking on the blue number under the Rotate option and entering **`-15`**. Alternately, the Free Transform tool should snap to –15 degrees if you rotate the field by hand while the tool's Snap to Objects option is on (in the context-sensitive area at the bottom of the Tools panel).

6. With the text field selected, use the Properties panel to position the field at approximately point 100, 65. Don't worry too much about the width and height of the field, as these properties will vary depending on your chosen typeface. Just make sure that the entire phrase "My Portfolio" is visible.

7. Use the Line tool to draw a line anywhere on the stage. Select the line with the Selection tool, then use the Properties panel to set the width and height of the line to **`200`** and **`1`**, respectively. Set a Stroke color of **`black`**, Stroke of **`4`** points, and Style of **`Stippled`**.

8. Use the Transform panel (or Free Transform tool with Snap to Objects turned on) to rotate the line to **`-15`** degrees. With the line selected, use the Properties panel to position the line at approximately point **`115, 110`**. Adjust the position based on the font you chose so that the line appears under the title text.

9. Now that you have the title and underline in place, remove the placeholder artwork by deleting the layer with the placeholder content. Lock the *title* layer by clicking the dot across from the layer name and under the lock icon to prevent unwanted editing in future sections.

10. Save your file and check your work against the sample file, *home_page_01.fla*. If desired, feel free to start with this file when moving on to the next section.

NOTE

Notice that after rotating the line, its width and height are approximately 190 pixels and 50 pixels, respectively. These settings have changed from the original values of approximately 200 and 1, respectively, due to the rotation.

Adding Skills Text and Underline

Before moving on to the primary graphical element of the page, you need to repeat the text exercise to add a small list of skills professed by the portfolio owner. Because the process is nearly identical to what you've just accomplished, this segment stresses only variances in the tasks outlined previously. From a practical standpoint this is more concise, but, more importantly, it's also good practice because not every step is described.

1. Add a new layer called `skills` to the *HomePage* movie clip Timeline, and place all art created in this section into this layer.
2. Repeat the title creation process, but populate the field with the word `Skills`. Use a font style that is not bold (`Plain`, `Regular`, `Roman`, `Book`, or equivalent), a type size of `24`, and set Paragraph Format to `Align right`. The field should be approximately 75 pixels wide × 36 pixels tall, or make adjustments to the size of the field or any relevant property values specified herein. Remember, the result of your efforts should please you, not adhere rigidly to these values.
3. Do not apply any rotation, and position the text at approximately point `460, 225`.
4. Create a second underline at approximately point `460, 260` that is `75` pixels wide with no rotation.
5. Create another Static text field, and populate it with a handful of one- to two-word skills. Use a `bold` style of the same font you used previously, at a size of `14` at approximately point `445, 265`. Set the Paragraph Format to `Align right` again.
6. Holding down the Shift key, click on all three elements added in this section—the title, underline, and list—and open the Align panel (Windows→Align). With the *To stage* option `off`, click on the *Align right edge* button (the last of the second group of buttons in the panel). All three elements should now be right-aligned and tidy.
7. Now that you have the remaining text in place, lock the *skills* layer to prevent unwanted editing in future sections.
8. Save your file and check your work against the sample file, *home_page_02.fla*. Again, if you prefer, start with this file when moving on to the next section.

Cascading Rectangles Primitives

With the page's text elements behind you, it's time to move on to the graphic centerpiece.

1. If you're not still working inside the *HomePage* movie clip, double-click the title text you created in the previous segment to open the file for editing.
2. Add a new layer called **cascade** to the Timeline. You will place all art created in this section into this layer.
3. Select the Rectangle tool from the Tools panel and select Object Drawing mode from the context-sensitive options at the bottom of the panel.
4. Use the Properties panel to set a Fill color of **black**. To choose no color for Stroke, select the red diagonal line (the universal "no" symbol) found near the upper-right corner of the pop-up color palette, accessed from the Tools, Properties, or Color panels.
5. Draw a rectangle approximately the length of the title you created previously (the source file rectangle is approximately 220 pixels wide), and approximately **40** pixels in height.
6. With the rectangle selected, use the Transform panel (or Free Transform tool with snapping enabled) to rotate the rectangle to **-15** degrees.
7. Position the rectangle under the line beneath the title, at approximately point **120, 100**.
8. Repeat steps 4 and 5, but vary the size and rotation of the rectangle each time to create several rectangles that appear to cascade down from the title. Create enough rectangles so they eventually exceed the lower bounds of the framed area shown in the guide layer on the main Timeline. Most importantly, overlap some portion of every rectangle so they appear to form a contiguous shape (at least one portion of every rectangle touches some other rectangle so there are no isolated shapes in the mix). Although you have not yet applied the gradient, refer back to Figure 2-23 for an example of what you're trying to achieve.
9. Check to make sure you are happy with your cascade, and that at least some portion of every rectangle overlaps some other rectangle, connected in one contiguous piece. After the next step, you won't be able to edit this layout.

NOTE

If breaks in the cascade still occur (not all rectangles appear part of one merged shape), you may have neglected to remove the stroke from your Rectangle tool before creating the shapes. If this is the case, select the entire layer, and select the Modify→Shapes→Convert Lines to Fills menu command. This is a handy way of changing any line to a fill if that will make it easier for you to edit.

10. Select all the shapes in the layer. You can either confirm that the only other layer (*title*) is locked and select all (Edit→Select All), or click on the *cascade* layer name in the Timeline to select all of the layer's contents. With all the rectangles selected, break them apart (Modify→Break Apart) so they convert from Drawing Objects to shapes that can merge together. To be sure you did this correctly, click anywhere on the Stage or Pasteboard to deselect the shapes, and then click anywhere in the jumble of rectangles again. All of the shapes should select as one. If not, you can either connect or delete any lone rectangles.
11. Lock the *cascade* layer to prevent unwanted edits, then save your file. Check your work against the sample file, *home_page_03.fla* or, if desired, start the next with this file.

Applying a Gradient to the Cascade

To reduce the weight of the cascade but still draw the viewer's eye all the way to the navigation panel, change the fill from solid black to a linear gradient ending in an alpha value of 0. This fade out to transparency will soften the transition to the bottom of the screen.

1. Return to your *HomePage* movie clip and unlock the *cascade* layer. Select the shape cascade using the Selection tool and open the Color panel (Windows→Color). You should be able to select any color, and all the shapes should change to that color. If this is not the case, review all the steps in the previous section carefully. Know, however, that erring in this segment will not affect any other segment, as these results are purely cosmetic.
2. In the Color panel, select the Fill color chip (which looks like a paint bucket) if it is not already selected. To create the linear gradient you need, select `Linear` from the Type menu.
3. Select `Extend` from the Overflow menu to make sure that your gradient continues seamlessly if enough display area is available. For example, the gradient you've created in this exercise will continue with 100% black above and 100% transparent below.
4. Create a gradient with two colors at the extreme ends of the gradient-editing bar. Make the left color `black` with an alpha value of `100` and the right color `black` with an alpha value of `0`. Refer back to the "Creating gradients" section of this chapter for more information.

5. If your shape cascade is not already fading, select the Paint Bucket tool from the Tools panel and click in the shape. Don't worry about the placement or size of the fade yet. It may default to the correct black-to-transparent appearance but orient horizontally, or it may look like a different gradient or even a solid if the scale of the gradient is not correct. If the shape is selected, deselect it by clicking in any unoccupied area of the Stage or Pasteboard. This will give you a better look at the gradient. The best indication that the gradient has been applied is to look in the Fill color chip in the Tools panel while the shape is selected. The chip should resemble the gradient you created, even if the shape does not.
6. Once you're sure that the gradient has been applied, use the Gradient Transform tool to set its appearance. As discussed previously, this tool is found in the Transform tool menu below the Subselection tool. With the tool selected, click on the shape. Scale the gradient to fit the shape (refer back to the "Gradient Transform Tool" section of this chapter for more information). Using the same Gradient Transform tool, rotate the gradient to orient it vertically, with black at the top.
7. When you are happy with your gradient, relock the *cascade* layer and save your file. Check your work against the sample file, *home_page_04.fla*. If the files don't match to your satisfaction, look over your work or start with the provided file when moving on to the next section.

Adding Grime

Tying the stippled underlines in with the overall distressed look of the portfolio, it's time for grime. Use the Spray Brush tool to quickly and easily add black shapes of random size along the bottom of the screen:

1. Create a new layer at the top of the Timeline and name it `spray`.
2. Select the Spray Brush tool (found in the Brush menu in the Tools panel) and configure its settings. Maintain the default settings of `black` color, Scale of `100`, Brush Width and Height of `96`, and Brush angle of `0`. Enable the Random scaling option to randomly adjust shape size during spray.
3. In the *spray* layer, spray the brush along the bottom edge of the content area so that grime is deposited both inside and outside the frame area. This will make the grime appear more natural within the frame. After trying these settings, you may want to adjust the brush size, through trial and error, to affect the density of the grime.
4. Once you are happy with your filth, lock the *spray* layer to prevent unwanted future changes.
5. Save your file and check your work against the sample file, *home_page_05.fla*.

The Project Continues...

In the next chapter, you'll practice with the Deco tool and get an early taste of ActionScript (just for fun). The example will provide an entrée into the use of Flash symbols, the topic of the next chapter.

You'll also practice editing vectors using both the Selection and Subselection tools, and you'll create your first mask—a layer that will only make parts of specific underlying content visible to the user.

USING SYMBOLS

Introduction

IN THIS CHAPTER

When Peter Sellers acted in Stanley Kubrick's *Dr. Strangelove or: How I Learned to Stop Worrying and Love the Bomb*, he played not one character, but three. Although there was only one Peter Sellers, costume, makeup, and acting brought Dr. Strangelove, President Merkin Muffley, and Group Captain Lionel Mandrake to life.

Flash assets known as *symbols* bring your Flash files to life much in the same way that actors contribute to films. Symbols are efficient Flash asset types that you can use again and again without materially contributing to file size, just as one actor can play multiple characters without increasing the size of a film's cast.

Furthermore, when a new *instance*, or unique occurrence, of a symbol is created, that instance can be transformed without leaving any lasting effect on the symbol. For example, you can add three instances of a symbol to the stage. The first can be rotated, the second can be scaled, and the third can be faded to 50% opacity. Despite all of these transformations, the symbol from which the instances were derived remains unchanged. Taken from another perspective, editing or replacing the symbol will immediately update all instances derived from it.

Likening this to film, one actor can play multiple characters, and costume and makeup do not permanently transform the actors into the characters they play. If an actor is replaced, all characters played by that actor reflect the change.

Reusability, global editing, and file efficiency, combined with the special features discussed in this chapter, make symbols the cornerstones of Flash assets.

Symbol Types

There are three types of Flash symbols: *button*, *movie clip*, and *graphic*. A button, as the name implies, is typically used when mouse interaction from the viewer is required. Buttons usually require ActionScript to be fully functional and are commonly used, for example, as interactive user interface elements. Movie clips are typically used for animation because each movie clip has its own timeline just like the main FLA. ActionScript can also control movie clips, making them popular assets for programming-driven tasks. Graphics, on the other hand, are essentially movie clips that cannot be manipulated by ActionScript. Instead, graphic symbols can be controlled in a few simple ways without ActionScript, making them useful to nonprogrammers.

Button

Button symbols have two distinct functions in Flash. First, a button can be controlled by ActionScript and is typically used as an input device of sorts. When a user clicks on a button, a script can be triggered to perform an action. Second, a button can automatically provide two kinds of feedback to the user when the mouse interacts with the button: a hand cursor is automatically displayed when rolling the mouse over the button, and optional *Up*, *Over*, and *Down states* can be shown during mouse interaction.

NOTE

Over and Down states are not required for buttons to work. It is not unusual for a button to only have an Up state if a project design does not require multistate buttons. This is because a designer can rely on the display of the hand cursor when the user rolls over the button with the mouse.

An *Up* state is the default button state, visible anytime the mouse cursor is not hovering over the button. An *Over* state is displayed when the mouse cursor is over the button but while the mouse is not clicked. Finally, a *Down* state is visible only when the mouse is clicked while over the button.

Creating your first button

Creating and editing buttons are the easiest ways to understand how they work. Although the ActionScript required to *use* a button won't be discussed until a little bit later, understanding the parts of a button, and seeing the visual feedback provided, will be a great first step. The following simple test file will give you some experience with your first button. It won't be used in the final project, so feel free to experiment.

1. Create a new file using the File→New menu command. Because this file will not be used in your project, you do not need to use the book template.
2. Enable Object Drawing mode for simplicity and practice (consult Chapter 2, if necessary).
3. Using the Rectangle tool, draw on the Stage a rectangle at the desired size of your button. Look at the Properties panel and see that you have created a Drawing Object (or Shape if you chose to use Merge Drawing mode).

NOTE

It's a good idea to follow the example using a Rectangle first, before experimenting with your own button art.

4. Test your Flash movie using the Control→Test Movie menu command. Your file will compile and display the SWF. Roll your mouse over the rectangle, and you will see no change in the cursor. Close the SWF and return to the FLA.

5. If needed, select the rectangle again (click or drag over it with the Selection tool) and choose the Modify→Convert to Symbol menu command (F8).

6. A Convert to Symbol dialog will appear (Figure 3-1). This dialog has basic and advanced options. Only the basic options are required for this exercise, so regardless of which is displayed, you only need to pay attention to the first section of the dialog.

Figure 3-1. The Convert to Symbol dialog

7. Name this symbol `MyFirstButton` and choose `Button` from the Type menu.

8. The Registration option lets you choose where the registration point will be placed when creating the symbol. This is the point that is used for things like positioning an object on the Stage. Click the upper-left corner in the nine-box grid to place the registration point in the upper-left corner of the button.

9. The Folder option determines where this symbol will be placed in your Library. For now, leave the default, Library root. This will place your newly created symbol in the root, or top, directory of the Library. In other words, it will not appear inside a specific folder in your Library. You'll learn more about using the Library later in this chapter.

10. When you've completed these steps, click OK. This will create the button symbol, close the dialog, and return your focus to the Stage. Save your file as *my_first_button.fla*. You can compare your work to the sample file, *my_first_button_01.fla*.

Your rectangle may not appear to have changed much, but if you look at the Properties panel, you will see that the object has changed from a Drawing Object (or Shape) to a Button, as shown in Figure 3-2. The icon and menu indicate that the symbol is a button.

Figure 3-2. A detail of the Properties panel with your button selected

The text at the bottom left of Figure 3-2 shows that the selected object (the button you clicked on the Stage) is an instance of the *MyFirstButton* symbol you created. If you look in your Library (Window→Library) you will see that the button symbol has been added to the Library. You can drag a few copies from the Library (dragging from either the button name in the list or the preview icon) to create additional instances of the symbol.

When you start adding ActionScript to your files, you will be able to distinguish one symbol instance from another by giving them instance names, as shown near the top of Figure 3-2. This isn't necessary until you begin programming, so for now you can leave the instance unnamed.

Finally, in the lower-right corner of the panel detail, you can see an icon named Swap. When an instance is selected on the Stage, you can click this icon to switch the symbol that the instance is derived from. This is handy when you want to swap one button for another because all attributes of the instance (such as transformations like x- and y-position, scale, rotation, alpha, and so on), will remain even after the symbol is swapped.

If you test your movie again (Control→Test Movie) and roll your mouse over your newly created button instance, the cursor will now change to the appearance of a hand with a pointing finger. This cursor feedback is automatic when using a button symbol.

Editing your button

WARNING

Editing a symbol is very different from transforming a symbol instance. Applying a transformation, such as rotation, to an instance is an isolated occurrence that affects only that instance. Such a transformation affects neither the symbol nor other instances derived from it. Editing the symbol, however, alters the symbol and every instance spawned from the symbol.

You've seen one visual aspect of button use (the cursor change), but you still need to examine the various button states to better understand how a button works. To do this, you must edit your symbol.

Double-clicking the button instance will allow you to edit the button symbol. If you look at the Edit Bar, near the left corner above the Stage, you will see that a breadcrumb trail has been started, showing that you are now inside the *MyFirstButton* symbol. The trail also shows that your button symbol is in Scene 1, or your main Timeline.

To leave the symbol-editing mode and return to your main Timeline, you can either double-click the Stage or use the Edit Bar (Figure 3-3). In the latter case, you can either click the Scene 1 button or click the left arrow to traverse back to the stage one object at a time. In this case, the two options yield the same results, but it's possible to have a symbol inside a symbol inside a symbol, and so on. Such a *nested* arrangement would cause the Edit Bar to include more objects through which you could navigate.

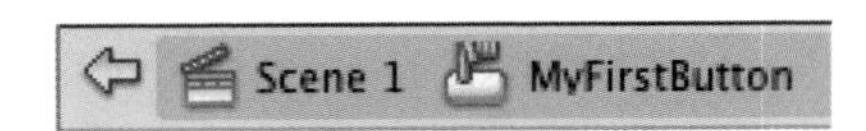

Figure 3-3. The Edit Bar after double-clicking to edit your button

In the Timeline, you will notice that your symbol has a unique set of frames, specific to buttons (Figure 3-4). They are labeled and represent the Up, Over, and Down button states, as well as a fourth state, called the *Hit* state. The Hit state is invisible at runtime and defines the areas of the button that will react to the mouse. Any nontransparent object will trigger button detection and mouse feedback. Semitransparent pixels are not supported. They will be seen as nontransparent and, therefore, as part of the button.

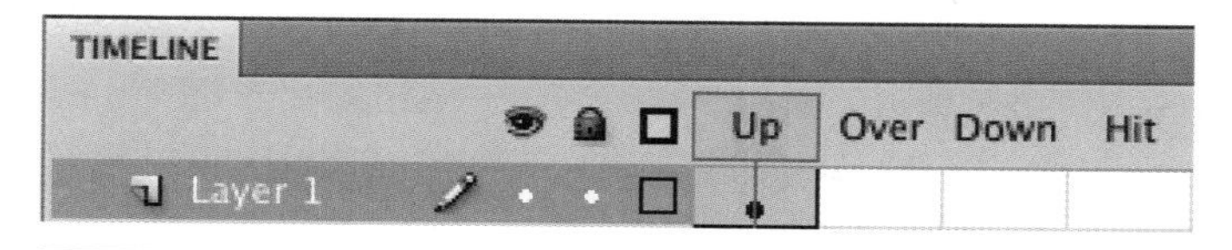

***Figure 3-4.** The button's Timeline*

Because the art in the Hit state defines the areas of the button that react to the mouse, it is possible to create hot areas that are larger (or smaller) than the button. For example, if a button must appear to be very small, and is hard to click with the mouse as a result, you can add area to the Hit state. This added area will not be visible, but will increase the area(s) with which the mouse can interact. Similarly, you could adjust the Hit state of a button that looked like a target by removing all but the bull's-eye from the target. This would prevent the outer rings of the target from being treated like a button, limiting mouse interaction solely to the bull's-eye. Finally, if no Hit state is used, Flash automatically uses the Up state to define the hot area(s) of the button.

To see a visual change in your button resulting from state changes (in addition to cursor changes), it is necessary to create one or more of the labeled Timeline frames and populate them with altered artwork. Follow these steps to give it a try:

11. Click in the frame labeled Over so the frame appears selected. Use the Insert→Timeline→Keyframe menu command (F6) to add a keyframe. This feature creates a new keyframe and duplicates the contents of the previous frame. Because you've created a keyframe rather than adding another frame to the existing frame span, you can transform the contents of each keyframe independently.
12. Create another keyframe by selecting the Down frame and inserting another keyframe. Leave the Hit frame empty for now.
13. Use your mouse to move the Timeline playhead back and forth over these three frames by clicking and dragging the red playhead over the words Up, Over, and Down (you won't be able to move the playhead over the Hit frame because there is no content in that frame). You will see the same artwork used in all three states.
14. Move the playhead to the Over frame and select the button artwork on the Stage. Change the fill color of the Drawing Object.

NOTE

Dragging the playhead through frames to interactively play segments of the timeline is called scrubbing.

WARNING

Remember that the Hit state is invisible at runtime. If the button's Up, Over, and Down states occupy the same area, it doesn't matter which state you duplicate in the Hit frame. On the other hand, if the size or location of any state changes, it is sometimes advantageous to use the union of these states in the Hit frame. Related examples can be found on the companion website.

15. Alter the Down state in a similar fashion and again scrub through the frames. You should now see a unique color for all three states.
16. Test your movie again. This time you will see art changes and cursor feedback when interacting with the button using the mouse. You can compare your work to the sample file, *my_first_button_02.fla*.

Once you've successfully tested your button, it's time to experiment with the Hit state. It's useful to see how adding to, or removing from, this frame affects Flash's recognition of the button. To start, add another keyframe to the Hit frame (F6), which will duplicate the Down state. When testing your movie, no change in behavior will occur because the prior example automatically used the Up state to define the button hot area(s).

Now edit the content in the Hit frame to remove the right half of the button. If you test your movie now, you'll see that only the left half of the button will respond to the mouse. By removing content from the Hit frame, you've reduced the area of the visible button with which the mouse can interact. Finally, go back into the Hit frame and double the size of the button. When testing your movie, you'll find that the mouse will appear to respond to the button even before you touch it. This is because all nontransparent pixels in the Hit frame will cause such a reaction—even when the art ends up being larger or smaller than the visible Up state.

Writing your first script

Just for fun, add your first ActionScript to this test file. You'll learn more about what this script does in later chapters, but trying this now will get your feet wet and give you some experience with the ActionScript-related interface elements.

1. If you haven't already, double-click the Stage or click the Scene 1 button in the Edit Bar to return to the main Timeline.
2. Select the button instance on the Stage and, at the top of the Properties panel, give the button an instance name of `myButton` (not only does spelling count in ActionScript, case-sensitivity counts, too, so be sure you type the instance name correctly). The button can now be referenced by this instance name when using ActionScript.
3. Using the mouse, click on the current frame (indicated by the red playhead line) in the Timeline. If you've followed these steps without any added experimentation, your file will have only one frame in it: frame 1 of layer 1. You will be adding your script to this frame, so you still cannot select the button for the next step.
4. Open the Actions Panel (Window→Actions). If you see the message "Current selection cannot have actions applied to it," look over step 3 again. If the Actions panel is empty and ready for typing, add the following script:

```
myButton.addEventListener(MouseEvent.CLICK, onClick);
function onClick(evt:MouseEvent):void {
    myButton.rotation += 10;
}
```

5. Test your movie and click the button. With each click, the button should rotate around the registration point you chose in an earlier exercise (the upper-left corner was recommended).

6. Although it will not be used in the project, you may wish to save your work for later review. You can compare your work to the sample file, *my_first_button_03.fla*.

Movie Clip

Movie clips are much simpler than buttons in that they are essentially self-contained timelines. That is, a movie clip contains a timeline just like the FLA documents you've been working with to date, but they are not additional external files. This means that you can create an animation inside a movie clip, and then separately animate the movie clip as a whole.

For example, certain types of character animation are simplified significantly by using what might be described as animations within animations. To animate a character walking across the stage, it is much more efficient to animate the task in two phases. First, animate the character walking in place inside a movie clip. This is often called a *walk cycle* because it is a full cycle of movement needed to give the appearance of walking. This might require 5 or 10 frames. Second, move the movie clip as a single element rather than copying and pasting the entire walk cycle over and over, updating the location of each frame every time.

To create a simple movie clip, follow these steps:

1. Create a new file using the File→New menu command. Because this file will not be used in your project, you do not need to use the book template.

2. As you did with the button exercise, create a symbol. Use the Rectangle tool to create a rectangle on the stage, then convert it to a symbol (Modify→Convert to Symbol or F8).

3. Choose Movie Clip as its type. Name the symbol `MyFirstMovieClip` and click OK.

4. Double-click the instance of the movie clip to edit it, and look at the Timeline. Instead of frames marked with Up, Over, Down, and Hit, the Timeline looks just like the main Timeline you've been using all along.

5. Again, as you did with the button exercise, create keyframes in frames 2 through 5 and change the color of the Drawing Object in each frame. Rather than using the color changes for button states, you are simply creating a five-frame animation of a rectangle changing colors.

6. Before you leave the Movie Clip Editing mode, scrub through the frames using the Timeline. The effect will be similar to scrubbing through the button frames, but this is a very important concept. Remember that you are inside the movie clip, editing its Timeline.

7. Double-click the Stage to return to the main Timeline. You should see that the main Timeline still has only one frame and that it's not possible to scrub through the animation any longer.

8. Save your file as *my_first_movie_clip.fla*, test your movie (Control→Test Movie), and watch your animation play.

Notice that your five-frame movie clip animates even though the main Timeline in which it resides only has one frame. You do not need more than one frame to play a movie clip, regardless of its length. However, short of manually previewing playback while editing your movie clip (as in this exercise), you must test your movie (Control→Test Movie) to see the animation play. You cannot otherwise preview movie clips in authoring mode.

Finally, notice that your animation loops forever. Movie clips are designed to loop by default and you must use ActionScript to alter this behavior. If you had any trouble with this stage of the exercise, compare your file with *my_first_movie_clip_01.fla* in the source code from the companion website. Then complete the exercise by adding some ActionScript:

1. Close your SWF to return to the FLA and, if not already in symbol-editing mode, double-click your movie clip to edit it (you can always tell where you are by looking at the breadcrumb trail in your Edit Bar).

2. Click the last frame in your animation and open the Actions panel. (Windows→Actions). If you are warned that you cannot apply actions to the selected object, be sure the last frame of the animation in the Timeline is selected—not the art (on the Stage) in that frame. Type the following script:

   ```
   stop();
   ```

3. Test your movie, and notice that the animation now plays through once and stops. The ActionScript you added told Flash Player to stop the movie clip animation on the frame in which you typed the relevant command (the final frame).

4. Save your work and compare it to the sample file, *my_first_movie_clip_02.fla*.

The most important thing to remember about movie clips is that you can control them with ActionScript. You used ActionScript inside a movie clip in the previous exercise, and you will control movie clips from the main Timeline in the upcoming "Project Progress" section of this chapter.

Another unique attribute of movie clips is that they are the only symbol type to which you can apply certain animation techniques and special effects. These points will be discussed in later chapters, but, for now, commit to memory the fact that movie clips are the best animation tools when ActionScript control is required.

Graphic

Graphic symbols are nearly identical to movie clips, but with three very significant differences:

- You cannot use ActionScript to control graphic symbols.
- The Properties panel provides three simple ways to control the playback of graphics in authoring mode.
- To fully play an animation in a graphic symbol, the Timeline in which the symbol instance resides must have at least as many frames as the graphic symbol itself.

For example, in the previous movie clip exercise, an instance of a five-frame movie clip played through all its frames while sitting in a main Timeline that contained only one frame. If that movie clip were a graphic, only the first frame of the instance would display. By contrast, if the main Timeline spanned five frames, the graphic symbol instance would have enough frames to play through its entire animation.

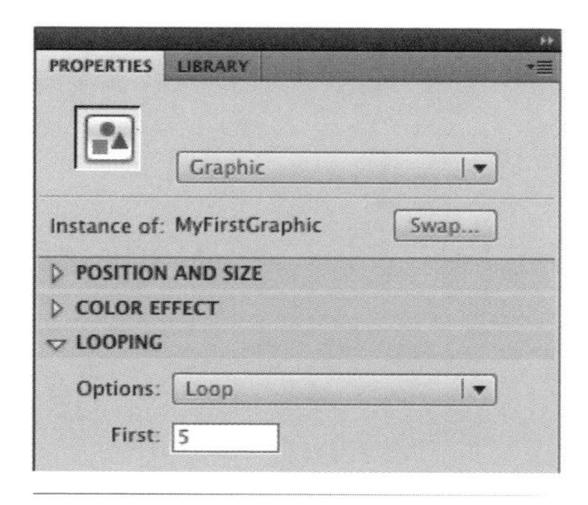

Figure 3-5. A detail of the Properties panel showing the properties of a graphic symbol instance

Graphic symbols were invented to serve as an animation asset that did not require ActionScript skills to use. As such, the Properties panel offers three playback options, found in the Looping section of the panel (Figure 3-5):

Loop

Plays all the frames in the graphic symbol for as many frames as the instance of the symbol occupies.

Play Once

Plays the animation from the first frame you specify in the accompanying text field, labeled *First*, to the end of the animation and then stops.

Single Frame

Displays a single frame, specified in the accompanying text field, labeled *First*.

NOTE

The number of frames in which the graphic symbol resides significantly affects the Loop and Play Once options. If you plan to use these options, test your work thoroughly. See the companion website for additional examples.

Converting Symbol Types

Creating a graphic symbol does not differ substantially from creating a movie clip. The only difference is specifying that the symbol is a Graphic instead of a Movie Clip during creation. So, to demonstrate the use of a graphic symbol, you'll learn how to convert a symbol from one type to another.

1. Open or continue with the file you created previously when building your first movie clip. Save the file with a new name, *my_first_graphic.fla*.

2. In the lower-left corner of the Library panel, click the Properties button, which looks like a lowercase *i* in a circle. The Symbol Properties dialog will open and will look almost identical to the dialog you saw when creating the movie clip. In this dialog, simply use the Type menu to change this symbol from Movie Clip to `Graphic`.
3. Drag a new copy of the symbol to the Stage somewhere near the copy that was already there.
4. Using the Selection tool, click on the new copy and look in the Properties panel. The symbol is now a graphic. Click on the previous copy and again look in the Properties panel. Despite changing its symbol type in the Library, the original instance is still a movie clip. This is because you can change the way an instance behaves just like you can change a symbol. You'll change the original instance to a graphic in a short while, but take a look at the differences and similarities between the two symbol types first.
5. Select the new instance of the graphic symbol again and make sure the Looping Options menu says Loop. Theoretically, this graphic should behave just like the movie clip, looping when it reaches the end of the Timeline.
6. Test your movie (Control→Test Movie) and see what happens. The movie clip plays all the frames and loops, but the graphic does not. This demonstrates the aforementioned need for as many frames in the host Timeline (the one in which the graphic is placed) as the graphic itself contains. Close the SWF and return to your FLA.
7. Because your movie clip had five frames, the graphic converted from that movie clip has five frames. In the Timeline, select frame 5 and use Insert→Timeline→Frame (F5) to add four new frames.
8. Test your movie again; the graphic now animates just like the movie clip.
9. Finally, use the Selection tool to select the graphic and, in the Properties panel, switch the Looping Options to `Single Frame` and enter `frame 3` in the text field.
10. Test your movie and notice that the graphic displays frame 3 only. A nice bonus is that this display change is also updated in authoring mode, meaning you don't have to test your movie to see the result as you did in the movie clip example.

Creating and Editing Symbols

There are two distinct approaches to creating and editing symbols, and the primary difference between the two is context. One way to create a new symbol is to start fresh with an empty canvas on which to draw. Alternately, you can convert an asset already on the Stage to a symbol. This is the more oft-used technique because you can take full advantage of surrounding art elements to inform your editing.

Convert to Symbol and Edit in Place

When creating a movie clip or button, it is usually most helpful to convert an existing asset into the new symbol type while you are working with the artwork. This allows you to work with the new symbol immediately, in the same location and context as the original art.

To demonstrate this approach, draw three separate rectangles on the Stage, then convert one to a symbol. It doesn't matter whether you create a movie clip, button, or graphic. The point is, after creating the symbol, you will see the Stage again and the shape you converted is now a symbol. This typically means that the location and appearance of the art remains intact after making the symbol, and you can immediately begin working with your creation.

You can also edit a symbol with the benefit of this graphical context. Double-click the symbol you just created and look at the surrounding rectangles. You can still see them, albeit slightly faded in appearance (Figure 3-6). This faded look tells you that you're in editing mode, but you can still see all the other stage-bound elements in the movie to assist your editing efforts. This approach to editing is called *Edit in Place* and can also be initiated using the Edit→Edit in Place menu command.

Figure 3-6. Using Edit in Place to edit one of three copies of a rectangle; surrounding elements remain visible but dimmed for context

Insert Symbol and Edit Selected

When surrounding context is not necessary, you can create a new symbol using Insert→New Symbol. This creates the button, movie clip, or graphic, but changes the interface to display only the empty new symbol in Editing mode. In other words, you can't see any other artwork while editing the symbol. Furthermore, when you're finished editing, the newly created symbol is not in the Timeline and you must drag it from the Library to the Stage.

This approach is used less often than converting an existing asset to a symbol, but allows you to edit a symbol without any surrounding distractions. After selecting a symbol, you can use the Edit→Edit Selected menu command to enter Editing mode. However, you will not be able to see anything other than the contents of your symbol.

NOTE

You can also edit symbols without visual distraction by selecting the symbol from the Edit Symbols menu in the Edit Bar above the Stage, or by double-clicking the symbol in the Library panel.

Transforming symbol instances does not permanently alter a symbol. Chapter 2 discusses several ways to transform symbol instances using features such as the Free Transform tool, Transform panel, and Properties panel.

Reusing Symbols

Reuse is one of the main benefits of symbols. In addition to using multiple instances of a symbol within a single document without contributing noticeably to file size, you can also reuse symbols by copying them from one FLA to another.

Within a FLA, adding instances of a symbol to your project is as easy as dragging the symbol from the Library to the Stage:

1. Create a new file using File→New. Save the file temporarily as *rectangle.fla*. Saving this file is just for convenience. It will not be used in the project.
2. Use the Rectangle tool to draw a rectangle anywhere on the Stage and then convert it to a movie clip called `rect`.
3. Drag the *rect* symbol to the stage more than once. Use the Free Transform tool to transform one of the symbol instances. Note that no other instance is affected, nor is the symbol in the Library changed.
4. Save this file and leave it open to continue with the next exercise.

Between FLA files, you can use exactly the same process of adding symbol instances to the Stage:

1. Create another new file using File→New. You will not need to save this file, nor will it be used in the project.
2. Using the Oval tool, draw an oval anywhere on the Stage and then convert it to a movie clip called `oval`. Delete the instance of this symbol from the stage so only the symbol remains in the library.
3. In the Library panel, use the menu near the top to switch from the library of the current file (displayed as some variant of *untitled.fla* if you haven't saved it) to the library of *rectangle.fla* (don't switch documents, just switch libraries). After switching libraries, you will see the *rectangle* symbol from the other document, and you can drag it from the Library panel to the Stage of the new document.
4. Leave this file open and continue with the exercise.

It is also possible to swap one symbol for another that is already in use:

1. If the Library from *rectangle.fla* is still visible, select the Library of the new untitled document again. In the new document, you should now see the *rectangle* movie clip somewhere on the Stage, and both the *oval* and *rectangle* movie clips in the Library.

2. Using the Selection tool, select the rectangle and use the Modify→Symbol→Swap Symbol menu command. A dialog displays a listing of all existing symbols in the current FLA. Pick the *oval* movie clip and click OK to close the dialog. On the Stage, the rectangle is replaced by the oval. All properties of the symbol instance, such as the x and y location, scale, rotation, and so on, are preserved.

The more you use symbols, the more you will appreciate the power of reuse. The next section of this chapter discusses the Deco tool, which takes great advantage of reusing symbols to create art with a little assist from your friend, math.

NOTE

The Swap Symbol feature is also accessible from the Properties panel when a symbol instance is selected. A button called Swap appears near the top of the panel, which will function the same way as the Swap Symbol menu item.

Using the Deco Tool

Chapter 2 briefly discussed the application of symbols while using the Spray Brush tool. When spraying a swash of stars, the Spray Brush reused the same symbol again and again, not only reducing file size, but also making it possible for you to edit all instances of the stars just by editing the original symbol.

Another tool new to Flash CS4 Professional that makes similar use of symbols is the Deco tool. The Deco tool automatically applies instances of a symbol, or default shape, to the stage, but uses three distinct effects.

The Deco tool is the last in the drawing section of the Tools panel, and looks like a pencil drawing a series of dots. Activate this tool and experiment along with the text.

Vine Fill

The first Deco tool drawing effect is *Vine Fill*. Similar to the way the Paint Bucket fills an area with color, Vine Fill fills a contiguous area in your drawing with a vine drawn by a computer algorithm. For example, Figure 3-7 shows a rectangle, originally filled with black and a large yellow flower, to which the Vine Fill effect has been applied. The origin point of the vine is the location clicked by the user, indicated by the star. Simply by clicking at that location with the Deco tool, the vine grew, branching and sprouting leaves and flowers. When it reached an obstacle, such as the edge of the shape or the large flower, the vine detected the collision and stopped growing in that direction.

Figure 3-7. Deco tool Vine Fill results

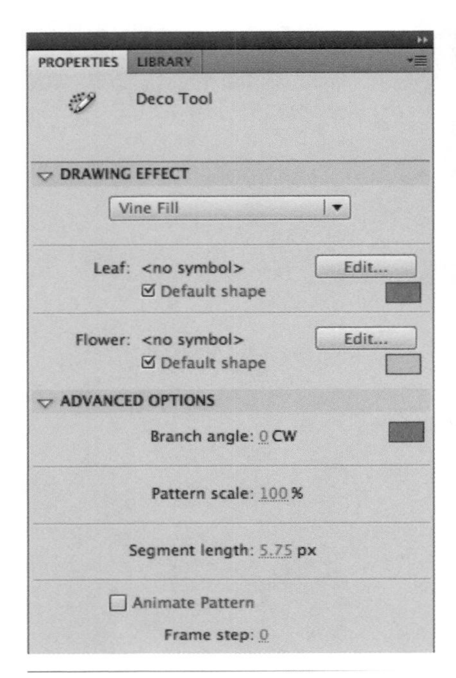

Figure 3-8. Deco tool Vine properties

Figure 3-8 shows the Properties panel with the Deco tool selected, illustrating the properties for the Vine Fill effect. You can choose to use the default leaf and flower shapes, using any color, or substitute your own symbols. You can also set the angle of the sprouting branches, the length of each branch segment, and the overall pattern scale (creating bigger or smaller leaves and flowers). You can even animate the process at runtime by enabling the *Animate Pattern* option; to determine how much of the vine is drawn each frame, specify the *Frame step* value. Increasing the step size will draw more of the vine in each frame, requiring fewer frames to complete the animation.

Grid Fill

The Grid Fill drawing effect fills an area in your drawing with a grid. The properties of the Grid Fill effect can be seen in Figure 3-9. Again, you can use a default shape (a small square) of any color or substitute your own symbol. You can also dictate the horizontal and vertical spacing between grid elements and the scale of the overall pattern. Figure 3-10 shows the result of the Grid Fill effect using the default shape.

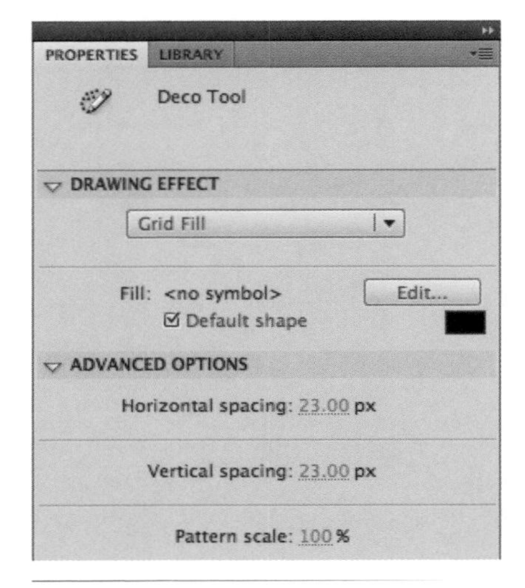

Figure 3-9. Deco tool Grid Fill properties

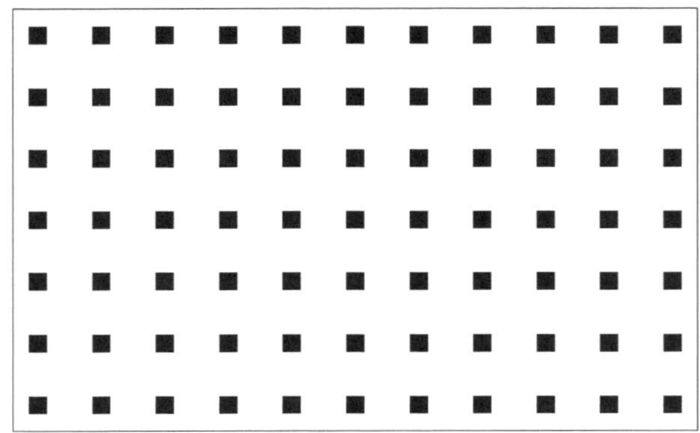

Figure 3-10. Deco tool Grid Fill results

Symmetry Brush

The last Deco tool drawing effect is the Symmetry Brush, which places shapes or symbols on the Stage in symmetrical patterns. As shown in Figure 3-11, the effect's only basic option is to use a default shape (again, a small square) of any color, or substitute your own symbol. The Symmetry Brush, however, has four possible advanced options—*Reflect Across Line*, *Reflect Across Point*, *Rotate Around*, and *Grid Translation*—all of which control how the shape or symbol is placed on the stage. In all four cases, the symmetrical patterns can check for collisions between symbols to prevent overlapping, or place the symbols without concern for one another.

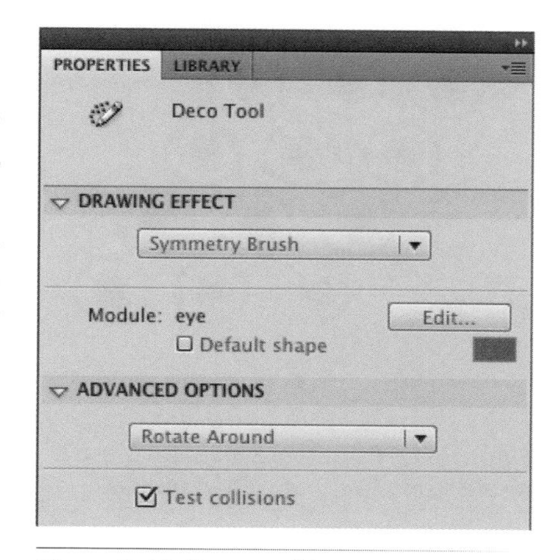

Figure 3-11. Deco tool Symmetry Brush properties

With every mouse click, *Reflect Across Line* positions two instances of your symbol reflected anywhere across a single line. For example, if you position your Deco tool in the center of the Stage, orienting the line of reflection vertically (the default behavior), every click on the right half of the Stage will be reflected on the left half of the Stage. You can also click and drag (before releasing the mouse) to see a live preview of the position of your objects. Control points allow you to rotate the line around which elements are reflected (effectively rotating the entire collection of shapes or symbols around the control point), or move the collection of shapes or symbols as a single object.

Reflect Across Point again creates two objects with every click, but reflects the desired shape or symbol across a center point. This is similar to Reflect Across Line, but is not limited to a single axis of reflection. That is, if you click and then drag to preview position, you can rotate the objects' positions in a complete circle, reflecting around the center point. A single control point allows you to move the collection of shapes or symbols as a whole.

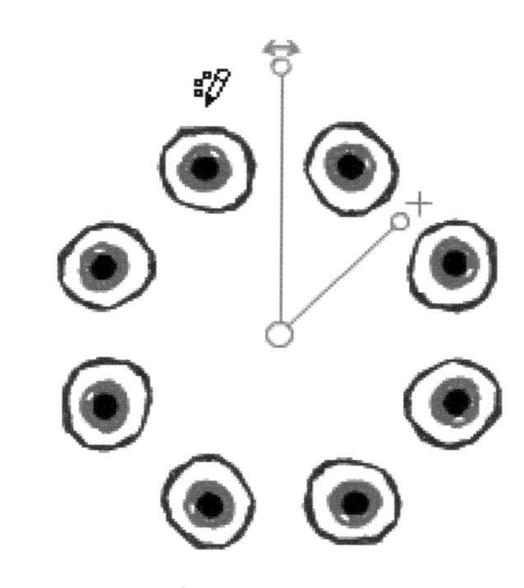

Figure 3-12. Deco tool Symmetry Brush results using Rotate Around setting

Rotate Around, shown in Figure 3-12, adds multiple instances of your shape or symbol along the circumference of a circle around your chosen center point. The distance from the center point determines how close to one another the shapes or symbols will be placed.

Three control points are provided to help you adjust the placement of the shapes or symbols. A center control point allows you to move the collection of objects as a whole. The control point with an accompanying plus (+) symbol adjusts the angle between objects along the unseen circular path you are describing. Dragging the control point at the end of this smaller axis (shown at an angle in Figure 3-12) closer to the other axis tightens this distance. Dragging it further away from the other axis increases this distance. Finally, dragging the control point at the end of the longer axis (shown as vertical in Figure 3-12) rotates the entire collection of shapes or symbols.

Grid Translation adds a grid of symbols, much like the Grid Fill Deco tool option. With Grid Translation, however, the grid is treated as an adjustable object and can be translated in several ways:

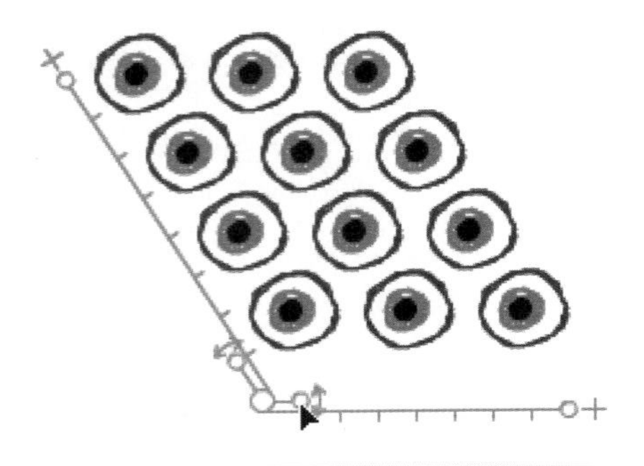

Figure 3-13. Deco tool Symmetry Brush results using Grid Translation setting

Grid size

You can reduce or enlarge the size of the grid (number of shapes or symbols added) along columns, rows, or both, by dragging the control handles at the end of each long axis, as shown in Figure 3-13.

Grid gaps

You can reduce or enlarge the gaps between columns, rows, or both, by dragging the handles of the shorter axes. Figure 3-13 shows the cursor manipulating the control point that affects column gaps. If you drag the cursor to the right, additional space is added between columns.

Angle between grid elements

You can reduce or enlarge the angle used to place grid elements relative to one another, causing the grid to appear more acute (between 0 and 90 degrees) or more obtuse (between 90 and 180 degrees).

To do this, drag the handle near the corner control point, which is on the *inside* of the axes. Figure 3-13 shows the cursor manipulating this control point. Instead of dragging this point left and right to affect the column gap value, you can drag it around the corner point to change the angle of object placement. If you drag the cursor that appears in Figure 3-13 clockwise around the corner point, the angle between the grid elements will open up even further than depicted. If you drag the point counter-clockwise around the corner point, the grid will begin to square off and eventually become more and more acute.

Rotation of grid as a whole

You can rotate the entire grid around its corner control point. This is akin to rotating a shape or symbol, but this feature allows you to preview the results while still editing the grid. To do this, drag the control point near the corner that is on the *outside* of the axes in a clockwise or counter-clockwise direction. In Figure 3-13, this point is outside the leftmost axis, approximately at the bottom of the lowest eye.

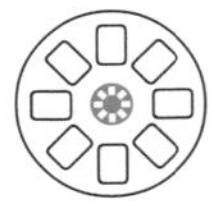

Project Progress

In this chapter you will focus your project work on symbols and gain additional experience with the Timeline, creating a sound control widget that you will eventually add to the portfolio's main interface. In addition, you'll use the Deco tool to create the first interactive element of the project using ActionScript.

Don't worry about your scripting abilities yet. The Deco tool exercise will be just for fun. Typing in a provided script will give you a little more experience with the Actions panel, and you'll learn how the code works in Chapter 6, where ActionScript is discussed in greater detail. Chapter 13 will give you the necessary information to program the sound controller so, for now, you'll only need to build the required parts.

Creating the Sound Controller

The portfolio project will include a sound controller that viewers can use to turn soundtrack music on and off. In addition, the controller will visualize the music during playback by manipulating volume meters for the left and right stereo channels.

These meters are often comprised of bars, oriented vertically or horizontally, that grow in length with the volume of the sound. Advanced meters of this type also display colors that indicate the relative "temperature" of a sound. Sound playing at a volume that is under a specific level—usually about two-thirds or three-quarters of maximum volume—typically contain less distortion and are less likely to cause hearing problems. These levels are represented by "cool" colors such as greens and blues. Volumes above this threshold run the risk of degrading the quality of the sound and damaging your ears or sound equipment. These levels are represented by "hot" colors like oranges and reds.

Left Channel Louder

Right Channel Hot!

Figure 3-14. Sound control in use

Figure 3-14 shows two states of the horizontal sound controller that you will be creating. The top image shows a point at which the left channel (top bar) is louder than the right channel (bottom bar), but both channels are at acceptable volumes. The bottom image shows that the right channel is louder than the left channel and that both are above the desired volume threshold. In particular, the right channel is "hot," or at maximum volume and distortion.

Using layer masks

Building a basic volume meter with a solid-color bar requires only that you set the width of the bar to the volume of the sound. However, if you used this simple approach with gradient colors, as described previously, all colors would be visible during the scaling. This would not expose the hot colors only at high volumes, as desired. Figure 3-15 shows the result of scaling the color bars. Note that the individual bar segments squash horizontally and all colors remain visible at all times.

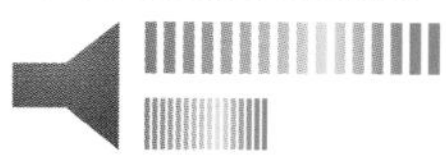

Right Channel Hot!

Figure 3-15. Errantly scaling meter color bars

The way around this problem is to use *layer masks*. A layer mask is a special kind of layer in the Timeline that serves as a window through which you can see underlying masked layers. Any nontransparent pixel in a layer mask will act like a hole cut out of a piece of paper. In other words, when adding to a layer mask, you are drawing the hole.

You will use masks twice in the sound controller. First, you'll create a static mask that will divide a continuous gradient into vertical bars. Second, you'll create a dynamic mask that will be controlled by ActionScript. By using ActionScript to control the width of the color bar's *mask*, not the bar itself, you can expose varying degrees of color in sync with the audio volume, rather than distorting the entire color bar.

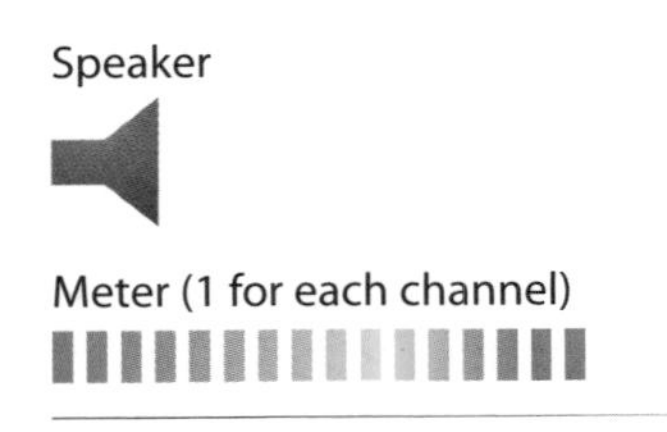

Figure 3-16. Sound controller parts

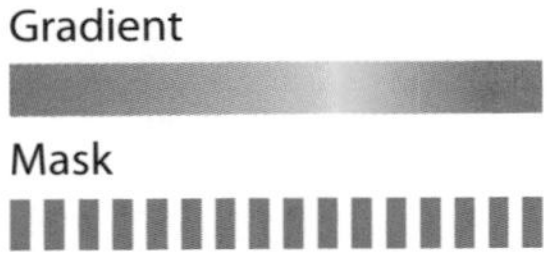

Gradient

Mask

Figure 3-17. Gradient and mask used to create color stripes

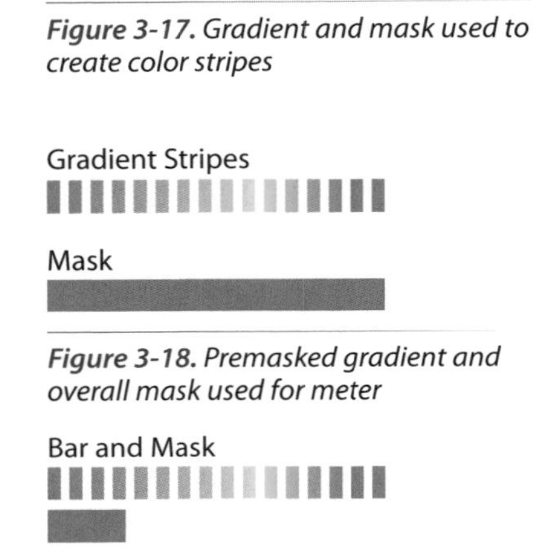

Figure 3-18. Premasked gradient and overall mask used for meter

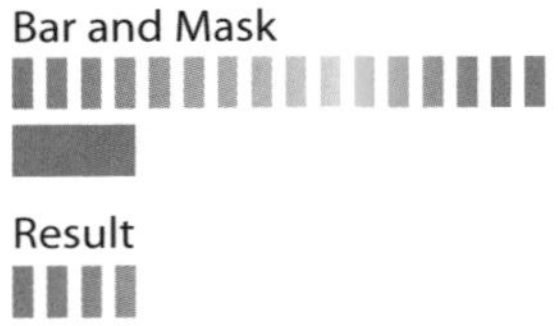

Figure 3-19. Result of mask in use

The controller parts

Before you begin drawing assets for the sound controller, it helps to have an inventory of the required parts. Ultimately, the entire controller will be a movie clip symbol, which will act like a button that the user can click on to turn the sound on and off. At the top level, the controller will consist of the two main parts shown in Figure 3-16: a speaker shape and (two instances of) a movie clip for the volume meter. The speaker requires no further explanation, but the meter consists of additional parts.

Before the meter can reveal a subset of color segments, you must create a bar that is divided into the stripes shown in Figure 3-16. Two parts are required: an underlying gradient with a range of cool to hot colors, and a mask of the stripes through which the colors will be visible. Shown in Figure 3-17, both are static parts and are combined to create the color bar that will be dynamically revealed during sound playback.

Once you have masked the gradient so that only individual segments of the color can be seen, you will add a mask to the entire meter. Unlike the mask in the previous step, this is a movie clip and will be dynamically controlled by ActionScript at runtime. Figure 3-18 shows the two meter parts.

At runtime, the horizontal scale of the bar mask will be synchronized to the volume of the sound. At full volume, the bar will be at full width, showing all the colors in the meter. When the volume is at 0, or when the sound is not playing, the width of the bar will be 0, showing none of the meter colors. Figure 3-19 shows one possible result of the meter when the sound is at approximately 25% of maximum volume. The horizontal scale of the mask is set to approximately .25 and the mask, therefore, reveals only cool colors.

Assembling the pieces

The first step in putting the sound controller together is to create the speaker icon:

1. Create a new file using the File→New menu command. You don't need to use the template because the sound controller will be relocated in the final file. Instead, create a new file and leave the document properties at their default settings. You should have a white Stage that is 550×400 pixels.

2. Use the Rectangle tool with Object Drawing mode off, and draw a rectangle anywhere on the Stage. Using the Properties panel, set its width and height to `20` and `25`, respectively, and position it at point (`210`, `185`).

3. Use the Properties panel to give the rectangle a thin stroke, or no stroke at all, and a blue fill to contrast the eventual background. Try to use a gradient to practice with the Color panel.

4. Use the Pen tool menu to select the Add Anchor Point tool. (Zoom in for easier editing.) Click twice along the left edge of the rectangle, dividing the edge into thirds, as shown in Figure 3-20. The sample file uses distances approximately 8 and 17 pixels down from the top, dividing the edge into 8-, 9-, and 8-pixel segments. You can check your work by opening the Info panel and hovering over the points with your mouse. The y-values along the left edge should be approximately 185, 193, 202, and 210.

Figure 3-20. Adding points to the rectangle

5. Switch to the Subselection tool. Select the upper-left corner of the rectangle and drag it down to about 10 pixels directly opposite the top point that you added to the left edge. Use the Info panel as in the previous step to confirm that the new point is at approximately (220, 193). Drag the lower-left corner up and over to a similar location opposite the second point you added to the left edge, as shown in Figure 3-21. The new location of this point should be at approximately (220, 202).

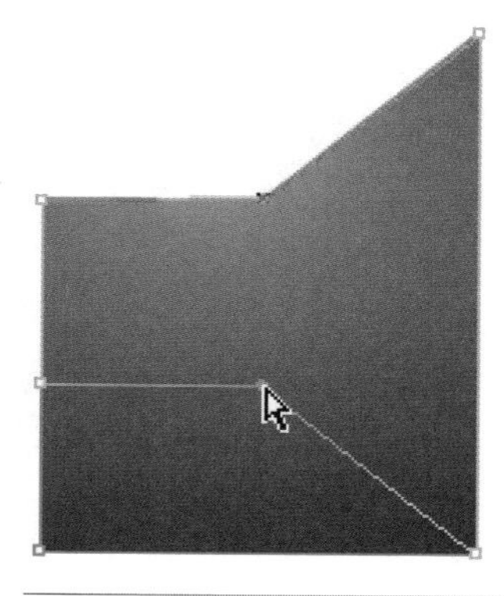

Figure 3-21. Dragging points to form a speaker

6. Your speaker icon is now complete, and because the icon itself won't be interactive it can remain a shape. To avoid damaging the icon with other shapes, select it and group it using the Modify→Group menu command. Finally, save your work as *sound_control.fla*.

Now it's time to create the meter:

1. Using the Rectangle tool, draw a horizontal bar anywhere on the Stage and use the Properties panel to set its width and height to `100` and `10`, respectively, and position it at point (`235, 185`).

2. Use the Selection tool to drag over the new bar to select it and open the Colors panel. Use the Stroke chip in the upper-left corner of the panel to select no stroke. You should see a red line through the chip.

3. Select the Fill chip in the Colors panel and switch the Type menu to `Linear` to create a linear gradient. Create a three-color gradient moving from green to yellow to red, approximating what you see in Figure 3-22. See the "Color Panel" section in Chapter 2 if you need any review on creating gradients.

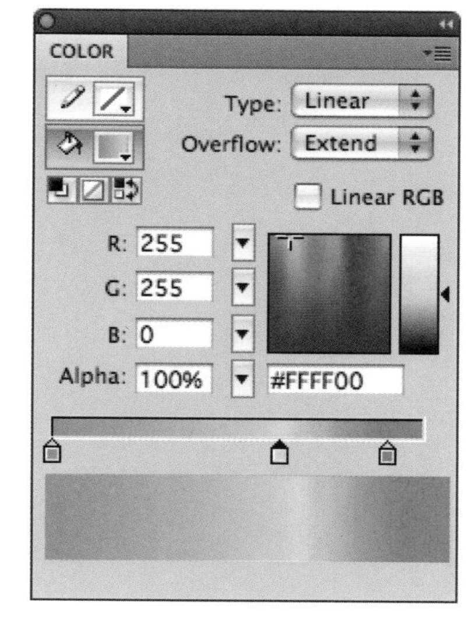

Figure 3-22. Adding points to the rectangle

4. If you've been following along, your horizontal bar is now filled with the gradient shown in Figure 3-22 because the bar was selected while you were using the Color panel. If the bar fill is not the desired gradient, switch to the Swatches panel and click in the bottom bar to add your new gradient to the panel. Choose the new gradient with the Fill color in the Tools panel and then use the Paint Bucket tool to fill the bar with the gradient. If the shape still has a stroke, use the Selection tool to double-click the stroke and delete it.

5. Use the Selection tool to select the bar and use the Modify→Convert to Symbol menu command. In the resulting dialog, choose `Movie Clip` and an upper-left corner registration point. Name the symbol `bar` and click OK.

Save your work and compare your file with the source file *sound_control_01.fla*. If preferred, continue with the provided file.

You will now add a layer mask to create the color segments of the meter:

1. Double-click the new movie clip to edit its contents.
2. Rename the only layer in the movie clip to `color`.
3. Add a new layer and name it `mask`. Double-click the layer icon of the *mask* layer to access the Layer Properties dialog. Change the layer type to `Mask`.
4. Double-click the layer icon for the *color* layer and change its type to `Masked`. Lock this layer to prevent future changes.
5. In the *mask* layer, use the Rectangle tool to draw a rectangle anywhere on the Stage. Use the Properties panel to set the rectangle's width to `4` and height to `10` and position it at point (`0`, `0`). (You are editing the contents of a movie clip with a registration point of (0, 0), so the stage coordinates you used to draw the original gradient bar no longer apply.)
6. Create 15 copies of this shape, so that you end up with 16 shapes evenly spaced horizontally between 0 and 96 pixels, all with a y coordinate of 0. The Align panel can help with this task and features both alignment and spacing options. Position the last copy you made at point (`0`, `96`), select all the shapes, and use the *Align top edge* and *Space evenly horizontally* options.
7. Lock the mask layer to see your finished color bar. Save your work and compare your results with the source file, *sound_control_02.fla*. If preferred, continue with the provided file.

Now that your color bar is finished, you need to assemble the meter in a single movie clip and create a mask that will expose the appropriate colors based on sound volume:

1. Using the Selection tool, select the instance of the *bar* movie clip on the Stage and convert it to a movie clip following the same process used earlier in the exercise. Select the upper-left corner for the registration point and name the new clip `meter`.
2. Double-click to edit the new movie clip and rename its only layer to `bar`. Lock this layer to prevent any unwanted changes.
3. As you did previously, create a new layer of type `Mask` (naming it `mask` again) and change the *bar* layer to a type of `Masked`.
4. In the *mask* layer, draw a rectangle anywhere on the Stage. Set its width to `100` and height to `10` and position it at point (`0`, `0`).
5. Use the Selection tool to select the entire rectangle and convert it to a movie clip. Again, select the upper-left corner as its registration point and name it `maskBar`. Later on, you'll need to control this movie clip with

ActionScript, so you need to give it an instance name. While the *maskBar* movie clip instance is still selected, give it a name of `barMask` at the top of the Properties panel.

6. Lock the mask layer to see your finished meter. If you sized your mask correctly, you should see the entire color bar. Save your work and compare your results with the source file, *sound_control_03.fla*. If preferred, continue with the provided file.

Before tying all the parts together, you must now duplicate the first meter you created and give both meters instance names so they can be controlled with ActionScript. Two meters are required because you will be visualizing stereo sound.

1. Using the Selection tool, select the instance of the *meter* movie clip on the Stage and copy it. Paste the copy onto the Stage and use the Properties panel to position the copy below the first meter at point (`235`, `200`).
2. Select the first meter and use the Properties panel to give it an instance name of `lPeak`. Select the second meter and name it `rPeak`. These names are derived from the left and right peak volume values that will later be used to visualize the volume of the sound.
3. Now wrap everything up into one controller. Using the Selection tool, drag over everything to select the speaker icon and both meters. Use Modify→Convert to Symbol to convert everything into a movie clip. Select an upper-left corner registration point and name the symbol `soundController`.
4. You are almost finished. Create a folder in your Library called `soundControllerAssets` and place the *bar*, *maskBar*, and *meter* movie clips in it. This will keep things nice and tidy when you transfer your controller to your master file later on.
5. Save your work and compare your results to the source file, *sound_control_final.fla*. In Chapter 13, you'll write the code necessary to put the controller into action.

Adding ActionScript-Controlled Animation

This exercise is strictly to increase your enthusiasm and get you creating interactive elements right away. You won't spend any time learning ActionScript prematurely. Instead, you'll revisit this code in Chapter 6 and learn everything you need to see it through. For now, it's just time to plug and play. After using the Deco tool to create a symbol in a new file, you'll paste a provided script into the Timeline and the symbol will animate.

1. Open the file you created in Chapter 1 for your home page. Double-click anywhere on the *HomePage* movie clip's Stage to edit its contents.

2. Create a new layer at the top of the timeline called `deco` and place all visual assets from this exercise in that layer.

3. Using the Oval tool, draw an oval anywhere on the Stage. Using the Properties panel, use a `black` fill and no stroke, and make the oval `6` pixels wide and `12` pixels tall.

4. Use the Selection tool to select the oval and convert the shape to a symbol (Modify→Convert to Symbol). Choose the `Movie Clip` type, name it `oval`, and use a center registration point. Click OK to create the symbol, then delete the instance from the Stage. You will use the movie clip in the Deco tool instead of manually placing instances on the Stage.

5. Select the Deco tool and drag the center point of the tool's interface to the center of the vacant area in your *Home* screen. This area is constrained to the top and right by the frame implied in the guide layer, to the left by the rectangle cascade, and to the bottom by the list of skills. Figure 3-23 shows a detail of the screen and the work in progress, and shows approximately where to position the tool interface. This need not be exact, but remember that the frame shown in the guide layer will cover anything that appears outside the framed area.

Figure 3-23. A detail of the HomePage movie clip, showing the location and appearance of the Deco tool creation

6. As described previously in the "Using the Deco Tool" section of this chapter, use the Properties panel to choose the tool's Symmetry Brush drawing effect. For the Module setting, select the *oval* movie clip you created and, for the Advanced Options, choose Rotate Around and Test Collisions.

7. Click and drag on the Stage to position a series of ovals along a circular path around the Deco tool's center point. Start close to the center and repeat the click and drag process four times to end up with four rings of ovals, as shown in Figure 3-23. The number of ovals added is not important, but this will approximate the example provided in the sample file.

8. When you are done, switch to the Selection tool. The Deco tool will deselect, so click on the group of ovals to select them. Convert this group to a movie clip, naming it `decoOvals` and choosing a center registration point. Click OK to create the symbol.

9. Using the Selection tool, click again on the ovals to select your new movie clip. In the Properties panel, give the movie clip an instance name of `ovals`. You are now done with the Deco tool and can add ActionScript to animate the ovals.

10. Add a layer to the top of the timeline, and name it `actions`. Select the frame in the *actions* layer and open the Actions panel (Window→Actions). Add the following script to the Actions panel:

```
1   ovals.addEventListener(Event.ENTER_FRAME, onEnter);
2   function onEnter(evt:Event):void {
3       var numOvals:int = ovals.numChildren;
4       for (var i:int = 0; i < numOvals; i++) {
5           ovals.getChildAt(i).rotation += 10;
6       }
7       ovals.rotation = mouseX;
8   }
```

11. Test your movie (Control→Test Movie). You should only see the *Home* screen assets (not the guide layer), and each of the ovals should be spinning. Also, if you move your mouse left and right across the stage, the entire set of ovals will rotate with your mouse. You'll learn more about this script in Chapter 6 when ActionScript is introduced.

12. When you are finished, lock both the *deco* and *actions* layers and save your file. Check your work against the sample file, *home_page_final.fla*. If your script isn't working the way you expect it to, feel free to use this sample file moving forward.

The Project Continues...

In the next chapter, you'll import assets and make use of powerful import features from native Photoshop and Illustrator files. You'll create the main user interface shell and a background for your content pages, and begin work on the Lab page of your portfolio. Finally, you'll bring in your first completed content: the home page and sound control!

CHAPTER 4

IMPORTING GRAPHICS

IN THIS CHAPTER

Introduction

Chapter 2 focused on drawing vector graphics within Flash, a common skill you will need for many projects. In many cases, you may also require bitmap graphics, such as those created in Adobe Photoshop, or you may need to import vector graphics that originated in Adobe Illustrator or another drawing application. This chapter describes how you can use these assets, and provides a few related workflow tips. You'll also learn a few techniques for creating animations from imported assets.

Importing Graphics

As the chapter progresses, you'll see that Flash supports the importing of a wide variety of graphic file formats. For simplicity, however, the import process is initiated the same way for all formats. Regardless of file format, you need only invoke a menu command and choose one or more files from a standard operating system file browser. Flash then presents the appropriate import dialog based on the chosen file type.

You will most likely want to import the file directly to the Stage (File→Import→Import to Stage). With an unlocked layer selected in the Timeline, importing a standard graphic file places the imported asset on the Stage, adds it to the active layer and frame, and places it in the Library. Flash support for native file formats, such as Adobe Photoshop PSD files, includes additional import options that allow the creation of layers and frames based on the structure of the imported file.

If you don't plan to use the asset immediately, you can import it to your file's Library (File→Import→Import to Library). This can be very helpful if you intend to batch-import several assets at a time.

Importing Nonnative Formats

Flash supports most standard graphics file formats, as well as several formats typically associated with specific operating systems. Table 4-1 shows the most common file formats available for import into Flash on both Mac and Windows operating systems. Table 4-2 shows additional file formats that are supported when QuickTime 4.0 or later is installed on your Mac or PC.

Table 4-1. *Graphics file formats supported for import into Flash*

Format	Extension	Additional notes
JPEG	.jpg, .jpeg	
PNG	.png	If the file includes transparent areas, transparency is preserved; see the "Importing from Other Native Formats" section later in this chapter.
GIF	.gif	If the file includes transparent areas, transparency is preserved; static and animated GIFs are supported; animated GIFs are converted to keyframes.
AutoCAD DXF	.dxf	3D is not supported and fills are not maintained; best for 2D line drawing, such as floorplans.
Bitmap	.bmp	
Windows Metafile	.wmf	Grouped when imported to the Stage.
Enhanced Windows Metafile	.emf	Grouped when imported to the Stage.
Flash SWF	.swf	SWF must not be protected; flattened and grouped when imported to the Stage; supported animation is converted to keyframe sequence. See the "Importing from SWF" section, later in this chapter.

Table 4-2. *Graphics file formats supported when QuickTime 4.0 or later is installed*

Format	Extension	Additional notes
TIFF	.tif, .tif	Flattened when imported.
Targa	.tga	Embedded alpha channels are supported.
PICT	.pct, .pic, .pict	Imports as a flattened bitmap on Windows.
Silicon Graphics Image	.sgi	
QuickTime Image	.qti	
MacPaint	.pntg	

Importing from SWF

The SWF ("swiff") file Flash creates when you test or publish a FLA is in a compiled proprietary format. Nevertheless, you can import graphics—and, to some degree, animations—from a SWF as if they were self-contained graphics files. When you import to the Stage, Flash adds graphical assets to the Library and approximates animations in the main Timeline as a sequence of keyframes. When you import to the Library, Flash imports graphical assets, but ignores animations and does not add keyframes to the Timeline.

Importing from a SWF is possible only if the file was not protected when compiled. Protecting your assets from importing is an optional feature that prevents others from appropriating your content. This feature is found in the file-specific Publish Settings (File→Publish Settings) in the Flash section under Advanced→Trace and debug (Figure 4-1). You will learn more about publishing your files for distribution in later chapters, but this feature, off by default, can be enabled to prevent an FLA from importing SWF content.

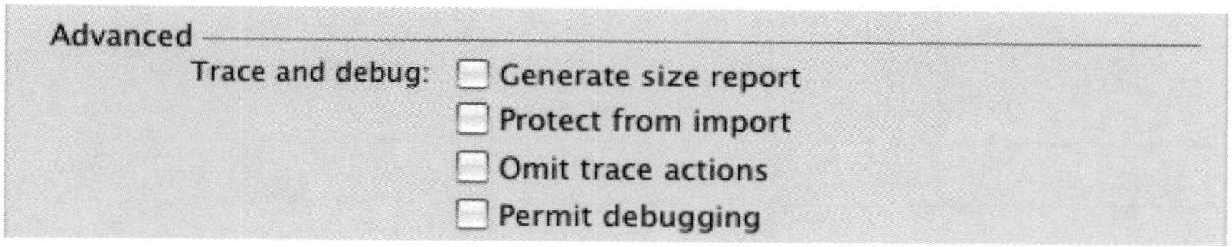

NOTE

When you import a SWF, Flash does not import the file as a whole. That is, you cannot embed a SWF inside another SWF. You will learn how to load an external SWF at runtime when reading about components in Chapter 9. For now it's enough to know that you cannot import a functional SWF during authoring.

Advanced
Trace and debug: Generate size report
Protect from import
Omit trace actions
Permit debugging

Figure 4-1. The Protect from import feature is disabled in Publish Settings to allow import from a SWF

NOTE

You can experiment with SWF imports by using the companion source code, which includes example SWFs with and without animation.

Importing from Adobe Photoshop

For Photoshop and Flash users, design productivity dramatically improved with the introduction of Flash's PSD importer, which can import multilayered PSD files in seconds. Flash not only understands the PSD file architecture, but can also convert select elements of a PSD to Flash assets (like movie clips) during the import process.

An average application or website interface created in Photoshop used to take hours to bring into Flash. You had to save every layer to individual JPEG or PNG files, import them all, reassemble everything on stage, create any necessary movie clips, reapply filter effects, and so on. Using the PSD importer, the same process is nearly automatic and can be completed in seconds. In addition, a useful preferences setting can preconfigure the importer to use your most commonly requested options, speeding up the process even more.

Supported Photoshop features

Flash supports many of Photoshop's layer features when importing a PSD. Table 4-3 shows a concise rundown of supported features, including known limitations.

Table 4-3. PSD features supported during Flash import

Feature	Notes
Color Mode	Only RGB is supported. CMYK will be converted to RGB.
Image Layers	Rasterized (converted to bitmap) to Flash bitmaps, transparency is preserved when converted to *Bitmap image with editable layer style*; when imported as a flattened bitmap, underlying layers seen through the transparency will be merged with the image layer.
Text Layers	Can be rasterized, converted to vector outlines, or left as an editable text element; text on a path must be rasterized to preserve the appearance.
Shape Layers	Can be rasterized or left as editable paths.
Merged Layers	Can be separated or converted to movie clips; adjacent layers can also be merged during import.
Layer Groups	Converted to Flash layer folders.
Smart Objects	Editability is not supported; the objects will be rasterized.
Layer Styles	Are rasterized.
Blend Modes	Compatible blend modes such as Normal, Darken, Multiply, Lighten, Screen, Hard Light, Difference, and Overlay, for example, are converted to Flash blend modes when *Bitmap image with editable layer style* is selected; incompatible blend modes are not applied, but their appearance can be preserved if rasterizing the layer to bitmap.
Adjustment Layers	Not supported; will be applied if the PSD is imported as a flattened bitmap.

Using the Import dialog

Initiating the import process for PSD files is no different than when importing a standard JPG. You can import to the Stage (File→Import→Import to Stage) or to the Library (File→Import→Import to Library). When you select a PSD file, however, Flash displays a special dialog for accessing the PSD layers (Figure 4-2).

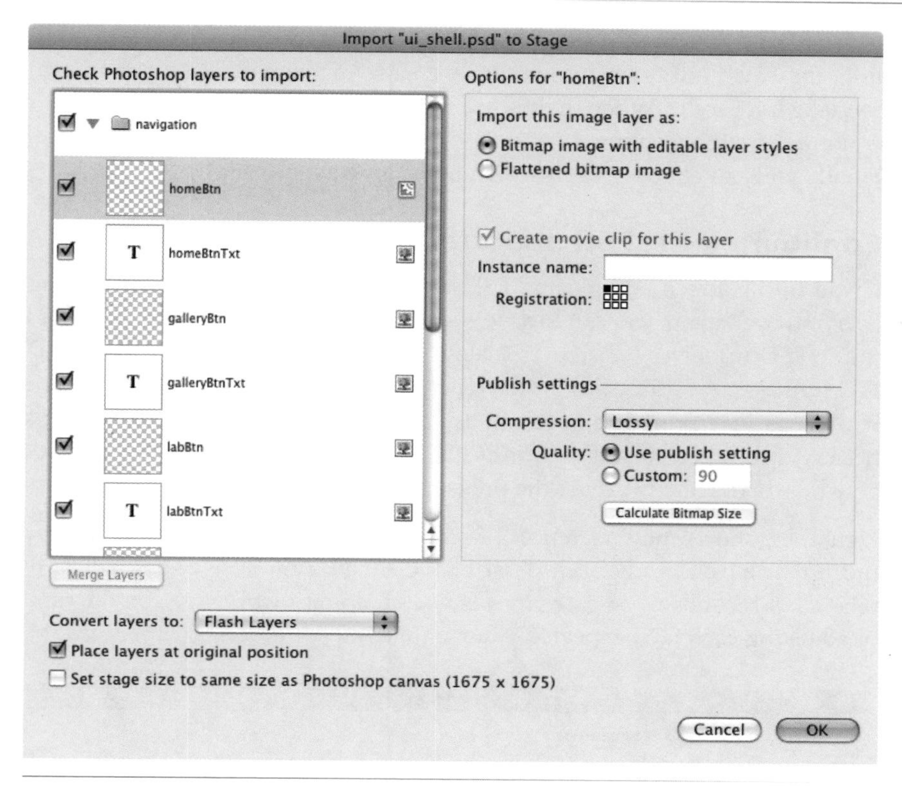

Figure 4-2. The PSD import dialog, setting properties for a layer of button art

When configuring the Import dialog, the first thing to decide is how you want the assets to flow into Flash. In the lower-left corner of Figure 4-2, the *Convert layers to* option specifies that PSD layers are converted to Flash layers or keyframes. Typically, you will convert to layers, but the keyframe option makes it possible to create an animation sequence in PSD layers and automatically import that sequence to a frame-by-frame animation. *Place layers at original position* preserves the location of layer elements during the import. *Set stage size to same size as Photoshop canvas* adjusts the Flash Stage size to match the dimensions of the PSD. This is handy when creating a Flash file from a PSD interface mockup.

After you determine the destination of your assets, it's time to configure each layer's import settings. You can configure each layer type according to the supported features described in Table 4-3. When either configuring image layers or rasterizing another layer type (converting to a bitmap image), you can set compression settings on a layer-by-layer basis. Compression is discussed in greater detail in the "Working with Bitmaps" section, later in this chapter.

In addition to choosing whether to maintain editability for each layer type or to rasterize to a bitmap, you will need to choose whether or not to create a movie clip symbol for the layer during the import process. A movie clip is necessary if you want to exert ActionScript control, animate the object using

NOTE

As an added convenience, dragging and dropping a PSD file from your operating system onto the Flash Stage invokes the PSD importer. This is not recommended for other file formats, as it may affect how transparency is rendered. In these cases, it is wisest to import the asset.

tweens (described in the next chapter), apply color effects, and more. In addition, the Flash importer automatically creates movie clips if you choose to preserve layer styles. When in doubt, encapsulate the layer in a movie clip, but you need to supply an instance name only if ActionScript control is required. Finally, you can choose a registration point for each movie clip.

Configuring the PSD Importer preferences

If you find yourself configuring PSD layers for import in a similar manner over and over again, you can save yourself a lot of time by setting these values in the PSD Importer Preferences dialog (on a Mac, Flash→Preferences→PSD File Importer, on a PC, Edit→Preferences→PSD File Importer). In fact, even if you end up having to import an asset more than a time or two (perhaps because you notice mistakes in your PSD file after importing it), it will probably be worthwhile to adjust the preferences for your immediate needs.

Figure 4-3 shows the Preferences dialog. As you can see, the settings mirror those that have already been described. Changing these settingss will only take a few seconds to do each time. However, doing so will save you the time of adjusting each layer repeatedly during the import process.

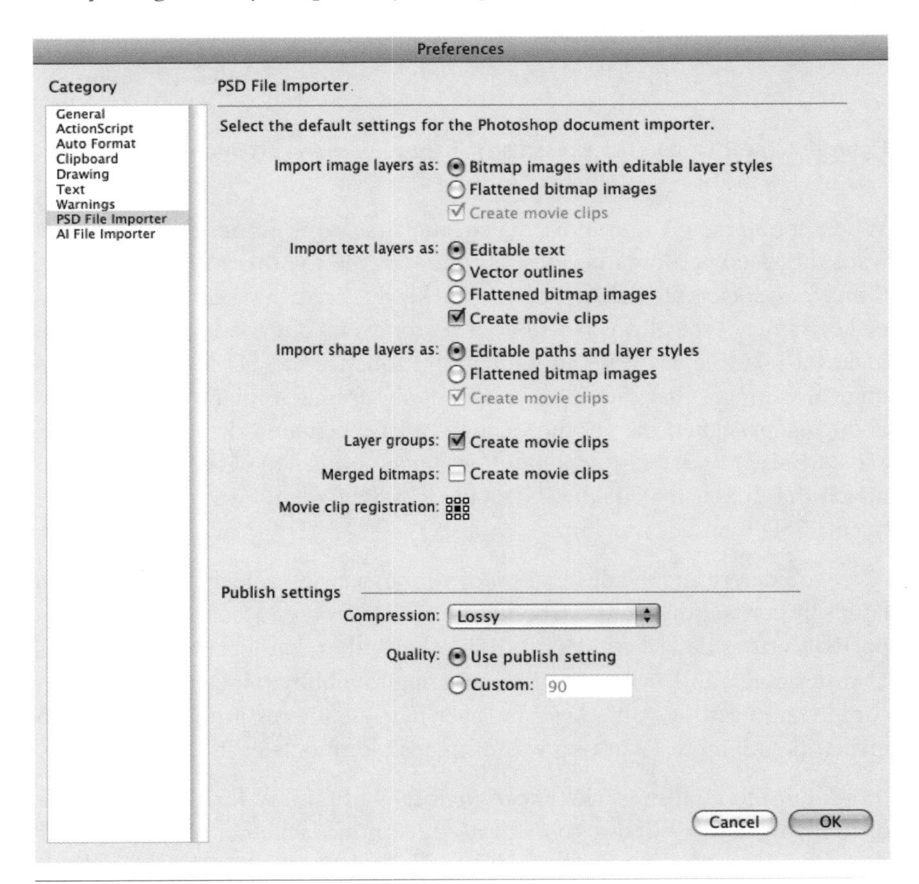

Figure 4-3. The PSD File Importer preferences

Importing from Adobe Illustrator

Importing Adobe Illustrator AI files is very similar to importing PSD files. It is, in fact, easier for two reasons. First, Flash and Illustrator have more in common from a vector-editing standpoint, so many of the features are more compatible from the outset. Second, Flash-specific features have been added to Illustrator that make it possible to create Flash symbols and text elements in Illustrator documents.

Supported Illustrator features

Translating Illustrator assets to the Flash world is relatively straightforward. Table 4-4 lists the most common Illustrator features you may wish to preserve and how Flash handles them.

Table 4-4. AI features supported during Flash import

Feature	Notes
Color Mode	Only RGB is supported. CMYK will be converted to RGB.
Transparency	Editability is preserved.
Clip Masks	Editability is preserved.
Gradient Fills	Editability is preserved.
Pattern Strokes and Fills	Editability is preserved.
Effects	Compatible effects such as Drop Shadow, Blur, Glow, and Bevel, for example, are converted to Flash filter effects; incompatible styles are not applied.
Blend Modes	Compatible blend modes such as Normal, Darken, Multiply, Lighten, Screen, Hard Light, Difference, and Overlay, for example, are converted to Flash blend modes; incompatible blend modes are not applied.

The real workflow enhancement, however, comes from the ability of any Illustrator user to originate select Flash assets right within Illustrator. Illustrator can create Flash movie clip and graphic symbols, as well as Flash static, dynamic, and input text fields. For more information, see the sidebar "Creating Flash Symbols and Text in Adobe Illustrator," next.

Creating Flash Symbols and Text in Adobe Illustrator

While Flash features the workflow improvements of the AI File Importer, Adobe Illustrator has a few improvements of its own. As of Illustrator CS3, you can also configure object settings that are applied while the file is imported into Flash. You can actually precreate two native Flash asset types while still in Illustrator: symbols and text fields.

Illustrator has featured symbols for some time, but the symbol workflow has been enhanced to closely resemble Flash practices. You can even use the same Convert to Symbol command in Illustrator with a matching keyboard shortcut (F8).

When adjusting settings, the symbol's ***Name*** property still exists from prior versions of Illustrator. Now, however, the name is inherited by the corresponding symbol created when importing into Flash. Enhanced features, introduced with Illustrator CS3, include the ability to make the symbol a graphic or movie clip, specify a registration point, and enable guides for ***9-slice scaling*** (Figure 4-4).

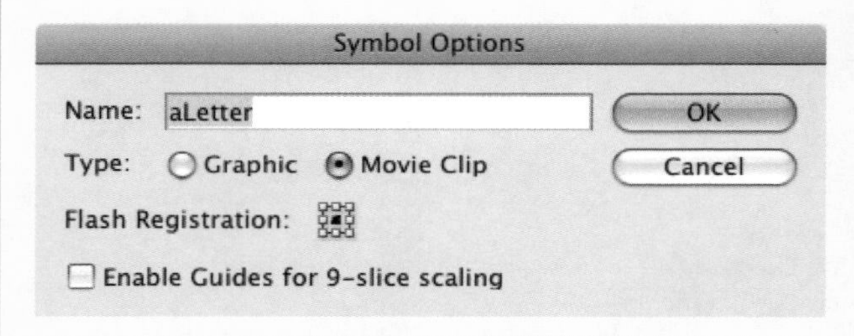

Figure 4-4. Adobe Illustrator's Symbol Options dialog

The 9-slice scaling feature helps prevent distortion when scaling vector objects. Behind the scenes, this feature slices an asset into an adjustable grid of nine pieces (the object remains intact for editing). Figure 4-5 shows an Illustrator symbol with 9-slice scaling guides enabled.

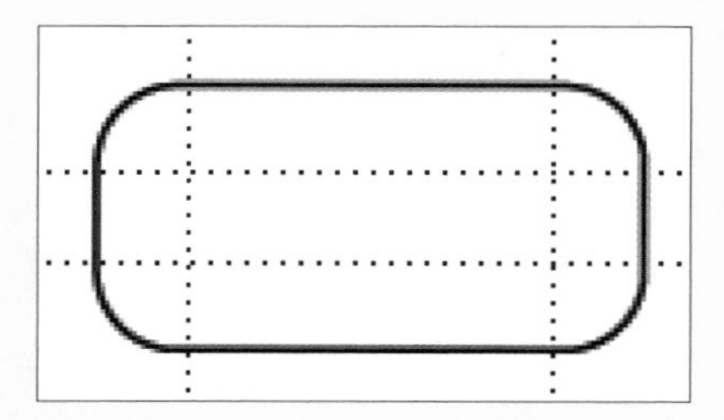

Figure 4-5. An Adobe Illustrator symbol with 9-slice scaling enabled

The corner slices are never scaled, the left and right slices are only scaled vertically, the top and bottom slices are only scaled horizontally, and the center slice is scaled in both directions. This is ideal for features like rounded corners where you don't want the corner radius to change if the object is scaled.

You can also create Flash text fields in Illustrator CS3 and later. With a text object selected in Illustrator, the Flash Text panel (Figure 4-6) becomes accessible in the Control bar above the artboard. Using this panel, you can configure three Flash text field types: ***Static***, ***Dynamic***, and ***Input***. In Chapter 2, you created Static text fields that are converted to graphics when publishing your SWF. In Chapter 11 you'll learn more about ActionScript-programmable Dynamic and Input text fields, including how to set antialiasing rendering options, background and border visibility, and text orientation. You can even create links and embed fonts to ensure consistent display on any computer.

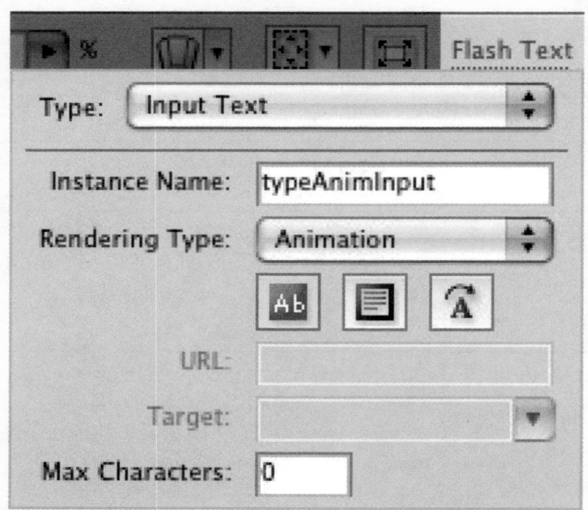

Figure 4-6. Adobe Illustrator's Flash Text panel

After configuring a text object's Flash Text properties, a corresponding Flash text field will be created when the asset is imported. This work prior to importing is relevant to all users, regardless of Illustrator experience, because you may collaborate with other designers that use Illustrator. Mirroring a real-life workgroup environment, these features are used in source files you will import later, in the "Project Progress" section of this chapter.

For more information on Flash asset creation in Illustrator, see this book's companion website.

Using the Import dialog

Flash's AI Importer dialog (Figure 4-7) is nearly identical to its PSD Importer dialog. The only appreciable differences are the fact that you can import unused symbols (symbols in the AI file's symbol library but not used on the artboard), and the fact that all Flash symbols and text elements created in Illustrator will be preconfigured when the import process begins. In addition, the importer can alert you to incompatibility issues, such as when an asset with a print color mode (CMYK) is imported into the web color (RGB) world of Flash. The Incompatibility Report button immediately below the list of Illustrator layers will generate a report describing any known issues.

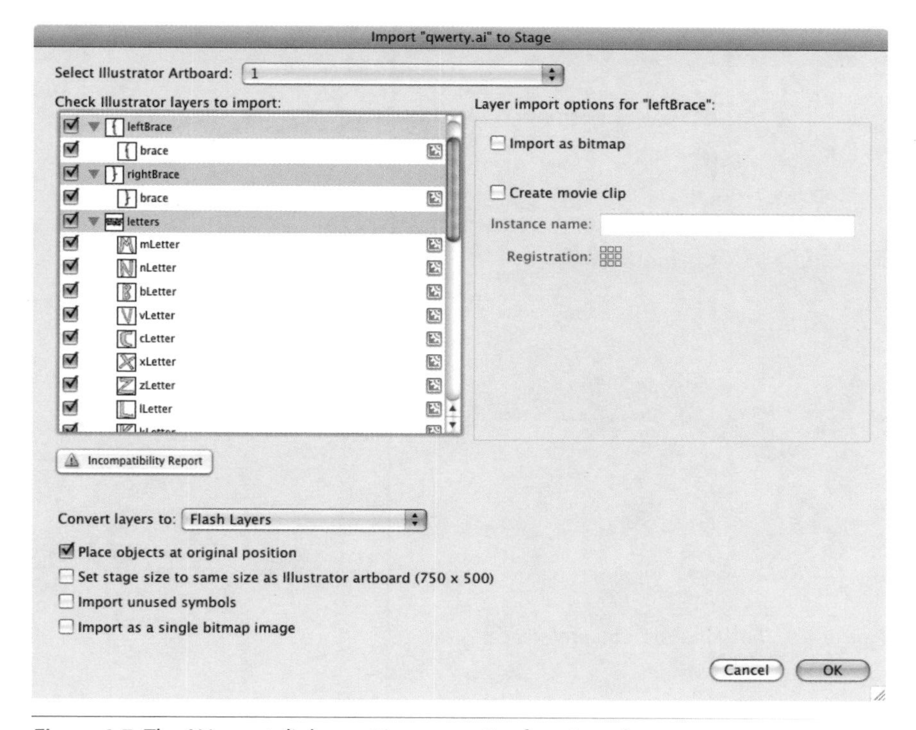

Figure 4-7. The AI Import dialog, setting properties for a text element

A nice, loosely related difference in the two workflows is that Flash has an improved copy and paste process for assets originating in Illustrator. Copying and pasting assets from Illustrator triggers a dialog (Figure 4-8) that prompts you for a method for handling the clipboard data. You can paste as bitmap or use the AI import preferences, and you can allow Flash to resolve conflicts and maintain layers.

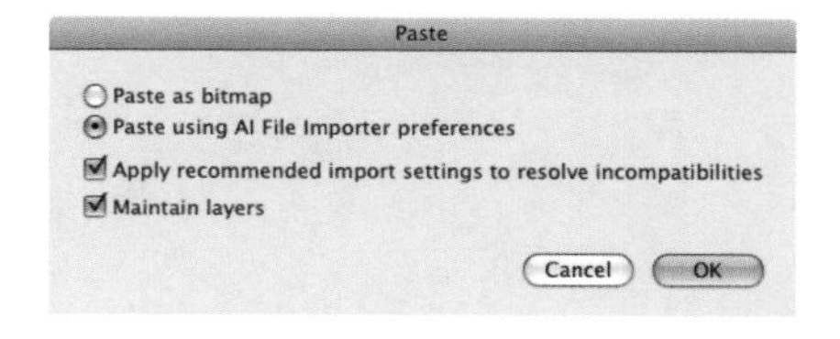

Figure 4-8. Adobe Illustrator's Paste options dialog

Configuring the AI Importer preferences

As with the PSD Importer, you can save a lot of time by configuring the preferences option for the AI Importer (Figure 4-9). Its Preferences dialog is nearly identical to the PSD Importer's, with one notable exception: you can exclude any objects outside the Illustrator artboard. This is convenient for quickly excluding assets during the import into Flash without deleting them in Illustrator.

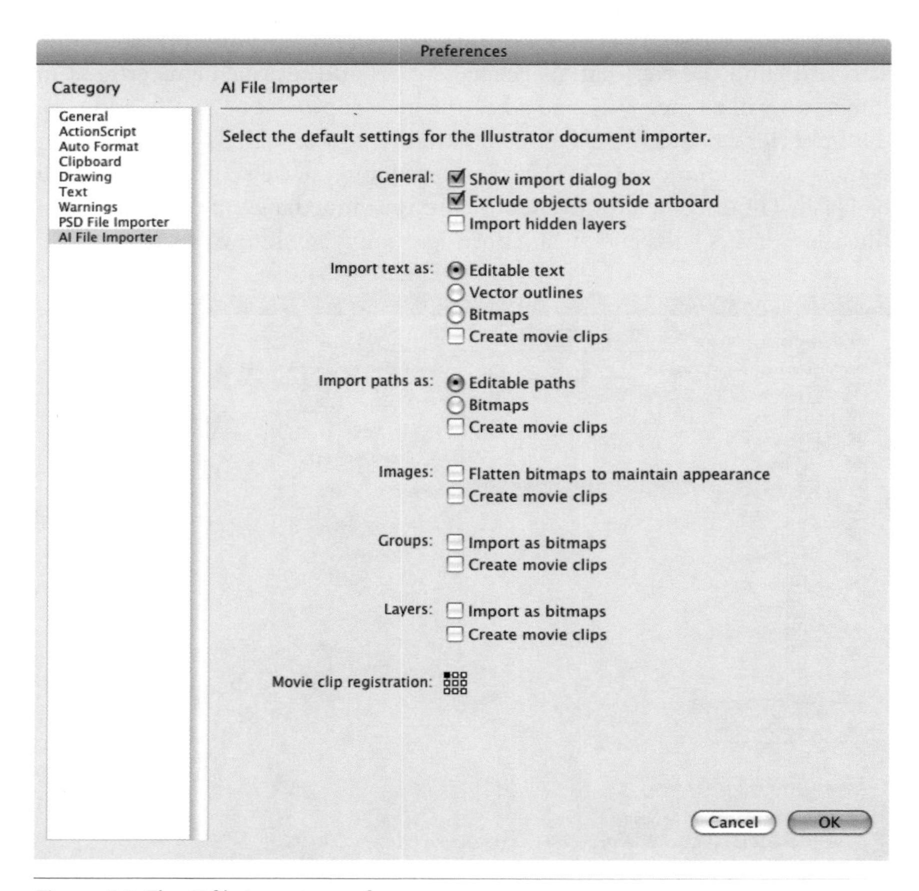

Figure 4-9. The AI file Importer preferences

Importing from Other Native Formats

Flash supports enhanced import options for two more native formats:

Fireworks

Importing Fireworks files preserves relative asset locations, editable paths and text, compatible blend modes and effects, and more. It will even convert multiple Fireworks pages into their own Flash scenes, if desired. If preferred, you can import a Fireworks asset as a flattened PNG and use Flash's round-trip editing features to edit changes in Fireworks. You can Control+click (Mac) or right-click (Windows) on the asset in the Library and choose the Edit in Fireworks option to open the file in Fireworks. After effecting the changes, the Flash file will be updated. Note that this only works on files that originated as Fireworks native files and were imported. It will not work on a flattened PNGs or files that were pasted into Flash.

Freehand

Although the application isn't still commercially available, legacy Freehand users will appreciate the fact that Freehand import support continues in Flash CS4 Professional. Freehand pages can be converted to Flash scenes or keyframes, and Freehand layers can be converted to Flash layers or keyframes, both of which make it possible to create animations from Freehand assets and to simply import artwork.

NOTE

The companion website has additional information and sample files for importing assets from Fireworks and Freehand. Even if you don't have Fireworks or Freehand, you can experiment with these features by importing files provided with the companion source files.

Importing an Image Sequence

Most of the animations you will create in Flash will originate in the Timeline, as discussed in the next chapter. However, it's also possible to create animations simply by importing external assets. For example, you can import an image sequence by filling a folder with several sequentially numbered graphics (*image_01.jpg*, *image02.jpg*, and so on). Importing the first one prompts Flash to ask if you want to import a sequence; if you consent, Flash automatically imports all the images and creates sequential keyframes for each. Instant animation!

You can even reverse the sequence automatically by selecting all of the frames (select the first frame of the sequence, hold down the Shift key, and select the last frame) and using the Modify→Timeline→Reverse Frames menu command. This is great for creating animation cycles. For example, you could import an image sequence of a squirrel peeking out of a hollow tree. You could then copy and paste the frames, reverse the sequence, and see the squirrel retreat back into its den.

Another nice feature helps when you want to animate more than one discrete element from a single layer. You will learn in the next chapter that timeline animations require that only one element be in each layer, so you must distribute each element to a layer of its own. Select the entire frame in which your multiple objects reside, and use the Modify→Timeline→Distribute to Layers feature. You can then apply the animation techniques that you'll learn in the next chapter to animate each layer.

Working with Bitmaps

Care for bitmaps doesn't end at the import stage. Flash includes a few features designed to help you work with pixel-based artwork in authoring mode. Options like manipulating compression settings, converting an image to vectors, and using a bitmap as a fill allow you to do more than just display static bitmaps.

Publish Settings

Publish Settings (File→Publish Settings) is a file-wide set of preferences that allows you to control many ways in which your SWF will behave. One such feature is the JPEG compression quality applied to certain embedded bitmaps (Figure 4-10). You can set the quality of the image and enable JPEG deblocking to help smooth out some of the visual artifacts caused by compression. This global setting applies to images pasted into Flash or imported as part of a larger file by way of one of the native file-format importers (such as the PSD Importer). It doesn't apply to individually imported bitmaps. To specify settings for the latter, use the Bitmap Properties feature described in the next section.

NOTE

To learn more about Publish Settings, see Chapter 14.

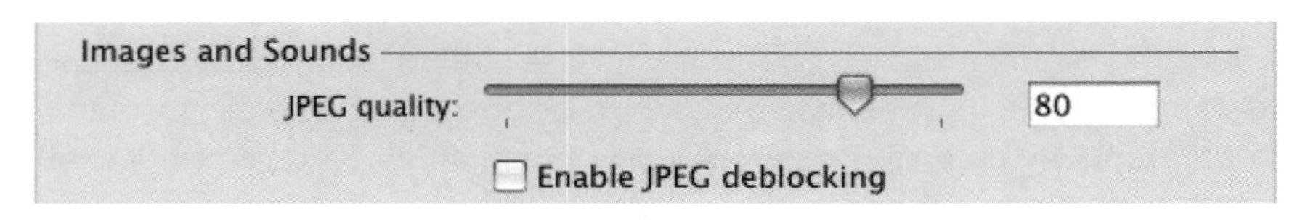

Figure 4-10. The Publish Settings JPEG compression option

Bitmap Properties

You can also vary compression settings on an asset-by-asset basis. If you select a bitmap in the Library and click the Properties button in the lower-left corner of the panel, you will open the Bitmap Properties dialog. Shown in Figure 4-11, the available options follow.

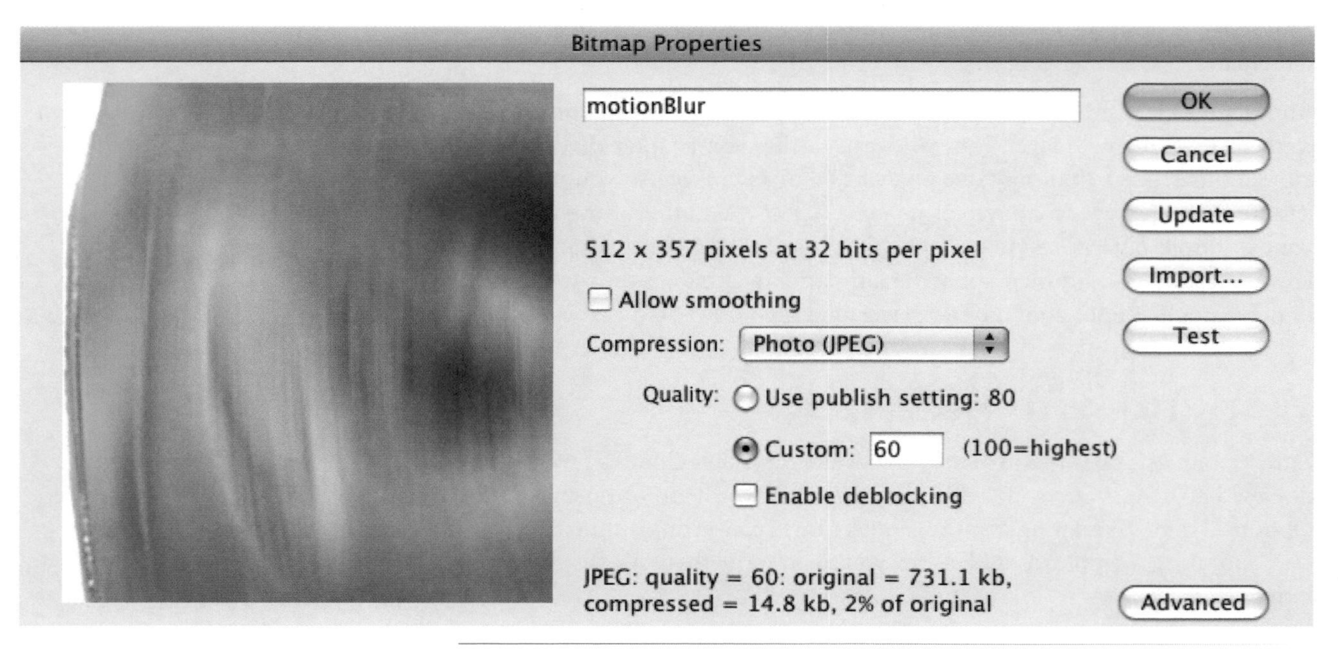

Figure 4-11. The Bitmap Properties dialog

Update and Import

When a bitmap is imported, clicking Update replaces the internal asset with the external source. This is a simple way to update a bitmap with a more current version without having to import a new file and replace all instances derived from it. If the bitmap was pasted into Flash or imported as part of a larger file by one of the native file-format import assistants, Update serves the same purpose as Import. Clicking Import opens a file browser dialog and allows you to import a replacement asset. All instances created from the prior bitmap are updated automatically.

Compression

This setting lets you choose between a lossy compression algorithm, called *Photo (JPEG)*, and a lossless option, called *Lossless (PNG/GIF)*. You need not select an algorithm based on the file type. For example, you can choose the lossless compression option for a bitmap imported from a JPEG.

NOTE

Historically, JPEG compression is best for continuous-tone images (such as photos and gradients), PNG compression is best for transparency, and GIF compression is best for large areas of solid colors. However, because Flash allows you to switch algorithms after an image has been imported, the choice of lossy versus lossless compression ultimately comes down to a balance of visual quality and file size.

Quality

If the *Lossless (PNG/GIF)* compression option is selected, no quality options are available. If the *Photo (JPEG)* option is selected, you can dictate whether Flash uses a preset quality value or a custom setting. The custom setting allows you to specify any quality value and enable deblocking. There are two types of preset settings. If the bitmap was imported directly, you can use the imported JPEG data (not pictured in Figure 4-11). That is, if you customize the compression values in Photoshop, this setting will use that data. If the file was pasted in or imported by a native file importer, you can tell Flash to use the file-wide publish settings quality value.

Test

Clicking this button test-compresses the bitmap using the compression option you specify. It also displays text at the bottom of the dialog showing the before and after size of the file and the percent of file size saved. This is handy for experimenting with quality sizes to see how much it will affect the size of the final SWF.

NOTE

The advanced bitmap properties options not pictured in Figure 4-11 are related to ActionScript and will be discussed in Chapter 6.

Allow Smoothing

This option smoothes the edges of images and is helpful when scaling bitmaps, particularly during animation.

Tracing a Bitmap

Occasionally, you may have a need or desire to convert a bitmap to a collection of vectors. The most compelling reason to do this is to edit or animate the graphic. Another useful reason to trace a bitmap is to scale the image to a much larger size. Enlarging a bitmap can cause blurriness and make the individual pixels too pronounced—an effect called *pixelation*. When you enlarge a vector, however, it remains crisp at any size because it is composed of lines and points that are redrawn each time the vector is scaled.

To trace a bitmap, select the bitmap and use the Modify→Bitmap→Trace Bitmap menu command. Use the dialog that appears (Figure 4-12) to set tolerances used when tracing. *Color threshold* determines when a new color in the bitmap is identified for tracing. The lower the threshold, the more colors will be seen as unique, and the more vectors will be created. *Minimum area* sets a minimum limit for the number of adjacent pixels within the same color threshold that must be found to create a new vector. The *Curve fit* menu contains six presets that prompt the trace routine to fit curves tighter or looser to the original image color areas. Tighter curves are more accurate and looser curves "round out" the color areas. Finally, the *Corner threshold* works much the same way, allowing you to choose more or fewer corners in the final trace. More corners make the trace closer to the original.

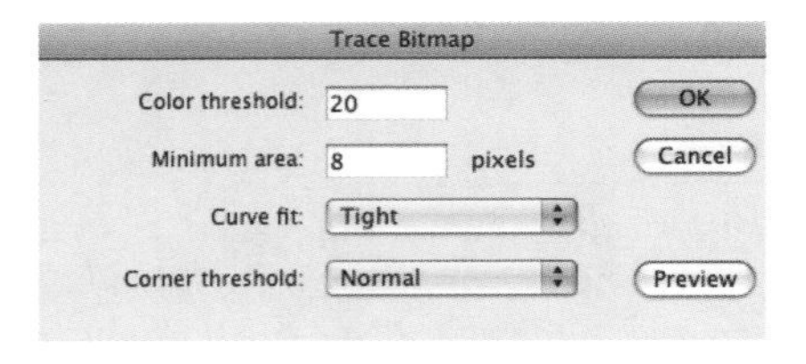

Figure 4-12. The Trace Bitmap dialog

NOTE

Look to the companion website for an interesting example of tracing bitmaps that includes ActionScript animation and a round-trip editing workflow from Flash to Illustrator and back again.

As the trace's visual fidelity improves, the resulting vectors become more complex. Smaller Color Threshold and Minimum Area settings, tighter curves, and more corners create many more vectors that must be drawn by the viewer's computer at runtime. The respective default values of 100, 8, Normal, and Normal produce a coarser tracing with fewer vectors. Figure 4-13 shows before and after trace images of a crumpled piece of paper, which appears as a background in the portfolio project. To achieve this level of accuracy, the values in Figure 4-12 were used, increasing the number of curves used in the trace.

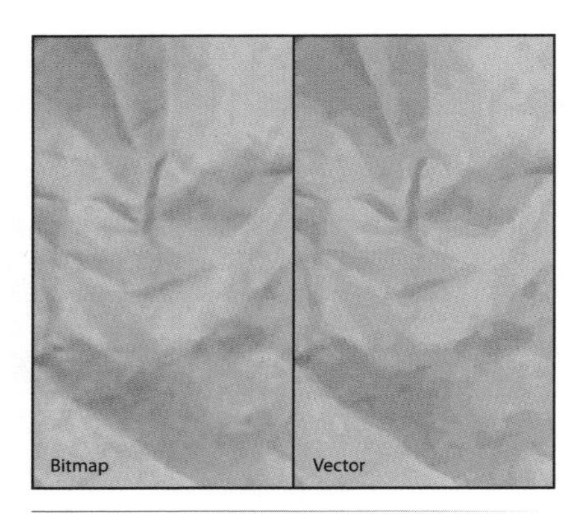

Figure 4-13. Before and after tracing a bitmap

Flash provides an Optimize Curves feature (Modify→Shape→Optimize) for reducing the complexity of a vector shape and, thus, the file size (Figure 4-14). You can vary the optimization strength from 0 to 100, preview the effect of the setting, and receive an optimization report. The vector image in Figure 4-13 was optimized with a setting of 100 and resulted in a 17% reduction in curves, but still maintained acceptable visual fidelity.

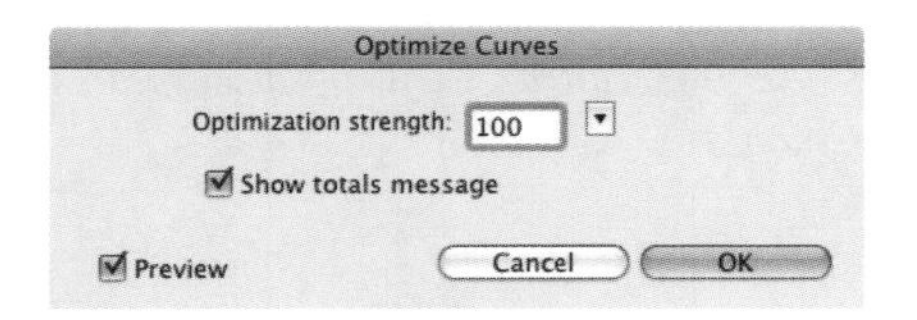

Figure 4-14. Optimizing curves to reduce file size

WARNING

Use bitmap tracing only when needed and follow the optimization techniques discussed in this chapter. It is tempting to include only vector-based assets in Flash, but a vector equivalent of a bitmap can be bigger and reduce performance at runtime.

Using a Bitmap As a Tile

Another bitmap manipulation technique is to use the bitmap as a tiled fill. This is most common when using photographic textures as fills or when filling large shapes where an equivalent bitmap would increase file size too significantly. Often, *seamless tiles* are used for fills, which are rectangles optimized to hide any seams when the tiles are joined. Figure 4-15 shows a seamless tile and a bitmap fill created from that tile.

To use a bitmap fill, open the Color panel, select the fill chip, and set the *Type* menu to Bitmap. This will show a series of thumbnails of all the bitmaps in the current file's Library panel. Select the bitmap you prefer and then apply the fill with the Paint Bucket just as you would apply solid or gradient color fills. Any new shape that you draw while the bitmap fill is active will also use that tile, and you can even apply tiles to strokes.

NOTE

The Gradient Transform tool can multitask when it comes to bitmap fills. The tool can rotate, skew, and scale the fill for interesting effects. The companion website contains more information about this technique.

Figure 4-15. A seamless tile bitmap (top) and a shape filled with a bitmap fill (bottom)

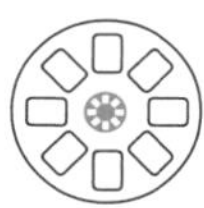

Project Progress

So far, you've created component parts of your portfolio. While that will continue, it's now time to create the master file into which all other assets will be placed. This file will contain the user interface elements common to all screens, including navigation buttons, logo, and so on.

Importing the Interface Shell

Before you import the native Photoshop document (PSD), it helps to understand how the file is constructed and what the portfolio will look like. In particular, the PSD uses an optimization technique that improves working with large bitmaps, and warrants scrutiny. Figure 4-16 shows the artwork in the PSD, and Figure 4-17 shows its layer configuration.

The design for the portfolio is based on a large viewing wheel that reveals each screen as it rotates. Figure 4-16 shows the wheel in the center of the document and the remainder of the interface in the upper-left corner. This arrangement allows you to import the entire interface in one process. The PSD is actually 1,675×1,675 pixels. The upper-left corner (outlined by a red reference outline that doesn't appear in the artwork) is 750×500 pixels, the size of the Flash document.

The portfolio's four navigation buttons are in the lower-left corner of the highlighted area. In Chapter 7, you will add filter effects to the Up, Over, and Down states of these buttons to give them their final appearances. The upper-right corner of this area will contain a logo you will import from Adobe Illustrator later in this chapter. The lower-right corner of the region is the eventual location of the sound controller you created in Chapter 3. Finally, the center of the highlighted area contains a graphic used to improve the experience of switching screens.

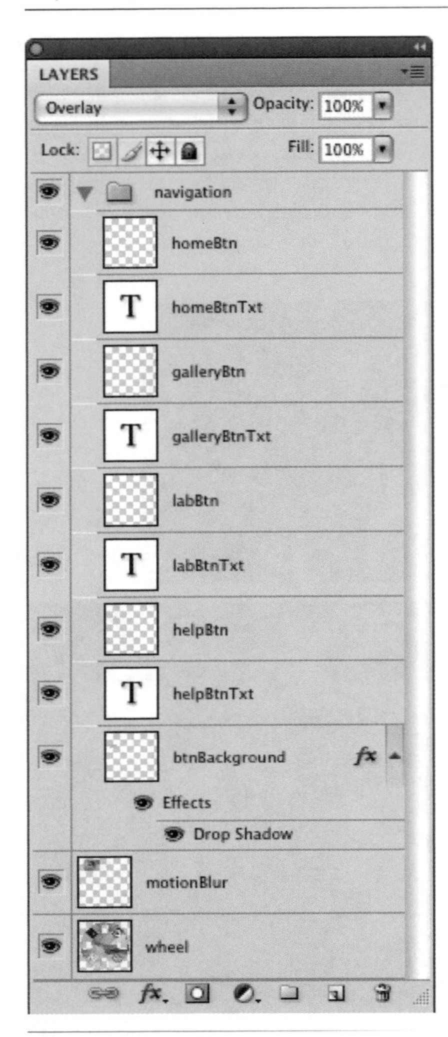

Figure 4-17. The project user interface PSD Layers panel

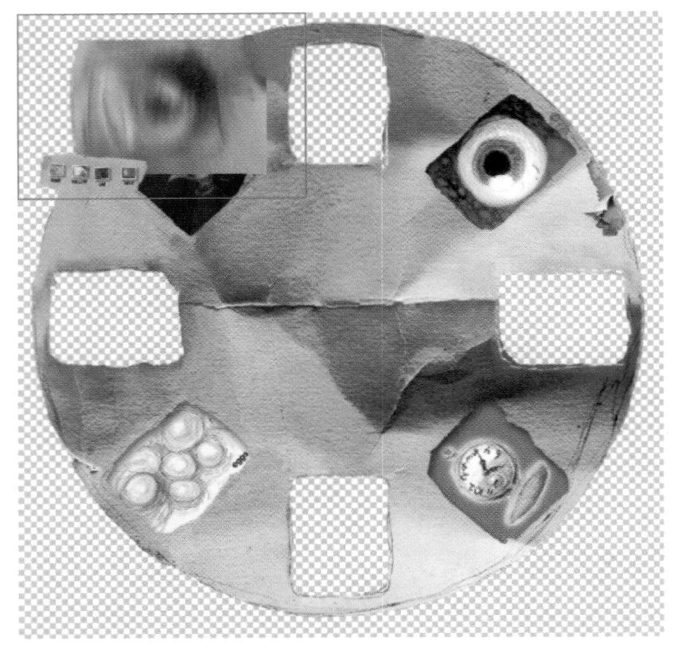

Figure 4-16. The project user interface PSD; the red outline, for reference only, indicates Flash file dimensions

When navigating to a new screen, the wheel rotates a few times and then slows down and stops at the chosen screen. To make the rotation time fast and consistent, the portfolio uses a *motion blur* illusion. Motion blur is a directional blur that fools the eye into thinking an object is in motion—a trick that has served animators for decades. When the wheel is rotating fast enough, the project fades in a graphic on top of the wheel. The graphic is a composite of the art in the wheel and is blurred to simulate circular motion. This reinforces the movement of the wheel in the viewer's mind. When the wheel begins to slow down, the motion blur fades away. This trick lets you move from any frame to any other frame in the same amount of time without appearing clunky or reducing the impact of the rotating wheel.

Figure 4-17 shows the layers in the PSD. The top nine rows of the document comprise the navigation system. Four pairs of button art and text sit atop a background, and all of these layers are contained in a layer group called *navigation*. Each button text layer is editable text, and a drop shadow layer style is applied to the background behind the buttons.

The motion blur and wheel reside in their own layers. Both are reduced to approximately 70% of their eventual size. This is a trick to reduce the file size of the final project SWF. Bitmaps can add a lot to file size, so it's sometimes helpful to reduce them prior to importing. However, if you reduce them too much, they will appear blurry when enlarged to full size. By reducing the images to this degree in the PSD, you will only need to enlarge them 140% in the FLA to reach their desired sizes.

1. Create a new FLA using the File→New menu command. This file will serve as the master project file into which all other assets will be placed, so you don't need to use the content template you created in Chapter 1. Save your file as *portfolio.fla*.
2. Change the document settings (Modify→Document) to apply the correct dimensions and Stage color. Set the size of the file to `750×500` pixels. Uncheck the *Adjust 3D Perspective Angle to preserve current stage projection* checkbox (you'll learn more about this and other 3D properties in Chapter 8). Set the Stage color to `black`.
3. Use the Import to Stage menu item (File→Import→Import to Stage) and, in the file browser dialog, select the provided source file, *ui_shell.psd*.
4. In the lower-left corner of the dialog that appears, set the *Convert layers to* option to `Flash Layers`, and enable the *Place layers at original position* option. Be sure the *Set stage size to same size as Photoshop canvas option* is not enabled. You want to preserve your file size, rather than match the much larger size of the interface's wheel asset.

If you set your PSD import preferences as described earlier in the "Configuring the PSD Importer preferences" section of this chapter, most of your work will be done for you. This is because Flash will automatically preconfigure each layer according to your preferences, based on the layer type.

5. If all of the layers in the Import dialog are already set to import as movie clips, your preferences match the options discussed herein. Skip to step 10. If you've chosen to configure each layer, continue with step 6.
6. Select all layers (click the *navigation* layer group at the top of the scrolling list, hold down the Shift key, scroll down to the bottom of the list, and click the *wheel* image layer). Enable the *Create movie clips for these layers* option and choose a center registration point.
7. Select each image layer and set it to *Bitmap image with editable layer styles*. While doing this, set the Publish Settings of each image layer to `Lossy` compression and choose the `Use publish setting` quality.
8. Select each text layer and set its import option to `Editable text`.
9. One by one, select the *navigation* layer group, and the *motionBlur* and *wheel* image layers, and give each movie clip an instance name that matches its layer name.
10. All that remains is to give the navigation buttons instance names. Select the *homeBtn* layer and name its movie clip `home`. Follow suit so *galleryBtn* becomes `gallery`, *labBtn* becomes `lab`, and *helpBtn* becomes `help`.
11. Click OK to complete the import process. Save your work. Your file should look like Figure 4-18. Compare it in greater detail to the furnished source file *portfolio_01.fla*. If your file doesn't closely resemble the furnished source, reconfirm your steps or continue with the provided file.

NOTE

As a typical project progresses, you will usually add several assets to your library. In addition to creating symbols as you work, you will likely import, copy and paste, or drag and drop symbols, bitmaps, sounds, and more, from external sources.

In doing so, your library will become cluttered, so you may want to organize it now and then by grouping assets into folders. How you arrange your files is often a matter of preference, but the companion source FLAs will periodically organize the Library to show you different ways of managing your assets.

For more information on using the Library, see the companion website.

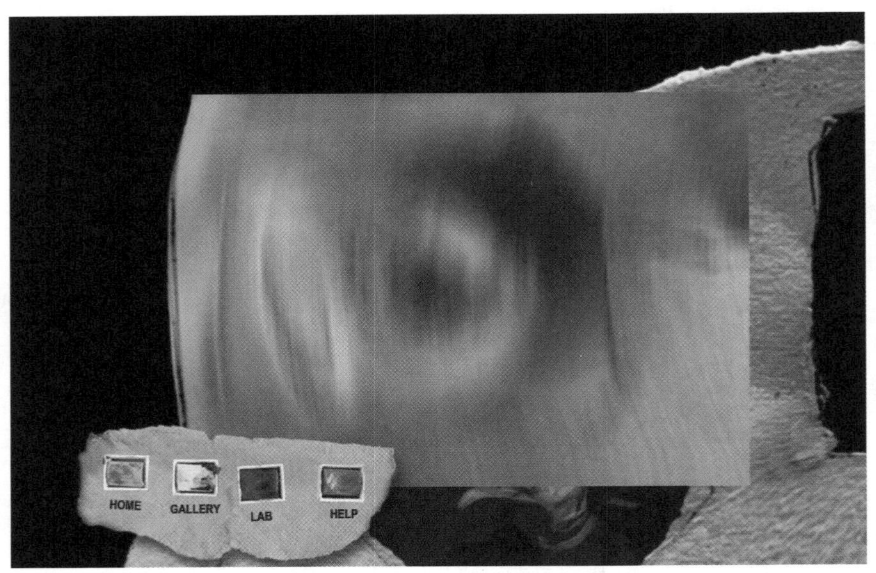

Figure 4-18. The user interface assets after importing into Flash from Adobe Photoshop

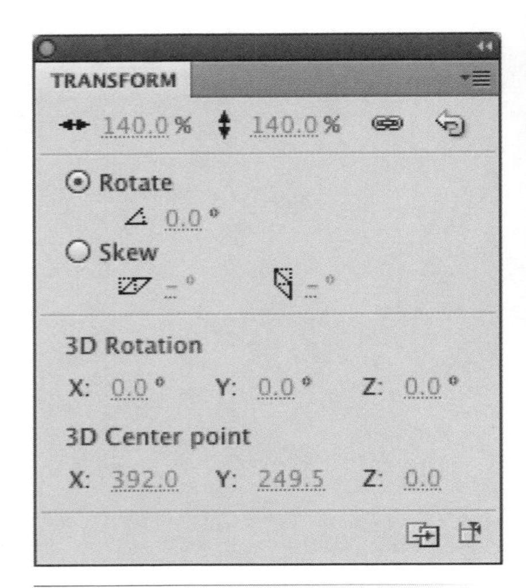

Figure 4-19. Transforming the motionBlur movie clip to a scale of 140%

12. Using the Transform panel (Figure 4-19), lock the link to the right of the scale properties and scale the *motionBlur* movie clip proportionally to **140%**. Wait to scale the wheel until the next chapter.

13. Save your work.

Importing the Background

When a page of the portfolio is displayed, its content will appear in one of the square holes punched into the viewing wheel. As such, you must use a contrasting background image so your assets don't disappear into the black Stage color. It would be trivial to change the Stage color, but that option is unappealing for two reasons. First, a flat-colored stage won't match the portfolio design and, second, the black stage color supports the illusion that the interface elements are free-floating and the wheel is spinning in place.

Another solution is to import a background graphic of crumpled paper to sit behind the wheel. This matches the design of the portfolio and, as the wheel spins, the empty frames reveal the background. When the wheel stops, your screen content animates in front of the texture.

1. In your *portfolio.fla* file, add a new layer at the top of the Timeline and name it **background**.

2. Select the frame in that layer and import (File→Import→Import to Stage) the provided source file *paper_background_small.jpg*. The smaller dimensions of this graphic will save on SWF file size later on, but the image must be scaled up to fill the area visible through the wheel frames.

Select the bitmap and, using the Properties panel, set its width and height to 625 and 500, respectively.

3. Create a movie clip from the bitmap. With the bitmap still selected, use Modify→Convert to Symbol. Name the movie clip paper and choose a center registration point from the nine-point grid.
4. Using the Properties panel, set the *x* and *y* of the movie clip to 395 and 250, respectively.
5. Click on the layer and drag it to the bottom of all other layers to make sure the background is below the wheel.

NOTE

Optionally, experiment with tracing bitmaps by tracing the crumpled background paper. Use an efficient setting, as described earlier in this chapter in the "Tracing a Bitmap" section, and compare the bitmap and vector versions of the asset. Determine which is more satisfactory: a bitmap scaled to 200% (with possible blurriness) or a traced bitmap (with a possibly coarser blending of colors).

Importing the Logo

Having imported the main interface assets from Adobe Photoshop, you'll now import two assets from Adobe Illustrator. The first is the portfolio logo, which you'll configure during the import process. The second is for the Lab screen you'll program later on, and that asset was preconfigured in Illustrator to smooth the import process. Before you get to that point, take a look at the logo (Figure 4-20) and the Illustrator file's layers (Figure 4-21).

Figure 4-20 shows a sneak peek at the logo after importing because its background and text appear white in Illustrator and are difficult to see. This is reflected in the layer thumbnails in Figure 4-21, in the *white* and *FOLIO* layers. The logo consists of four discrete elements: the word "Folio" in white text, a white background, a red center icon, and black outlines. All four parts are named appropriately and reside in a single layer named *logo*. The white logo background does not appear white in Figure 4-20 because a blend mode has been applied in Illustrator. You'll learn more about blend modes in Chapter 7 but, for now, you'll see that the blend mode is preserved during the AI import process and remains editable in Flash.

Figure 4-20. The portfolio logo after importing

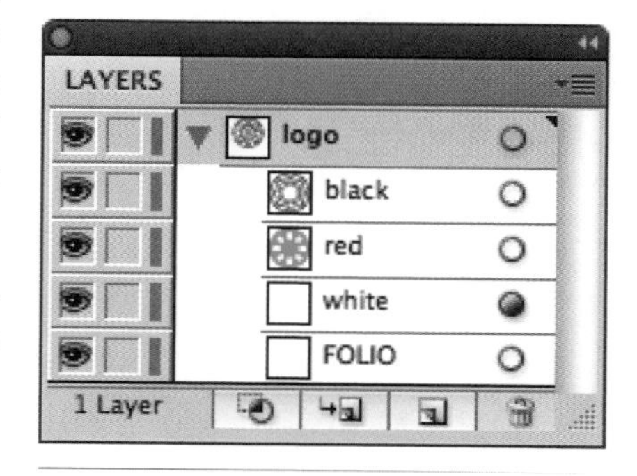

Figure 4-21. The layer structure of logo.ai

1. In your *portfolio.fla* file, click the *motionBlur* layer in the Timeline panel and then add a new layer. Don't worry about its name, as you'll change it later.
2. Select a frame in the new layer and import (File→Import→Import to Stage) the provided source file *logo.ai*.

The AI File Importer appears, and the layer structure of the file will be reflected in the dialog (Figure 4-22). The icons to the right of the element names indicate that the black and red elements are set to import as editable shapes, the text is set to import as editable text, and the white background element is set to import as a movie clip. If your icons don't match these options, revisit the "Configuring the AI Importer Preferences" section earlier in the chapter.

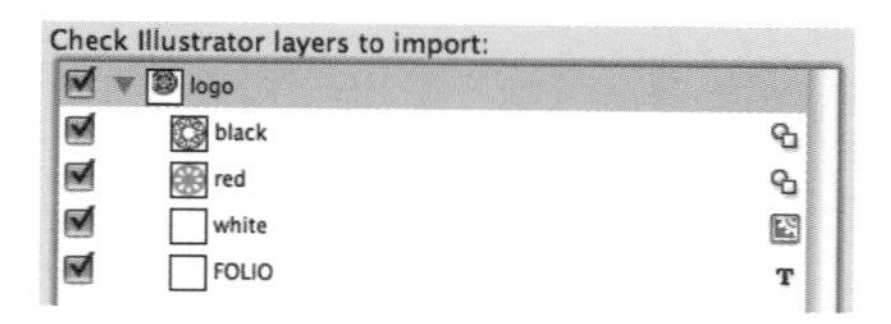

Figure 4-22. The logo parts in their unedited formats from the AI file

Even though the AI File Importer preferences discussed earlier in the chapter do not automatically create movie clips, the logo's white background element is set to import as a movie clip before configuring any further options. This is necessary to preserve the blend mode applied in Illustrator. If you alter this import option, the dialog will display a warning that an incompatibility exists, and the blend mode will not be translated into Flash.

3. Hold down your Shift key and click on the *logo*, *black*, *red*, and *white* elements and check the *Create movie clip* checkbox. Give the movie clips a center registration point by clicking the center box in the 9-box grid marked *Registration*. These options are shown in Figure 4-23.

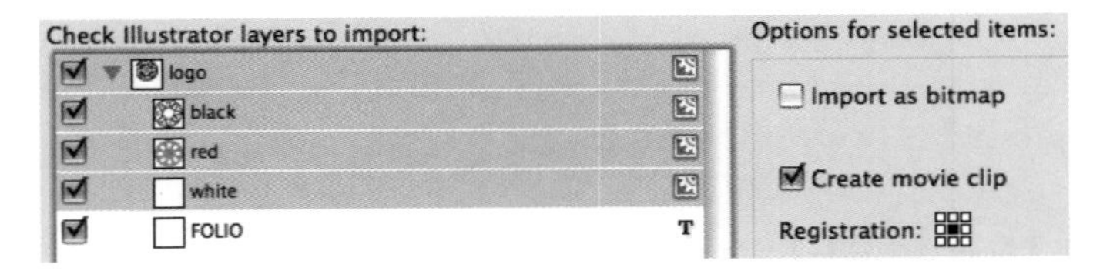

Figure 4-23. The logo parts after turning all into movie clips

4. Individually select the *logo*, *black*, *red*, and *white* elements and give them movie clip instance names that match their names in the dialog.
5. At the bottom of the dialog, choose to *Convert layers to* a Single Flash Layer, and to *Place objects at original position*. The Illustrator document is the same size as your portfolio FLA's stage, so the logo will appear in the correct position after import.
6. Click the OK button to import. You will notice that the layer name has assumed the name of the AI file. Rename it to *logo*.
7. Double-click on the movie clip on the stage, and you will see everything in a single layer. In Chapter 7, it will be easier to work with the logo as separate parts, so you a nice productivity tip to move each element to its own layer. Select everything by clicking once in the frame in the Timeline panel, and use the Modify→Timeline→Distribute to Layers menu command. A layer for each logo part will be created, along with the original, now empty default layer. Delete the latter, as you will no longer need it.
8. Return to the stage using Edit→Edit Document and save your file. You are now finished with the logo until you revisit it later in Chapter 7 to put blend modes to work.

Importing the Lab Screen

Lab, short for laboratory, is a name often given to an experimental section of a project. In the portfolio project, the *Lab* screen is a place to show design and coding experiments, and focuses on a text animation tool. In the finished project, the user types into a text field and triggers an animation in the corresponding on-screen letters.

Because the Flash assets for the Lab screen were prepared for you in Adobe Illustrator, the import process is nearly effortless. There is only one thing you need to do to compensate for a feature of the AI importer.

As you saw when importing the logo, when importing directly to the Stage, two options help control placement of assets. You can change the size of the Stage to match the dimensions of the Illustrator document, and you can place the imported assets at their original positions in the Illustrator document, relative to the upper-left corner of the Stage.

However, when importing into a movie clip, these options are not available. This behavior mimics the process of importing to the Library instead of the Stage, where position is not relevant. Unfortunately, you want to automatically position the elements in your movie clip, too. You're importing into a movie clip because you've prepared your workflow to ease asset integration in the future. When the time comes, you will transfer the Lab movie clip to the main project FLA, and your prep work will really pay off.

There are a few ways to work around the issue. First, you can import directly to the Stage and then create a new movie clip once the import is complete. In this case, you will need to manually adjust the location of the assets after the movie clip is created because the movie clip registration point will not be (0, 0). This will make transferring the movie clip to your main file more troublesome later on, so it is not ideal.

A better approach is to continue to use the content movie clip in your template. Import to the Stage, copy the newly imported assets to the clipboard, open the *content* movie clip, and then Paste in Place (Edit→Paste in Place) to preserve their prior locations. Because the registration point of your content movie clip is also (0, 0), the assets will be right where you want them. This is the approach you should use when confronted with a similar situation without control over your imported assets.

However, there's another trick that continues your workflow optimization by preparing for this issue while still in Illustrator. By simply adding a placeholder asset in the upper-left corner of your matching-size Illustrator document, the placeholder will correspond to point (0, 0) after import and all the assets will be correctly positioned. This trick has been used in the furnished source files, and all you need to do is remove the placeholder asset:

1. Create a new file using your *Content* template and double-click the *content* movie clip to prepare for importing.
2. Create a new layer and delete the placeholder layer. Your movie clip should now have only one layer with nothing in it.
3. Select the frame in that layer and import (File→Import→Import to Stage) the provided source file *qwerty.ai*.
4. All the work you would have put into the Import dialog has been done for you in Illustrator, so just click OK to continue with the import.
5. Once the import is complete, the assets will be correctly positioned and you will see a small pink square in the upper-left corner. Look at the

NOTE

Remember, importing AI files is nearly identical to importing PSD files, so you can easily experiment with the process using your own files. Rather than focusing on that process again, this portion of the exercise focuses on workflow and collaboration between Illustrator and Flash users.

Timeline panel and you will see a layer at the top of the Timeline called *placeholder*. Delete this layer, and you're done.

6. Save your file as *lab_page.fla* and compare it to the provided source file, *lab_page_final.fla*.

Figure 4-24 shows the finished movie clip after removing the placeholder layer. Poke around and see that the braces, letters, and the tabbed card at the bottom of the screen are all movie clips. Double-click the tabbed card and look inside to find a group. Grouping elements in Illustrator helps preserve their relative locations during the import process. Break this group apart (Edit→Break Apart), and you'll find that the text field already has an instance name. This is ready to program with ActionScript, which you'll do in Chapter 6.

Figure 4-24. The Lab movie clip after importing into Flash from Adobe Illustrator

The Project Continues...

In the next chapter, the heavy lifting begins. You'll position all the UI elements, apply some Timeline animations to introduce the portfolio when it loads, and add the content you've created to date.

ANIMATION

Introduction

For many users, animation is the cornerstone of Flash. Sure, there are large groups of users who choose Flash over other tools because of its video capabilities, or for use as an application development tool, or even to learn programming in a fun, visual way. A big part of the Flash user base, however, employs the application as the main tool, if not the exclusive tool, for animation.

For years, animators have used Flash to further web-based storytelling, produce content for animated television shows, and even contribute to feature films. In some ways, it's easy to see why Flash is a boon for animators. Using Flash, artists no longer have to draw every single frame of an animation by hand. Instead, animators can draw a few key poses by hand or build characters from many smaller posable parts, and then let the computer fill in the frames between poses. This process is called *tweening* because the computer calculates the frames between each pose.

Of course, Flash is not a wholesale substitute for hand-drawn animations. Highly expressive poses or sequences with rapidly changing poses still require a lot of manual illustrations. Even in these situations, however, Flash can lend a hand with backgrounds, transitions, and other elements that aren't the primary focus of attention. Deciding when to use Flash in your animations will usually be a matter of choosing the best tool for the job.

In this chapter, you'll learn how to create animations using a variety of techniques ranging from the Flash equivalent of traditional cel animation to the simple motion of symbols, and to morphing between shapes. Along the way you'll learn more about the Timeline and frame types, how to animate gradient and bitmap fills, how to use Flash CS4 Professional's new Motion Editor, and much more. This chapter will focus exclusively on interface-driven animations, and the next chapter will introduce some ActionScript basics for controlling animation with code. By the time you have finished reading these two chapters, you will probably recognize which kinds of animations Flash is most suited to create. Ideally, your animation tasks will be conquered more

IN THIS CHAPTER

often than not with a big computer assist, leaving you more quality time to spend on fewer hand-drawn sequences.

Knowing the Timeline

The single most important step in conquering noncoded animation in Flash is gaining an understanding of the Flash Timeline panel (Figure 5-1). Prior to Flash CS4 Professional, the Timeline panel was the primary interface element for all things animated. You'll learn later in this chapter that Flash CS4 Professional introduces a new tool for manipulating animations, the Motion Editor, but even when you're using this new tool, animations must originate in the Timeline panel.

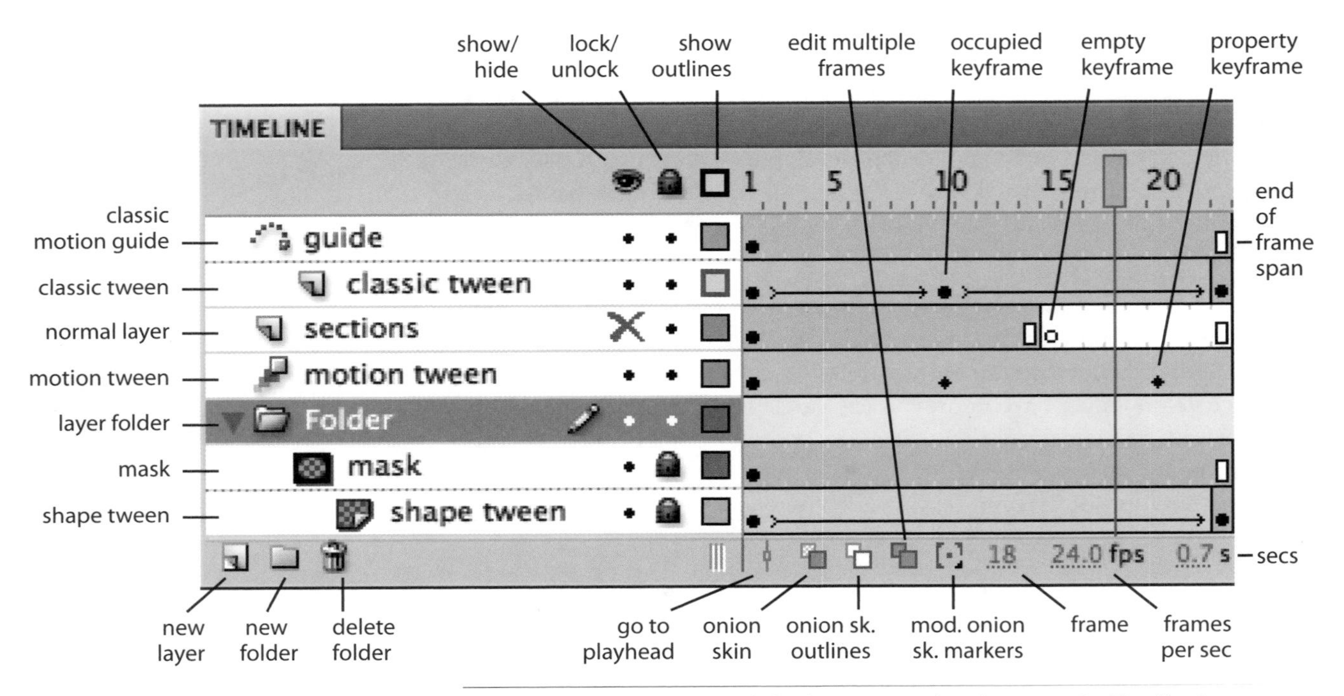

Figure 5-1. The Timeline panel, displaying examples of many major Timeline features

Layers

Figure 5-1 shows an example file viewed in the Timeline panel. This example uses many of the animation tools that will be explained in this chapter. As described in Chapter 1, there are two primary ways to organize assets in a Timeline: by vertical stacking order (layers) and by time (frames).

Horizontal spatial arrangement in a single frame is determined simply by where you place your assets on the Stage. You can set vertical stacking of overlapping assets by arranging them in a single frame and by bringing an asset forward or sending it backward (using Modify→Arrange) when they overlap.

However, to animate objects in the Timeline panel, you place assets into discrete layers so that only the desired asset is included in the animation. This visual stacking arrangement will be familiar to users of Adobe Illustrator and Adobe Photoshop.

There are several layer types, not the least of which are the *normal* layer and *guide* layer discussed in Chapter 1. Normal layers have no unique functionality, and guide layers display content only during authoring (guide layers are excluded when SWFs are compiled). You can also organize layers into *layer folders*. This is handy when files have a large number of layers, because you can collapse the folder using the arrow to the left of the folder icon and hide all the layers therein.

NOTE

In Chapter 10, you'll learn about a brand-new layer type, called an armature layer, which contains inverse kinematics (IK) armatures. IK is an animation tool that lets you join objects together with bones and joints like a skeleton or robotic arm.

While they are not specifically reserved layer types, two kinds of tweens reside in their own layers. *Shape tweens* are typically used to morph shapes, and *classic tweens* (formerly called motion tweens) are legacy animation types used to animate symbols, such as movie clips. The Flash CS4 Professional upgrade for the classic tween retains the name *motion tween* and has become a dedicated layer type. You will learn how to create both kinds of motion tweens in this chapter and even use both techniques in your ongoing portfolio project.

You can use a special kind of guide layer, called a *classic motion guide*, with classic tweens. Classic motion guide layers allow you to draw a path for an associated classic tween for the animated asset to follow. At the top of Figure 5-1, you can see the name and icon of the classic tween layer icon indented below the name and icon of the classic motion guide layer. This indicates a relationship between the two layers, and the guide is partially controlling the location of the asset in the classic tween.

For now, the final unique layer types you'll learn about are the *mask* and *masked* layers. A mask layer can be used to show only select content from an underlying masked layer. At the bottom of Figure 5-1, you will notice another pair of associated layers. This time the name and icon of a masked layer—in this case, a shape tween—are indented below a mask layer.

There are three buttons beneath the layer icons and names in Figure 5-1. From left to right, these buttons let you create a new layer, create a layer folder, or delete a selected layer. To the right of the layer icons and names are three columns. Clicking in these columns in a layer of interest will, from left to right, show or hide the layer's contents, lock layers to prevent further edits, and display all layer contents as colored outlines. For example, the red x indicates that the *sections* layer is hidden, the locks show that the *mask* and *shape tween* layers are locked, and the purple outline box indicates that the *classic tween* layer is displaying its contents in purple outlines.

NOTE

You can use the show/hide option to control the parts that are compiled into a SWF. If you disable Include hidden layers *in the Flash section of the File→Publish Settings dialog, hidden layers will be excluded from your SWF. Use this feature to hide assets, prevent scripts from executing, or speed up compiling by excluding sounds.*

Frames

The right half of Figure 5-1 is composed of the frames that make up the FLA's Timeline, much like the frames of a film, with each showing only the content of a single frame of the total animation to the viewer at a time. Chapter 1 outlines the principle differences between keyframes and computer interpolated frames, but, as animation techniques add complexity, a broader overview is in order.

NOTE

When a frame contains content, creating a new keyframe is usually based on the need to change a property of the frame's content (such as its location) or even replace the content altogether. However, it's also sometimes useful to add empty keyframes to a layer. For example, an empty keyframe can remove content from a layer only after a specific frame, or even serve as a visual cue to divide a Timeline into easily noticeable segments.

An ordinary frame is a frame in which no change occurs. Such frames might be empty or they may contain static, or unchanging, content. Within the Timeline panel, ordinary frames have no visual appearance other than a gray tint if the frame contains content. In Figure 5-1, for example, frame 18 of the *sections* layer (directly under the red playhead) is an empty frame, and frame 5 of the *sections* layer is an ordinary frame with static content.

A *keyframe* is a frame that contains animator-defined change. By definition, the first frame of every frame span is a keyframe, whether it contains content or not, because the animator has chosen to isolate that frame from its previous frame.

Traditional keyframes are marked with a circle. An empty circle represents an empty keyframe (frame 15 of the *sections* layer in Figure 5-1), while a filled black circle marks a keyframe that contains content (frame 1 of the same layer). A filled black diamond marks the new *property keyframe*, which is part of the new motion tween format. You'll learn more about property keyframes when you read about motion tweens, but you can see property keyframe icons in frames 10 and 20 of the *motion tween* layer.

NOTE

You cannot edit content in interpolated frames of shape tweens and classic tweens. You must create a keyframe to make any changes. Properties of content in new motion tweens, however, can be changed without first creating a keyframe. A property keyframe will automatically be created for you.

An *interpolated frame* is a frame in which content properties have been altered by the computer. Varying color tints indicate interpolated frames, depending on which kind of tween is at work. For example, frame 5 of the *shape tween* layer is tinted green to indicate that a shape is likely morphing from its original appearance in frame 1 to its final appearance in frame 23. Frame 5 of the *motion tween* layer is tinted light blue to show that it is part of the new Flash CS4 Professional motion tween. A movie clip in this layer may be moving between three positions set in frames 1, 10, and 20. Finally, frame 5 of the *classic tween* layer is tinted a purplish blue to show its inclusion in a legacy style motion tween. This layer may be tweening another property, such as *scale*, for example, between sizes set in frames 1, 10, and 23.

Finally, although not a keyframe, the end of a frame span is marked by a vertical rectangle. This serves no other purpose than as a visual indicator that an ordinary frame span has ended. Frames 14 and 23 of the *sections* layer mark the end of content-filled and empty frame spans, respectively.

Adding and removing frames

Prior to adding frames, the Timeline panel is restricted to working only with frame 1. You can't even move the playhead through the Timeline panel beyond the last frame. Although you can draw static illustrations in a single frame or control animations with ActionScript, Timeline-based animations require a minimum of two frames.

To add frames to a single layer, click in the layer where you want to insert the frame and press F5 (Insert→Timeline→Frame), shown in Figure 5-2a. If you click in an existing frame, a frame will be added at that point. If you click beyond all existing frames, additional frames will be added up to that point. For example, if you start with a new empty file containing only frame 1, and want the Timeline to contain 20 frames, click in frame 20 and press F5.

The process is similar for adding frames to more than one layer of the Timeline at once. Instead of clicking once in a single frame, drag through all the layers in which you will insert frames (Figure 5-2b). To insert new frames between existing frames in *all* layers, click in the number span atop the Timeline to position the playhead. By not clicking in a specific layer, frames will be added to all layers at the position of the playhead (Figure 5-2c). This technique will only work within existing frame spans, however, because the playhead can't move past the last frame in the Timeline.

Finally, to remove frames, select the frames you want to remove (by dragging your mouse over them or Shift-+clicking the first and last frames) and press Shift-+F5 (Edit→Timeline→Remove Frames).

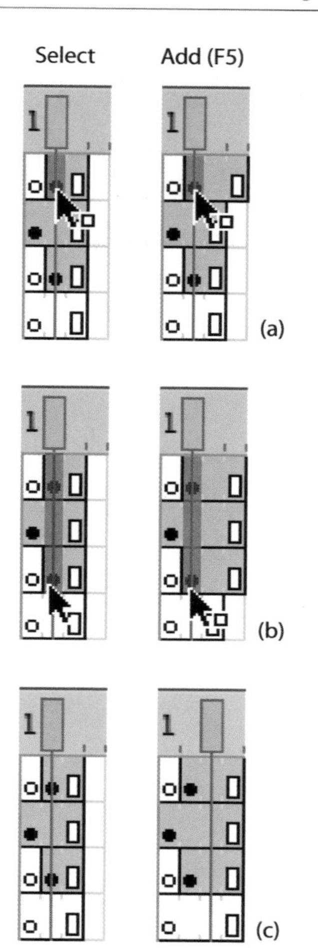

Figure 5-2. Adding frames to one layer (a), select layers (b), and all layers(c)

WARNING

Removing frames will literally remove the frames, not the contents of the frames, and will shorten the entire frame span. If you only want to remove the contents of a frame, click once to select the frame span in question and delete. If you want to remove the contents from a segment of a frame span, insert keyframes at the start and end of the segment so you can then delete only that portion of the content.

Creating and clearing keyframes

Creating and clearing keyframes is similar to adding and removing frames. Select the frame that you want to convert to a keyframe and press F6 (Insert→Timeline→Keyframe). To end a frame span by inserting a keyframe and deleting its contents, there's a more direct process: create an empty keyframe by using F7 (Insert→Timeline→Blank Keyframe) instead of F6.

To convert a keyframe back to a normal frame, select the keyframe and press Shift+F6 (Modify→Timeline→Clear Keyframe).

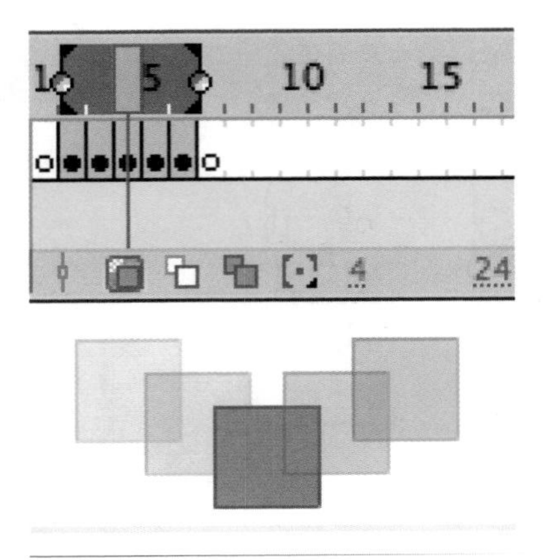

Figure 5-3. Onion skinning enabled, showing two frames behind and two frames ahead of the current frame

Frame Editing Controls

Below the frames, in the bottom of the Timeline panel shown in Figure 5-1, is a series of frame editing controls. The first button will scroll the Timeline panel to *center the frame* in which the playhead currently resides.

The next two buttons will turn on *onion skin* preview and show the onion skin frames in outline view. Onion skin preview allows you to display frames to the left and/or right of the current frame in a muted image or outline, depending on your setting (Figure 5-3). The additional frames are rendered right on the Stage, showing the current frame in the context of its surrounding frames.

This feature gets its name from the translucency of onion skin. The more layers of onion skin you must look through, the harder it is to see the content. Therefore, the frames farthest from the playhead are the least opaque. The onion skin feature is discussed further in Chapter 10 when working with inverse kinematics armatures.

The next control allows you to *edit multiple frames* at once and will be explained in greater detail later in this chapter.

The final button offers a quick setting to *modify the onion and edit multiple markers* that appear on top of the frame numbers. These markers change the number of frames that appear to the left and right of the playhead when using the onion skin or edit multiple frames features. You can manually drag the markers to show any number of frames, but this quick menu will show two, five, or all the frames surrounding the playhead. It also allows you to anchor the visible frames (even when you move the playhead) and show the markers all the time, even when the features are not in use.

The three editable values that conclude the array of controls are the current playhead position measured in frames, the frame rate (the speed at which the playhead moves through frames) measured in seconds, and the current playhead position measured in seconds. The playhead in Figure 5-1 is in frame 18 and the frame rate of the file is 24 frames per second, so the playhead is resting at 0.7 seconds of elapsed playback time.

WARNING

Animation speed is often affected by many factors of the end user's computer setup. These include processor speed, available RAM, other running applications, and even which tasks are being performed in Flash at any given moment. As such, don't rely too heavily on a frame-to-time correlation, especially when trying to synchronize

Creating Frame-by-Frame Animations

If you plan to create very stylistic animations with many unique art elements, it's likely that you won't be able to rely too heavily on tweening for assistance. In this scenario, you'll have to start from the beginning and use the time-tested technique of frame-by-frame animation.

There's nothing very special or difficult about frame-by-frame animation beyond the care and attention it requires. The method literally entails creating a keyframe in every frame and drawing or manipulating each frame's artwork manually. Figure 5-4 shows a layer with keyframes in every frame.

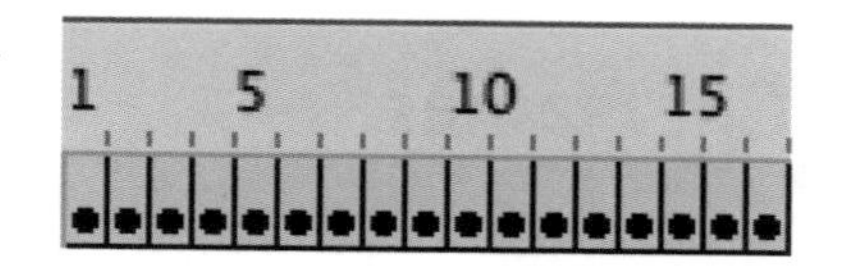

Figure 5-4. Keyframes in every frame allow precise manipulation of frame-by-frame animations

While this is the most straightforward of all animation techniques, there are a few tips and features that can make the process easier. For example, you should create keyframes as you go, rather than all at once at the outset of your effort. That is, avoid selecting a long span of frames and pressing F6 to create individual keyframes from the span. True, this is a quick process. However, if you add one keyframe at a time, each new frame will duplicate the previous frame. This allows you to adjust each new frame based on the prior frame's appearance, rather than starting from scratch.

Additional features can provide big animation assists. *Editing multiple frames*, *copying and pasting frames*, *swapping symbols*, and the aforementioned onion skin option not only improve frame-by-frame animations, but can be used with other animation techniques as well.

Editing Multiple Frames

In some cases, particularly when working with a sequence of several keyframes, you face the prospect of having to edit many frames. For example, assume you are working on a frame-by-frame animation and later decide that you want to move the art in every other frame down a bit on the stage. This would ordinarily require you to move the playhead to every other frame and repeatedly select the frame's contents and make the adjustments. What's more, you would have little context from the surrounding frames during the editing process.

Fortunately, the Edit Multiple Frames feature allows you to work in as many keyframes as you like, not only making it possible to edit the frames' content, but also to see each frame on the Stage at the same time. To do this, click the Edit Multiple Frames button at the bottom of the Timeline. Brackets will appear to the left and right of the playhead. You can drag these brackets to set the number of editable frames to the left and right of the playhead. Thereafter, you can work in any of the included frames.

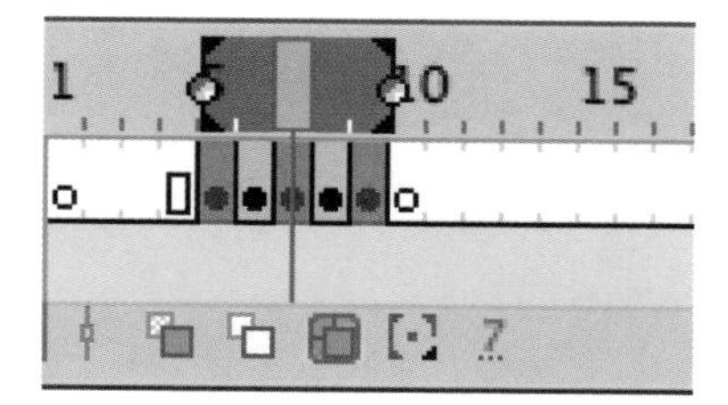

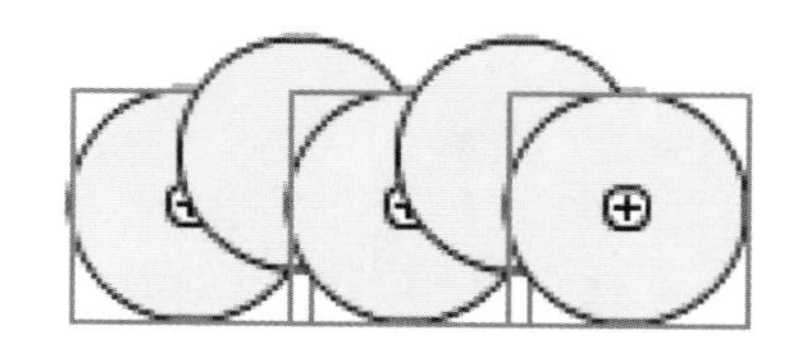

Figure 5-5. The Edit Multiple Frames feature allows simultaneous editing of more than one frame at a time—in this case, the two frames on either side of the current frame

The top of Figure 5-5 shows the feature enabled at the bottom of the Timeline panel detail (the button just to the right of the playhead, under frame 8, is pushed in, meaning the feature is active). The editable frames are set to two to the left and two to the right of the playhead. The selected movie clips on the Stage at the bottom of the figure, as well as the highlighted frames in the Timeline, show that content in frames 5, 7, and 9 are selected. They have been moved down 10 pixels, creating the desired effect.

Copying and Pasting Frames

In addition to copying and pasting content from a single frame, it is also possible to cut, copy, and paste the frames in their entirety. Doing so will preserve the frames' contents, but also keyframes, tweens, layer properties, and more.

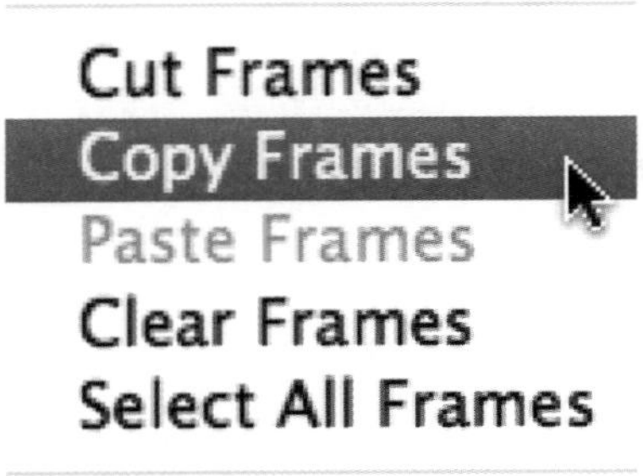

Cut Frames
Copy Frames
Paste Frames
Clear Frames
Select All Frames

Figure 5-6. Using the Copy Frames feature instead of the traditional Copy feature

For example, if you selected five frames and used the standard Copy feature, you would only copy the content from the current frame—the frame in which the playhead resides. However, if you selected those same five frames and used the Copy Frames feature, you could then paste five frames with all their content and attributes intact.

You can access the Copy Frames command through the application menu (Edit→Timeline→Copy Frames) or by right-clicking (Windows) or Control-clicking (Mac) the selection (Figure 5-6). You can then paste frames using the Paste Frames option found in the same menus.

This is *really* handy when you realize, after a lot of hard work, that you want all the animations you finished in the main Timeline to be in a movie clip instead. In that case, you can copy (or cut, if you prefer) all the frames, create a movie clip, double-click to edit that movie clip, and then paste all the frames into the movie clip. All of the layers, frames, and contents will be recreated inside the movie clip. Copying and pasting frames is also convenient when you want to create an animation of repeating frames.

The only tricky part of the process is when you want to *replace* frames with a paste. In the aforementioned examples, you are likely creating frames where none previously existed, such as when adding content to an empty movie clip. However, if you need to overwrite existing frames, you must first *select* the frames you want to replace. If you fail to do this, you will be adding to the existing frames, rather than replacing them, and your frame span will grow, possibly affecting frame synchronization later on.

Figure 5-7 shows the steps required to replace frames with a paste. The first step, (a), is to select the desired frames by dragging over them with your mouse and copying them. The next step, (b), is to select the frames you wish to overwrite using the same selection procedure. The final step, (c), is to paste the frames into the selection.

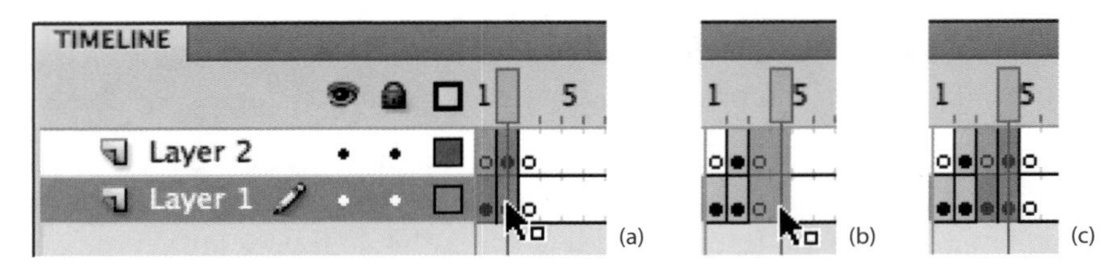

Figure 5-7. Copying and pasting frames to repeat a frame sequence; (a) select four frames and use the Copy Frames feature, (b) select frames to be replaced, and (c) use Paste Frames

Swap Symbols

When using symbols in animations, it can be very convenient to automatically replace a symbol without otherwise affecting animation properties. For example, if you had to replace a symbol by deleting it and adding a new symbol in its place, you would then have to reapply any properties such as location, rotation, scale, and so on, that defined the animation's characteristics. This is not only tedious, but you run the risk of not accurately reproducing all the original symbol's properties. Alternately, if you edited the symbol, any unrelated instance of the symbol would also be changed.

Enter the *Swap Symbols* feature to protect the integrity of your symbols and save you time and frustration. This feature does exactly what its name implies. After selecting a symbol on the Stage, click the Swap button at the top of the Properties panel (Figure 5-8) to swap the original symbol with a replacement from the Library.

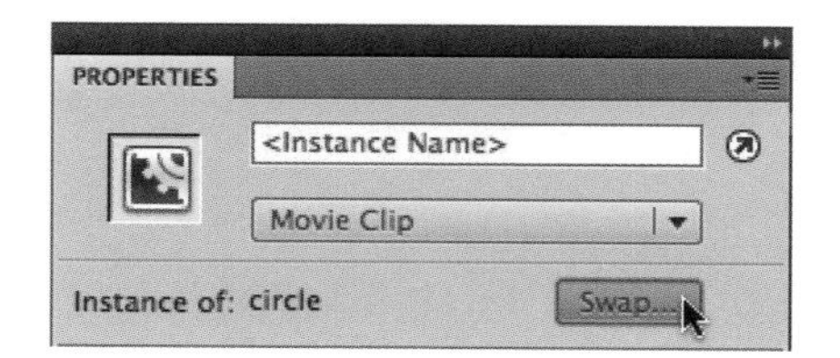

Figure 5-8. The Swap Symbols button in the Properties panel

The Swap Symbol dialog that appears (Figure 5-9) has a list and preview thumbnail of all symbols that you can swap for the original. The Duplicate Symbol button (at the bottom of the dialog) allows you to duplicate the original and swap the copy for the prior symbol instance, all in one step. This is useful when you want to edit the original symbol in one or more uses, but can't change it throughout the entire file.

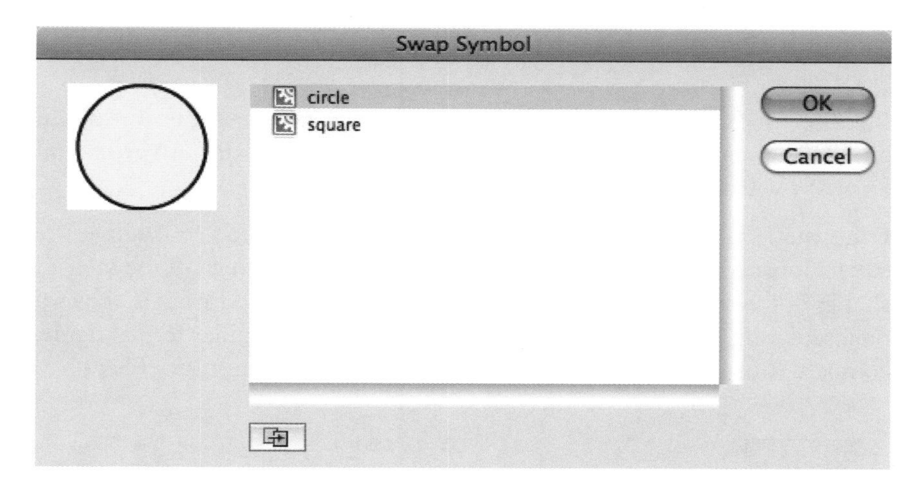

Figure 5-9. The Swap Symbol dialog shows available symbols to replace the original; the Duplicate Symbol button is located at the bottom of the dialog

Creating a Shape Tween

Now that you have some experience with a few workflows that will help you create frame-by-frame animations, it's time to get the computer more involved. Using one of Flash's tweening processes will allow you to set a few keyframes and turn the rest of the work over to the computer. Specifically, Flash will create the *interpolated*, or *in-between*, frames by calculating all the property values to further the animation from one keyframe to the next.

A shape tween is a tween that uses shapes (rather than symbols) as its assets, and is typically used to morph one shape into another. In this section, you will morph a square into a circle. To create a shape tween, you need two keyframes, each containing a shape. To try this in Flash, use the following steps:

1. Create a new file using File→New. This file will not be a part of the ongoing portfolio project.
2. Practice creating a keyframe by clicking in frame 5 and pressing F7 to create an empty keyframe.
3. Using the Rectangle tool, draw a rectangle in frame 5 at the left edge of the Stage.
4. Create an empty keyframe (F7) in Frame 15. This will create a gray frame span from frame 5 to 14 and an empty keyframe in frame 15.
5. Switch to the Oval tool and draw a circle in frame 15 at the right side of the Stage.
6. Click anywhere in the first frame span and use the Insert→Shape Tween menu command. Alternately, you can right-click (Windows) or Control-click (Mac) on the first keyframe (or even the shape on the Stage in the first keyframe), and select Create Shape Tween from the pop-up menu (Figure 5-10).

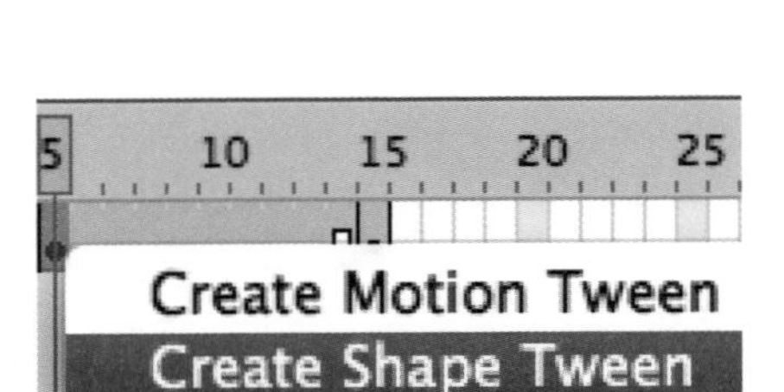

Figure 5-10. Creating a shape tween using the context-sensitive mouse menu

Because you created the tween in the first group of frames, the tween will span from the first keyframe to the next. The frame span will turn green and an arrow will appear between frames 5 and 15 (Figure 5-11).

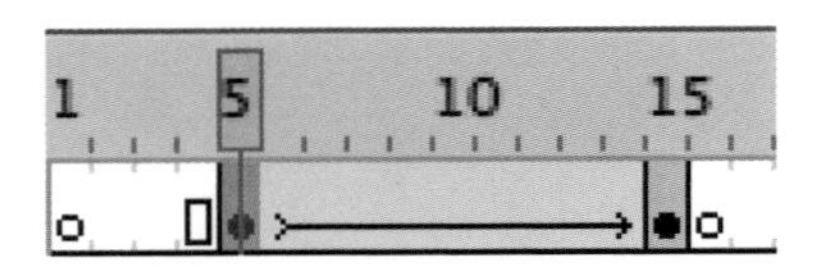

Figure 5-11. A green tint and arrow in the Timeline indicate a shape tween

Using the Timeline panel, scrub through frames 5 and 15, and you will see the square morph into a circle. Flash has calculated and drawn all of the interim shapes of the artwork required to change its appearance. Figure 5-12 shows a representation of the tween. If your tween doesn't look similar to this figure, compare your file to *shape_tween_01.fla* from the provided source files.

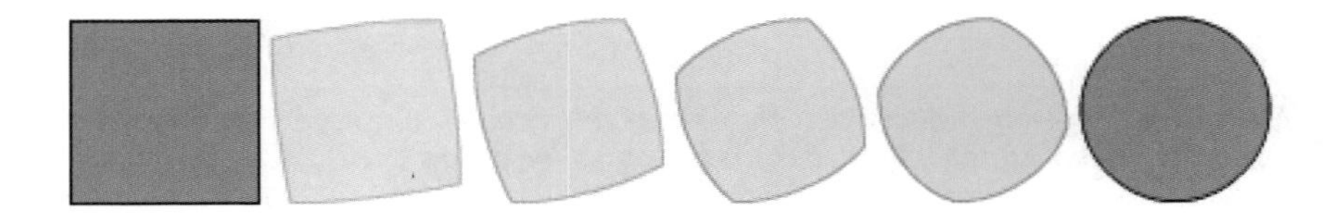

Figure 5-12. A visual representation of a shape tween on the stage; dark shapes are in keyframes, faded shapes are in interpolated frames

Although the ultimate effect is the same (changing one shape into another), shape tweens are not only useful for what clearly look like morphs. For example, you can also use them for transitions, as when covering up content with a shape that matches the Stage color, or for animating the look of drawing lines.

Shape Hints

If you look carefully at the shape tween in Figure 5-12, the square appears to rotate as it changes into a circle. However, the shape is not really rotating, it is simply distorting in an unexpected way during the morph. This is because the main control points for the square are at its corners, while the main control points for the circle are at the top, bottom, left, and right of the circle. As such, when the shape tween process attempts to change the square into the circle, it tries to move the control points about 45 degrees counterclockwise to meet up with the similar points on the circle.

This is a simple, easy-to-demonstrate example of distortions that occur during morphs, but the effects can actually be much worse. Flash often needs a little help—a few instructions to assist the morphing process—so the algorithm knows what kind of morph you had in mind. *Shape hints* are markers that you place on the original and final shapes to match up control points during the morph. For example, you can place a hint at the upper-left corner of the square and in the same approximate northwest position on the edge of the circle, and Flash will try to synchronize the position of the two points during the morph.

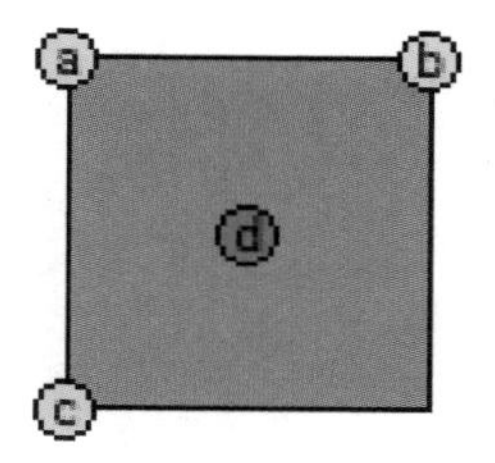

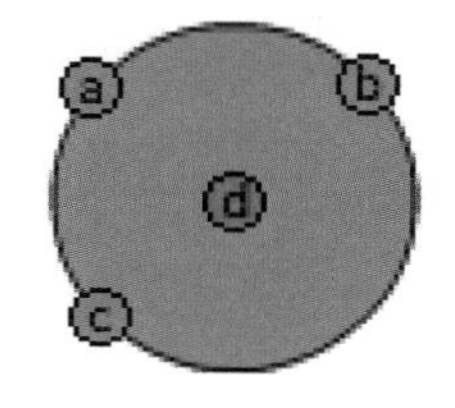

Figure 5-13. Shape hints in the first keyframe (top) and last keyframe (bottom) of a shape tween

Figure 5-13 shows four shape hints applied to the square-to-circle example discussed previously. A single hint can be referenced in both keyframes by noting the letter of the hint. In this case, the hints are marked a, b, c, and d in both frames.

When you place hints, they turn yellow in the first keyframe and green in the second keyframe. In Figure 5-13, the fourth hint has yet to be positioned, and remains red in both keyframes.

After successfully placing shape hints in both frames, Flash tries to match up each pair during the shape tween, resulting in a morph between square and circle that no longer appears to rotate (Figure 5-14). If desired, you can compare your work to *shape_tween_02.fla*.

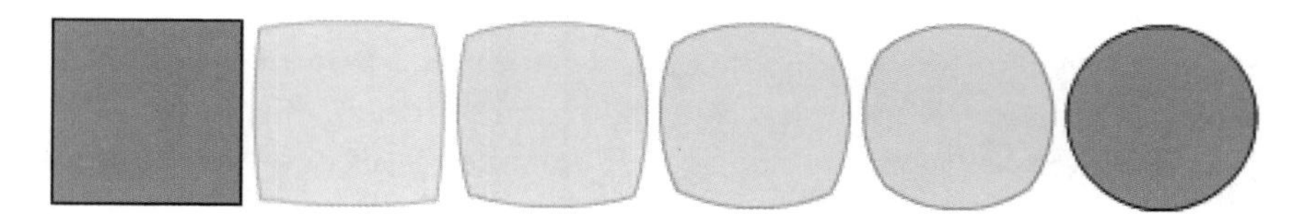

Figure 5-14. The shape tween, after using shape hints, with the apparent rotation distortion corrected

You can use up to 26 shape hints, remove shape hints by dragging them off the stage in either keyframe, and remove all shape hints by using the Modify→Shape→Remove All Hints menu command. You can also display shape hints that are no longer visible using the View→Show Shape Hints menu command. For these hints to be visible, you must unlock the shape tween layer and the frame in one of the keyframes using the hints.

WARNING

Even when using shape hints, Flash still sometimes has to make guesses when it morphs one shape into another. Sometimes only one or two shape hints are required, and sometimes too many shape hints can make a morph worse than when you started. When relying on shape hints to improve a shape tween, be prepared to test the tween often and start fresh if things go awry.

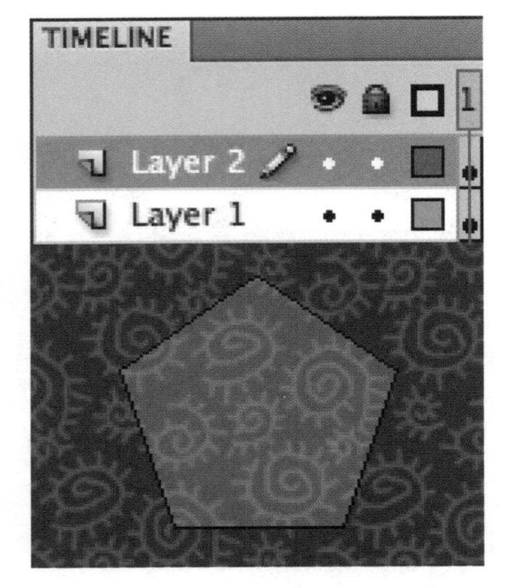

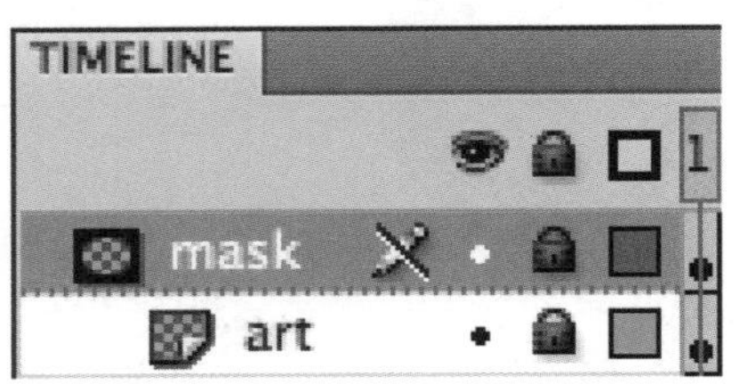

Figure 5-15. A shape layer over a bitmap before using the shape as a mask (top) and after (bottom)

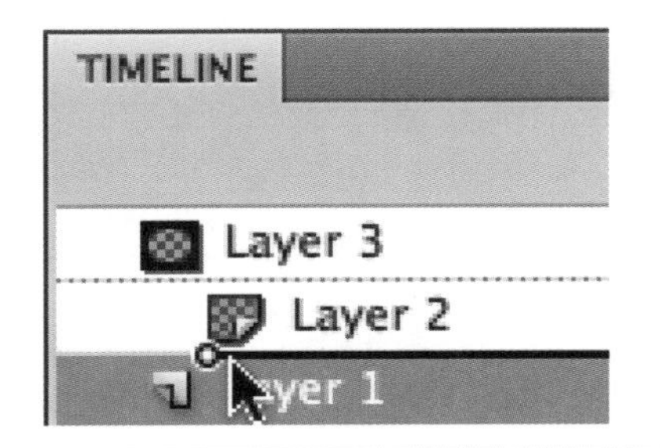

Figure 5-16. Dragging additional layer to mask

Adding a Layer Mask

Like visual compositing applications such as Adobe Photoshop and Adobe After Effects, Flash supports the use of *masks*. Masks behave like holes through which you can see the content in underlying layers.

To create a mask effect, you need a minimum of two layers: one layer will serve as the mask, and the other will contain the content seen through the mask. Figure 5-15 shows detail views of the Timeline and Stage of the *layer_mask.fla* source file. The top of the figure shows two layers prior to being converted to *mask* and *masked* layers. Layer 1 contains a bitmap and layer 2 contains a shape.

To convert these layers to a mask/masked pair, right-click (Windows) or Control-click (Mac) the top layer and select Mask from the pop-up menu. The top layer will be converted to a mask layer, the mask will affect the bottom layer, and both layers will lock to display the visual result of the masking process. The bottom of Figure 5-15 shows the end result, and the layers have been renamed to clearly indicate which purpose each layer is serving.

You can unlock either layer at any time, but when doing so, both the mask and the underlying content will become visible. To return to viewing the masked state, simply lock both layers again.

Unlike Photoshop, Flash does not allow you to choose whether dark or light colors serve as the mask. Furthermore, Flash layer masks do not support alpha values. Any nontransparent pixel will be understood as opaque and will become part of the mask. As a result, all edges in layer masks are hard edges, so feathered masks are not possible when using layers. To achieve a soft-edge mask, you will require ActionScript and bitmap compositing effects (this technique is discussed in Chapter 7).

A mask can affect multiple adjacent layers, if desired. For any normal layer that sits immediately below a masked layer, you can double-click its icon and select masked from its layer properties. You can also drag layers into a mask setup either by dragging them between the mask and masked layers (which may adversely affect layering and require further adjustment) or by dragging them below and to the right of a masked layer until the mouse is under the layer name (Figure 5-16). Visual feedback below the layer icons will give the appearance that the layer has been indented to match the left margin of the other masked layers.

Multiple mask layers, however, cannot apply to the same masked content. To achieve this effect, simply duplicate the mask/masked pair for as many masks as you wish to apply. If you are using symbols in both layers, this will not noticeably affect your file size. However, if you are using shapes, creating multiple layers will increase file size. In this scenario, you may want to consider ActionScript masks, discussed in Chapter 7.

Creating a Motion Tween

Where shape tweens morph from one shape to another, *motion tweens* animate symbols. Although the phrase "motion tween" has been part of the Flash vernacular for a long time, this is not your father's motion tween. Instead, this is a brand-new way of animating in Flash. The new tweens are easier to create and edit, provide better visual feedback to the animator during authoring, and are even supported by a new panel called the Motion Editor for exerting more granular control. You'll learn more about the Motion Editor panel later, but for now, see how easy it is to create and edit a tween right on the Stage:

1. Create a new file using File→New. This file will not be a part of the ongoing portfolio project.
2. Click in frame 5 and press F7 to create an empty keyframe.
3. Using the Oval tool, draw a circle in frame 5 at the left edge of the Stage. Select the shape and convert it to a movie clip (Modify→Convert to Symbol). Symbols are required for motion tweens.
4. Create an empty keyframe (F7) in Frame 15. This will create a gray frame span from frames 5 to 14 and an empty keyframe in frame 15.
5. Click once in the first keyframe and use the Insert→Motion Tween menu command. Alternately, you can right-click (Windows) or Control-click (Mac) the first keyframe (or even the shape on the Stage in the first keyframe) and select Create Motion Tween from the pop-up menu (Figure 5-17).

NOTE

If you try to create a motion tween without first creating a symbol, Flash will not only warn you about the problem, it will also offer to automatically convert the shape to a symbol for you!

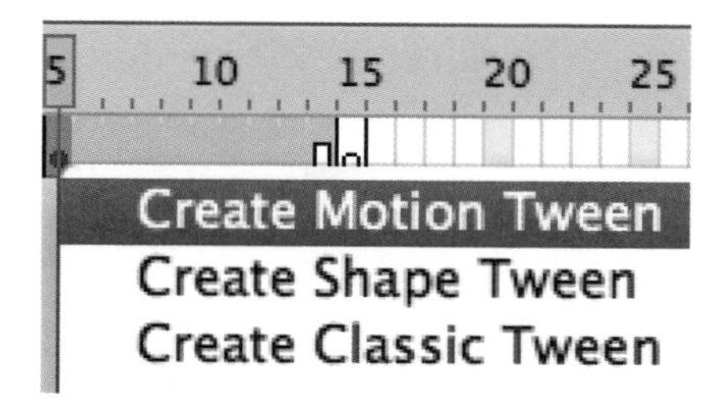

Figure 5-17. *Creating a motion tween using the context-sensitive mouse menu*

Notice that, unlike when working with shape tweens, you don't need a second keyframe to create a motion tween. The keyframe in frame 15 serves only as a terminus for the tween. Because you created the tween in the first group of frames, the tween will span from the first keyframe to the next and turn blue (Figure 5-18).

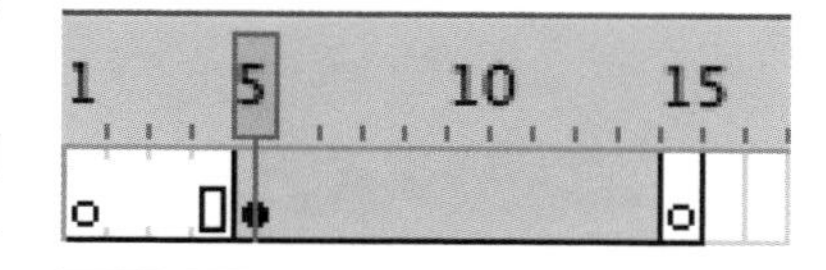

Figure 5-18. *A blue tint in the Timeline indicates a motion tween*

Scrubbing through the Timeline, however, you'll notice that the symbol doesn't move. In Flash CS4 Professional's new motion tween format, you create keyframes *after* creating the tween.

6. Move the playhead to frame 14 by clicking in the frame number bar above the Timeline.
7. Using the Selection tool, drag the circle to the right side of the stage.

This creates a property keyframe in frame 14, represented by the diamond icon (Figure 5-19). Using the new motion tween format, you can add property keyframes on the fly in the Timeline, just by changing a property value. In this case you changed the *x* and (possibly) *y* properties by moving the movie clip to the right side of the stage.

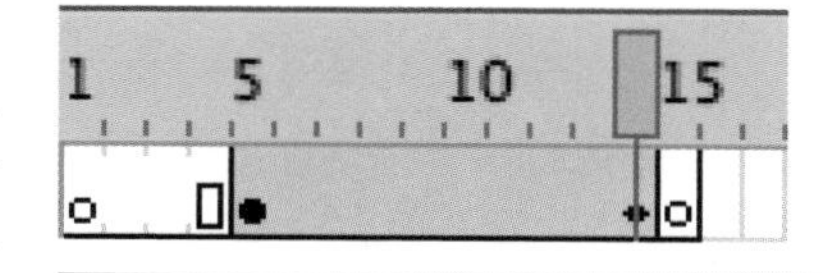

Figure 5-19. *A property keyframe added to the motion tween*

When you click any motion tween in the Timeline, the entire tween will be selected. To select a portion of the tween, such as single property keyframe or the first half of an animation, hold down the Ctrl (Windows) or Command (Mac) key while selecting. To clear a property keyframe, select only the keyframe and press F6. You can also right-click (Windows) or Control-click (Mac) a single selected keyframe and remove only certain properties using the Clear Keyframe menu.

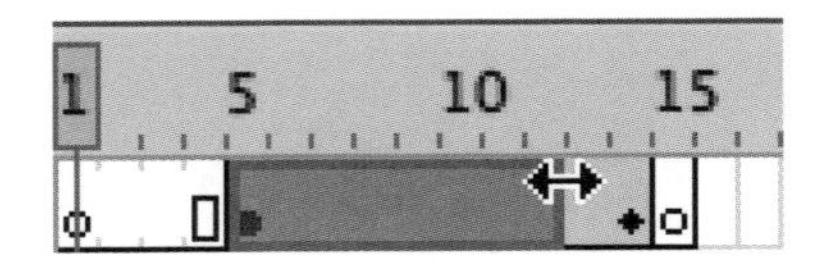

Figure 5-20. Shortening a tween span

If you grab the very end or beginning of a tween, the mouse pointer will turn into a horizontal pair of arrows, and you can slide it left or right to change the length of the tween (Figure 5-20). When doing so, the tween will expand or contract the time between keyframes. As you resize, the keyframes will move to retain their same relative positions within the tween. If you only want to add more frames to a tween, and don't wish the keyframe locations to change, hold down the Shift key before dragging to resize the tween.

Now if you scrub through the Timeline, you will see the circle move from left to right. Flash determined the distance between the first and second locations and calculated how far the circle would have to move to reach that distance. You'll also notice a line depicting the path the circle travels, and dots along that line. The large dots at the ends of the path depict keyframes, and the smaller dots along the path depict interpolated frames. Figure 5-21 shows a representation of the tween. If your tween doesn't look similar to this figure, compare your file to *motion_tween.fla* from the provided source files.

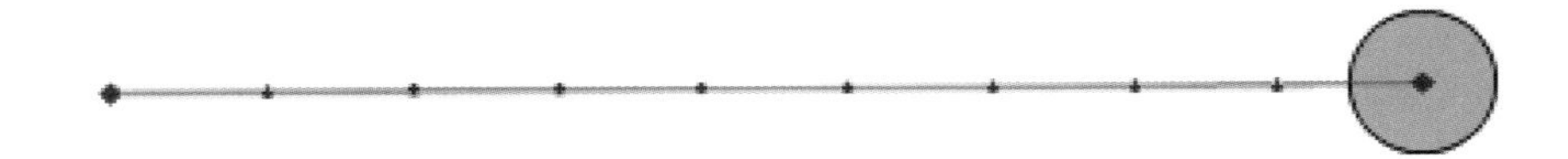

Figure 5-21. A motion path for a tween that moves a circle in a straight line from left to right

Editing the Motion Path

One of the nicest things about the new motion tween format is the degree to which you can dynamically alter the tween just by tinkering on the Stage. Previously, you inserted a property keyframe just by dragging the symbol in the last frame. You can edit the tween's motion path just as easily. In many respects, the tween's motion path behaves just like a regular stroke.

Figure 5-22 shows a motion path being dragged into an arc using the Selection tool and Flash's natural drawing techniques. You can grab a motion path anywhere along its length and distort it just as you would any other line.

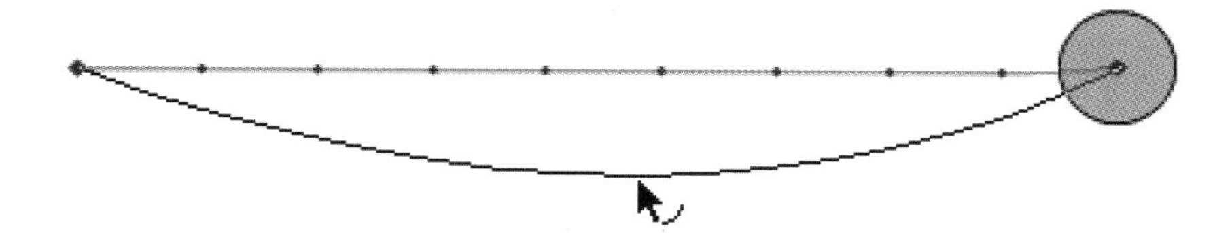

Figure 5-22. Dragging a straight motion path into an arc with the Selection tool

The Subselection tool is being used in Figure 5-23 to edit the motion path as if it were a Bézier curve created with the Pen tool. You can move each keyframe independently or drag its control handles to distort the shape of the curve. (In Figure 5-23, a third keyframe was added to the center of the tween to show additional control handles for demonstration purposes.)

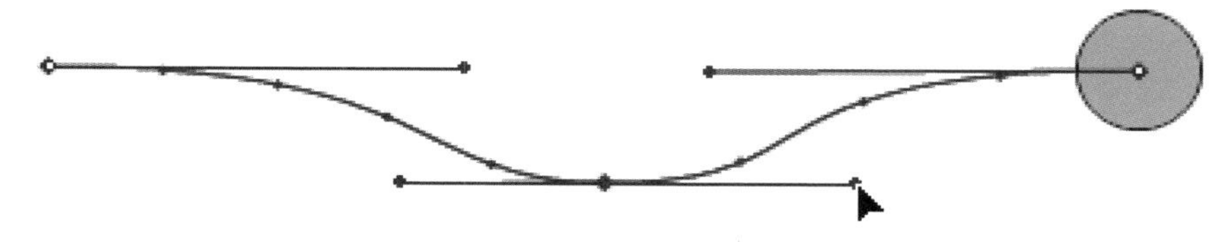

Figure 5-23. Using the Subselection tool to manipulate the motion path as a Bézier curve

As when working with regular strokes, if you click and release to select the path (rather than immediately dragging the path to reshape it), you can then drag the entire animation around on the Stage without creating or changing a property keyframe or further editing the path. Finally, you can draw motion paths from scratch, but not while editing the tween; you must create the path first, as described in the next section.

Drawing a Motion Guide

When working with a motion tween, you can use the Selection tool to drag the motion path and use the Subselection tools to manipulate the path's control handles. However, you can't actually draw in the motion tween layer. To apply a hand-drawn motion path, you must first draw a stroke in another layer as you ordinarily would, using any preferred drawing method. Figure 5-24 shows just such a stroke.

Figure 5-24. A standard stroke drawn in a tapering sine wave shape

Once the path is complete, you can simply copy the stroke and paste it into the motion tween layer. Flash may ask you if you want to replace the existing motion path, which you can decide on a case-by-case basis. If you proceed with the paste, the motion guide will take on the shape of the stroke you drew, and the frames will be distributed across its length (Figure 5-25).

Figure 5-25. The stroke from the previous figure pasted into a motion guide to become a custom motion path

If the stroke you drew is too complex to be recreated with the number of frames allotted to the tween, Flash will prompt you to simplify the curve by optimizing it or resampling it across the available frames. Optimizing the path is recommended as a first step, but experimentation with both options may yield different results.

Roving keyframes

When editing motion paths, watch out for signs of stuttering motion. This can be caused when the distances the object must travel between keyframes are not proportional to the number of frames allotted. Figure 5-26, for example, shows a tween in which half its distance must be covered in approximately one-third of the frames. This causes the first half of the tween to move quicker than the last half of the tween, with an abrupt change at frame 4.

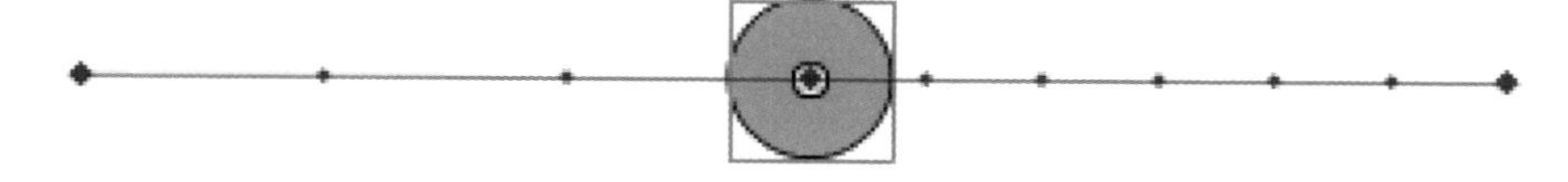

Figure 5-26. Irregular motion caused by the same amount of distance traveled in differing numbers of frames

A new Flash CS4 feature called *roving keyframes* solves this problem. Roving keyframes are specific to motion paths and are not linked to frames in the Timeline. Flash will disassociate roving keyframes from the Timeline and automatically adjust their positions to create more uniform motion throughout the tween.

Figure 5-27 shows the same tween depicted in Figure 5-26, but using roving keyframes. The motion has become much more consistent.

Figure 5-27. Overcoming irregular motion with roving keyframes

Roving keyframes only work with *x* and *y* properties, as well as the *z* property you will learn about in Chapter 8. To enable roving keyframes, right-click (Windows) or Control-click (Mac) a motion tween and choose Motion Path→*Switch keyframes to roving* from the pop-up menu (Figure 5-28). You can disable roving keyframes by selecting the *Switch keyframes to non-roving* option.

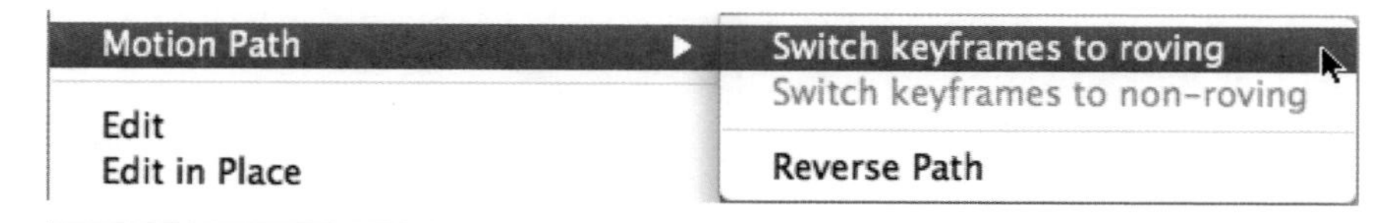

Figure 5-28. Enabling roving keyframes using the mouse's context-sensitive menu

Reversing a motion path

Another option in the Motion Path context-sensitive menu is the ability to reverse a motion tween. If, for example, your tween moves a movie clip across the Stage from left to right, you can automatically reverse this animation by using the Motion Path→Reverse Path menu option. The resulting animation will move across the stage from right to left.

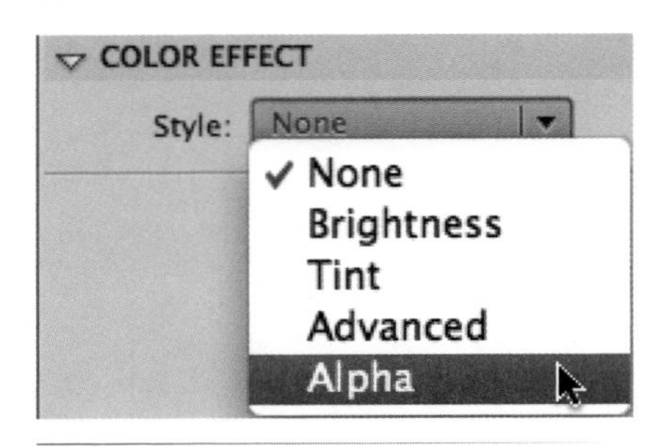

Figure 5-29. Applying an alpha color effect in the Properties panel

Color Effects

In addition to obvious properties such as *x* and *y* location, *scale*, and *rotation*, you can apply simple color effects to symbols and animate them using motion tweens. When you select a symbol, you can access color effects through the Properties panel (Figure 5-29). The Style menu in the Color Effect section of the panel contains four color effects:

Brightness

Brightness adds white or black to the color of a symbol instance. Values range from –100% (solid black) through 0 (no change) to 100% (solid white).

Tint

Tint sets the color of a symbol instance, ranging from 0% (no tint) to 100% (solid tint), regardless of the original symbol color.

Alpha

Alpha sets the transparency of a symbol, ranging from 0% (transparent) to 100% (opaque).

Advanced

Advanced separately adjusts the red, green, blue, and alpha values of a symbol instance. There are two controls for each channel. A percentage of the total value allows you to reduce the amount of any channel. This value is based on how much of any color or alpha channel originally appears in the symbol instance. If, for example, a color starts off with only 50% red, a setting of 100% for the first control allows you to use all of that available red, while a setting of 0% removes red. An offset value also exists for each channel, allowing you to offset the amount of color or opacity in a symbol instance from –255 to 255. This allows you to add more of the chosen color to the symbol instance, effectively changing the hue of each channel.

NOTE

Brightness and tint essentially cover a symbol instance with a white, black, or tinted overlay until the symbol disappears under that color at full value. The Advanced setting applies subtle changes to the actual color in the symbol instance, rather than covering it up. As such, Advanced is a better choice for altering the color bitmaps and for preserving, as much as possible, the impact of separate stroke and fill colors.

Just as you saw earlier when changing the *x* property in your first motion tween, changing a color effect during a motion tween will automatically create a property keyframe. Here, for example, is a simple process for fading in a movie clip, which also demonstrates a nice feature of motion tweens: automatic frame spans. If you need to review any of these steps, refer back to the beginning of this section.

1. Create a new file using File→New. This file will not be a part of the ongoing portfolio project.
2. Create a movie clip in frame 1.
3. Click frame 1 in the Timeline panel and create a motion tween.

Notice that frames were automatically added for you! When a frame span doesn't already exist to convert to a motion tween, Flash will add one second's worth of frames for the motion tween to occupy. So, if your frame rate is the default 24 frames per second, you will end up with 24 frames.

4. Move the playhead to frame 1 and select the movie clip. Choose *Alpha* from the Color Effect Style menu in the Properties inspector and give it a value of **0**.

5. Move the playhead to the last frame of the tween and select the movie clip. Choose *Alpha* from the Color Effect Style menu in the Properties inspector and give it a value of **100**.

Scrub through the Timeline, and you will see the movie clip fade in from transparent to opaque over the course of the animation. Compare your file to *alpha_tween.fla* from the companion source files.

Using Motion Presets

A fun, easy way to get started with motion tweens is to use *motion presets*. Motion presets are a collection of property tweens bundled into a single preset that you can apply to a tween to recreate a desired animation. Because motion tween values, and therefore motion presets, can be applied to any symbol, that symbol will take on the animated attributes defined by the tween.

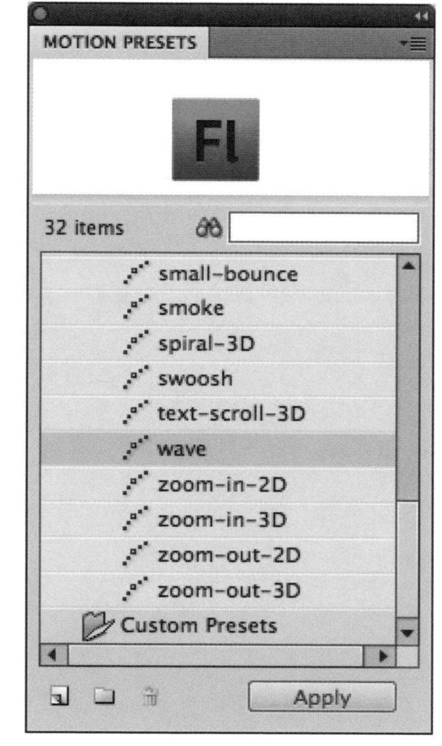

Figure 5-30. The Motion Presets panel, showing a preview of the selected Wave effect

Presets are stored in the Motion Presets panel (Figure 5-30). To apply a preset, simply select a motion tween, choose a preset from the panel, and click the Apply button. Flash will automatically insert the required frames in the Timeline layer and apply all the property value changes needed to create the desired animation.

Because using a motion preset is just an automated way of creating a motion tween, Flash will also help you apply motion presets even if the focus of your attention is not a movie clip. For example, you can create a new file, draw a shape on the Stage, select that shape, and apply a motion preset. Flash will do all the rest and, a second later, your former shape will be bouncing along or going up in smoke.

If you're thinking that canned effects like these have limited long-term value (a common sentiment), reserve judgment until you've used the feature a few times. Some of the effects, such as variants of fly in and out, zoom in and out, and spiral-3D, are useful for common tasks like presentations and intros.

Best of all, you can easily create your own presets and build a library of animations to reuse at any time. To create your own preset, right-click (Windows) or Control-click (Mac) the motion tween you want to archive and choose Save as Motion Preset from the pop-up menu. You can then name the preset and it will appear in the Custom Presets folder of the Motion Presets panel.

At the bottom left of the Motion Presets panel are three additional buttons to help you manage your own presets. From left to right, you can create a preset from your tween using the *Save selection as a preset* button instead of the mouse context menu, create folders within the Custom Presets folder to organize your presets, and delete a custom preset.

Copying Motion

If you want to reuse a motion tween on a whim or in a single case, you may not want to clutter up your custom Motion Presets menu. In this situation, you can quickly copy the motion from one tween and paste it into another.

Copy Motion
Copy Motion as ActionScript 3.0...
Paste Motion
Paste Motion Special...
Save as Motion Preset...

Figure 5-31. Accessing the Copy Motion command using the context-sensitive mouse menu

To see this process in action, create a file with two layers, with a unique movie clip in each layer. Create a motion tween in one layer and leave the other layer untouched. When you're finished, click the existing motion tween and use the Edit→Timeline→Copy Motion menu command, or right-click (Windows) or Control-click (Mac) a motion tween and choose Copy Motion from the pop-up menu (Figure 5-31).

This copies all the pertinent information from the tween, which can then be recreated with a paste. Click the movie clip in the second layer and revisit one of the same menus to invoke the Paste Motion command. The initial motion tween will be recreated and will feature another movie clip as its star player.

Using the Motion Editor Panel

Flash CS4 Professional introduces the Motion Editor panel (Figure 5-32), a new tool to help manage its new motion tween format. You can create and edit many basic tweens without using this panel, but some tasks, such as applying easing and editing property values by manipulating curves—both of which you'll learn how to do in a moment—require the Motion Editor panel.

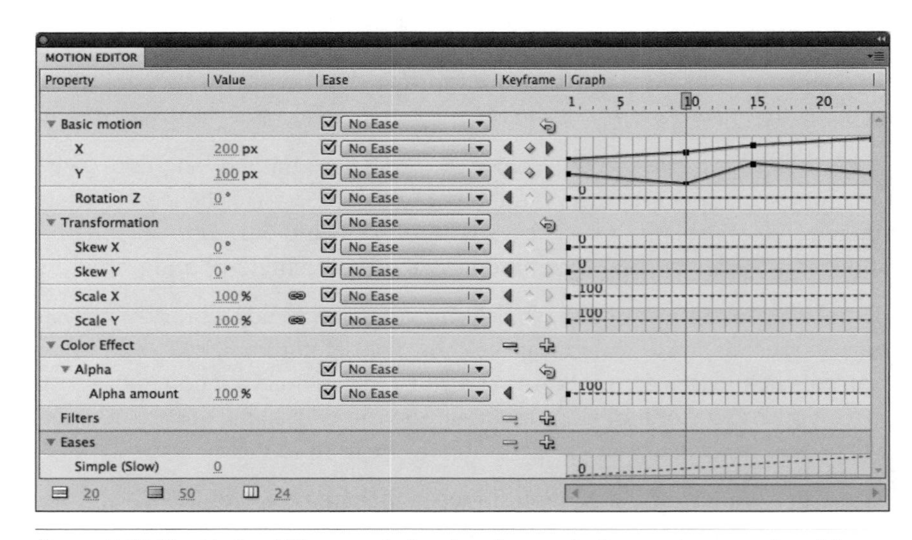

Figure 5-32. The Motion Editor panel, showing changes in the x and y properties of the tween

Think of the Motion Editor panel as a combination of the Timeline and Properties panels with a little Pen tool thrown in for good measure. If you have seen timeline-based video editing applications, particularly Adobe After Effects, the Motion Editor panel will feel familiar.

The most important thing to understand when using the Motion Editor for the first time is that it's not a replacement for the Timeline panel. Instead, you use the Motion Editor to edit property values for one tween at a time. For example, if no tween is selected in the Timeline panel, the Motion Editor panel will be empty. If the Timeline panel contains three tweens, you will need to select each tween in turn if you want to use the Motion Editor to change their values.

NOTE

Standard rotation *and* rotation z *are essentially the same.* Rotation z *bests standard* rotation *in one respect because it's more compatible with text fields. You'll learn more about that in Chapter 11, which discusses text.*

The x, y, and rotation z *properties are grouped together in the Motion Editor mainly because the panel can include 3D properties, which you'll learn about in Chapter 8. Using* rotation z *as the rotation of choice improves clarity and reduces redundancy when 3D properties*

After selecting a tween and opening the Motion Editor, you will find that properties appear by name in the first column and their corresponding values appear in the second column. Basic Motion and Transformation properties appear by default. The *rotation* property applied with the Transform tool, and edited numerically in the Properties and Transform panels, has been moved to the Basic Motion section and listed as *Rotation Z*.

For all intents and purposes, the Color Effects, Filters, and Eases sections start out empty (the Eases section does have a default ease setting available, but its value amounts to no easing). To add properties in these categories, click the plus (+) icon in the desired section and select the desired property from the pop-up list that appears. To remove effects, click the adjacent minus (–) icon.

Property keyframes are added using the Keyframe column and, in addition to the Value column, you can alter their values using graphs that appear in the Graph column. This process is covered in the next section.

Finally, at the bottom of the panel are three settings that control the row height and the width of the Graph column. Use these three settings to customize the view of the panel to improve your workflow. The first setting controls the height of a property row when not in use. Reducing this number allows you to see more rows without scrolling, while increasing the setting's value makes the graphs more legible.

The second setting controls the height of a row when in use. Splitting the height values into two settings makes it possible to maintain a panel as compact as you like, but then instantly increase the height of any row with a single click. Each time you work in a new row, that row expands and the others return to their inactive height.

The last setting controls how many frames are visible in the Graph column. This effectively allows you to zoom in on a portion of the graph or see as much of the graph as your panel size will allow. The wider your panel, the more frames can fit in the column.

Adding Property Keyframes in the Motion Editor

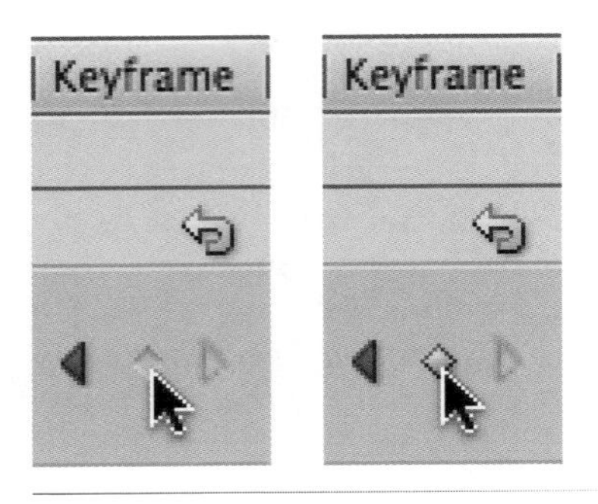

Figure 5-33. Manually adding a property keyframe in the Motion Editor

Any keyframe created in the Timeline panel will appear in the Motion Editor panel, and adding property keyframes in the Motion Editor is just as easy. The playhead at the top of the Graph column in Figure 5-33 corresponds to the playhead in the Timeline panel. You can use the playhead to scrub through the animation and stop at any frame.

At any point, you can add a keyframe by clicking the diamond shape in the center of the Keyframe column. The left image in Figure 5-33 indicates that there is a property keyframe only to the left of the playhead (see arrows), and that the current frame is not a property keyframe (no diamond icon is visible). After clicking the diamond icon (in the right image), the current frame becomes a property keyframe.

Another way to add property keyframes is to hold down the Ctrl key (Windows) or Command key (Mac) and click anywhere on the property curve. Similar to the Pen tool, you can add a keyframe where none existed before or delete a keyframe that already exists. Additional curve editing options are discussed in the next section.

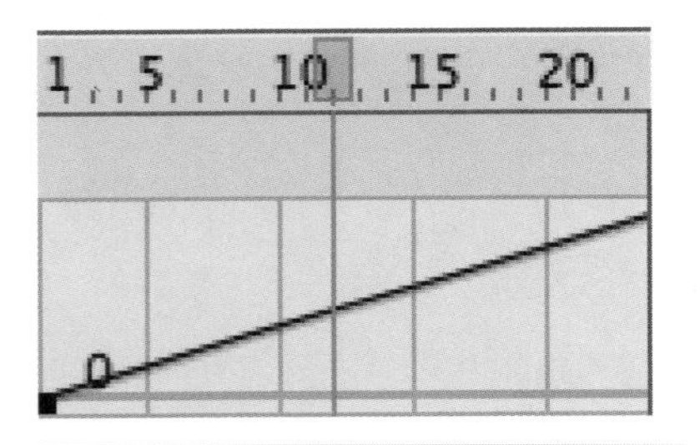

Figure 5-34. A Motion Editor property graph showing the x coordinate move from 0 to 300 in 24 frames

Editing Property Curves in the Motion Editor

The Graph column of the Motion Editor panel contains, well, a graph. The horizontal axis represents time, in frames. The vertical axis represents the value of the property, with higher values at the top of the axis, just like a normal graph. A straight dashed line and the value of the property represent an unedited property.

Consider, however, a one-second animation that lasts 24 frames with a frame rate of 24 frames per second, and changes only the *x* property to move a movie clip across the Stage from 1 to 500. The graph of this animation (Figure 5-34) starts at the lower-left corner (a time of 1 frame, or 0 seconds, and an *x* value of 0) and ends at the upper-right corner of the graph (a time of 24 frames, or 1 second, and an *x* value of 500).

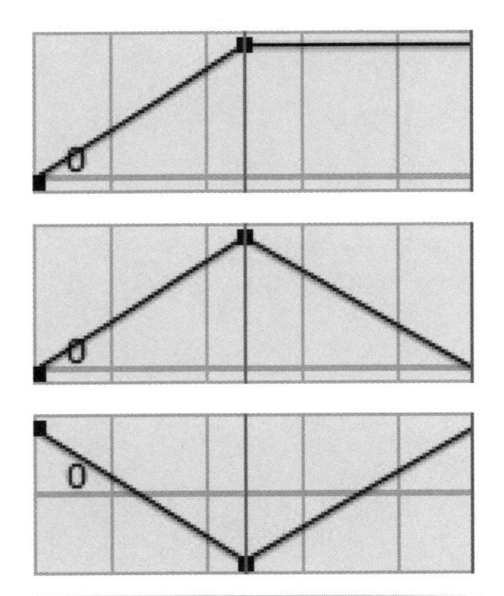

Figure 5-35. More Motion Editor graphs of x property change over 24 frames, with a property keyframe at frame 12; 0 to 300 to 300 (top), 0 to 300 to 0 (middle), 300 to –300 to 300 (bottom)

Using the property curve, you can control the value of the property in any frame. Figure 5-35 shows more examples of *x* property curves, including one that shows the *x* position offstage to the left in negative territory.

In addition to working directly with keyframes, you can edit the curve to edit the animation as a whole, including interpolated frames. To do so, edit the curve in the same way that you would use the Pen tool to edit strokes on the Stage (for more information, see Chapter 2). Each property keyframe is similar to a point on the curve, and the interpolated frames run along the rest of the curve.

Figure 5-36, for example, shows the default keyframe at the start of every frame span in frame 1, and a single property keyframe at frame 11. All frames between 1 and 11 and from 12 to 24 are interpolated frames. The curve in Figure 5-36 was created by holding down the Ctrl key (Windows) or Command key (Mac), clicking near the middle of the line, and dragging down and to the right forming the curve.

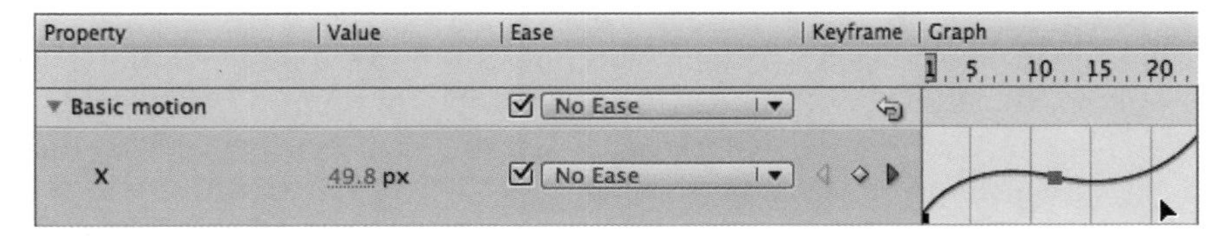

Figure 5-36. Manually editing a Motion Editor property graph

Cursor feedback and modifier-key use for editing property curves is pretty consistent with stroke editing on the Stage. Because you must select a motion tween for the Motion Editor to remain active, you will typically be using the Selection tool when you begin working with a property curve. However, the Motion Editor will handle the behavior of your mouse machinations, based on context.

When the cursor resembles the Subselection tool cursor (a hollow arrow), you can drag the entire curve or individual keyframes around. Hold the Shift key when dragging keyframes to constrain their movement in left-right or up-down directions. If you hold down the Ctrl (Windows) or Command (Mac) key, the cursor will resemble the Pen tool, allowing you to add new points and remove existing points. Dragging with this mouse/key configuration lets you create curves, as discussed previously and shown in Figure 5-36. Finally, you can hold down the Alt (Windows) or Option (Mac) key while clicking to change a keyframe marker to a corner point, straightening the curve surrounding the point, just as you would alter a point on a stroke with the Pen tool.

Right-clicking (Windows) or Control-clicking (Mac) on a point allows you to remove a keyframe or switch between roving keyframes (indicated on the graph by a circle) or the default nonroving keyframes (indicated on the graph by a square). You can also set the curve segment to the left or right of a point to a straight line.

If you right-click or Control-click in the area behind the graph, you can easily reset the property, reverse the keyframes, or copy and paste the curve from one property to another.

Easing

When you're just starting out with animation, one great way to add life to your work is to use easing. As the name implies, this is when an object eases into motion (like when you ease into traffic from a slowed or stopped state), or eases out of motion (like slowing gradually to a stop).

Typically, these movements are not perfectly consistent. If a character isn't the Road Runner of Warner Brothers cartoon fame, it doesn't run to a location and instantly stop. Nor does a character typically slow to a stop at a perfectly fixed rate. By applying easing, you can make your animations more expressive as they slow into or out of keyframes.

Figure 5-37 shows a movie clip easing into (top) and out of (bottom) a 10-frame animation. In both cases, the movie clip is at frame 5 of the animation. The clustered interpolated frame markers on the motion path convey the speed at which the movie clip is traveling.

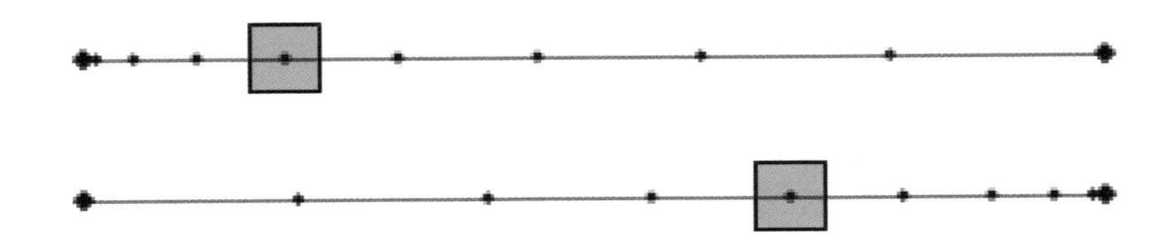

Figure 5-37. Visual representations of ease in (top) and ease out (bottom)

When the movie clip eases in, it initially moves slowly, traveling a shorter distance over more frames, and then speeds up to cover a greater distance over the remaining frames. Note that halfway through the animation, the movie clip is farther to the left of the path. When the movie clip eases out, it initially moves faster, traveling a longer distance over fewer frames, and then slows down, covering a shorter distance over the remaining frames. In contrast to the prior example, halfway through the animation, the movie clip is further to the right.

You can add easing to all tweens—shape, motion, and classic motion—through the Properties panel, and to motion tweens in the Motion Editor panel. Shape and classic motion tweens can have multiple easing settings on a per-keyframe basis, but motion tweens can only accept a single easing setting for the entire tween.

Adding Easing with the Motion Editor

In simple terms, easing is a slow acceleration or deceleration that lends realism to motion animation.

In this chapter, you'll look at two ways to apply easing: through the Motion Editor and Properties panels. Adding easing in the Motion Editor can be a bit of a challenge the first time. It will help you if you remember a few things:

- You can only apply one easing setting per property, and it applies to the entire animation. You will therefore repeat this action several times out of necessity, rather than redundancy.
- You can't edit the easing curve along with the property curve.

- You must edit the easing curve independently, and you can apply that setting to as many properties as desired through the pop-up menus in the Ease column.

Figure 5-38 shows an example of this workflow, and the sections that follow discuss how to apply easing curves in greater detail. Figure 5-38 shows the Motion Editor after applying the *wave* motion preset to a movie clip. Near the bottom of the panel you can see that the preset has added a new layer called *Sine Wave* in the Eases section.

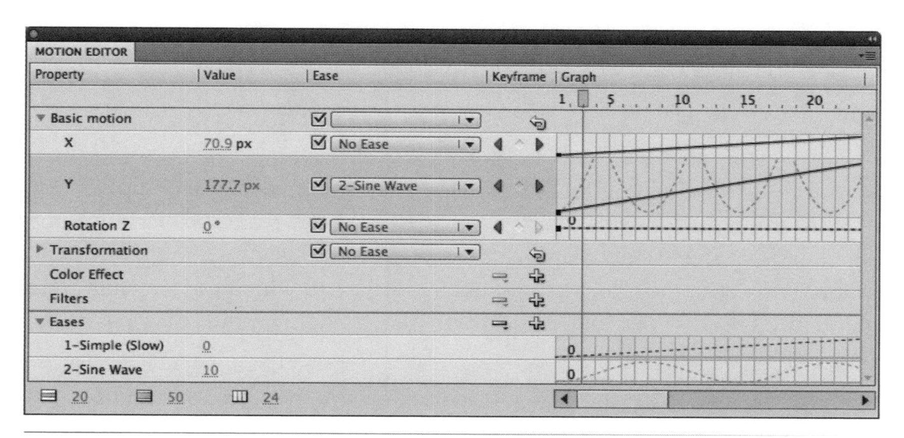

Figure 5-38. The Motion Editor after applying the Wave Motion Preset to a tween; note the Sine Wave easing preset at the bottom of the figure and its application to the y property of the tween

In the Ease column in Figure 5-38, the pop-up lists for each property show that no easing is used except with the *y* property, where the pop-up list shows the Sine Wave easing curve, and you can see a dotted line that runs behind the *y* property curve in the Graph column. The dotted line differentiates the easing curve from the property curve and further emphasizes the fact that you cannot edit the easing curve where it has been applied.

The easiest way to learn how easing curves work is to create your own, so that process is discussed in the next section. However, this is not required. You can always add any of the 16 existing easing presets to the Motion Editor menus and apply them to properties as described in the upcoming exercise. For now, familiarize yourself with creating your own curves, whether you intend to use them frequently or not.

Creating a Custom Easing Preset

Preparing an easing preset, whether it's custom or not, involves two steps: creating or using a curve, and determining the strength of the ease. In all cases, begin the process by adding a preset using the plus (+) menu in the Ease section of the Motion Editor. For a preset to be accessible to all properties, choose it from the menu by name. To create a custom preset, select Custom from the menu. In both cases, the curve will be added to the Ease section.

Editing your own custom easing curve requires the same process as editing a Bézier curve on the Stage and has been discussed in the "Editing Property Curves in the Motion Editor" section. Figure 5-39 shows a custom curve with two control handles.

Figure 5-39. Drawing a custom easing curve in the Motion Editor

In addition to being comfortable with Bézier curve editing techniques, you need to understand a small difference between easing and property graphs. The x-axis of the easing graph is measured in frames, just like the property graphs. However, the y-axis represents the degree to which the tween is complete, measured in percent, rather than the changing value of a particular property.

For example, the lower-left corner of the graph is 0% complete at the first frame of the tween. The upper-right corner of the graph is 100% complete at the final frame of the tween. The ease curve depicted in Figure 5-39 is based on the same 24-frame example used throughout the discussions on the Motion Editor. It will slowly ease into the animation, reaching approximately 20% complete at frame 10, move relatively quickly through the animation, reaching approximately 80% complete by frame 15, and then slowly ease out of the animation over the last 9 frames, reaching 100% complete at frame 24.

Once you're satisfied with your easing curve, you need to set the strength of the easing that will be applied to a property. As with properties, this value is set in the numeric entry field in the Values column of the panel.

For simple preset ease settings, this is a value between –100 and 100. Negative values ease into an animation, resulting in slower change at the beginning of the curve. Positive values ease out of an animation, increasing the easing applied to the end of a curve. For wave presets, such as Sine Wave and Sawtooth Wave, the value determines the number of half-cycles the wave makes during the duration of the tween. Finally, for custom ease settings, the value is the percent complete at any point throughout the curve.

Because you must create an easing preset and add it to a list of available settings for use in any property, and because you must set the strength value of the ease at its time of creation, it's not possible to vary the strength from use to use. The only way to accomplish this is to create multiple copies of the ease curve, each with different values. Unfortunately, it's not currently possible to rename any ease setting, even the custom settings you create, so you're stuck with remembering which setting is which by number.

After you have created the curve and set its strength, you can apply the setting to a property. Figure 5-40 shows the custom ease curve created in Figure 5-39 being applied to the *x* property. In the Ease column, access the custom setting from the pop-up menu; the dotted line showing the ease appears behind the property curve. In this example, the *x* value changes from 0 to 300, but does so at a pace based on the easing curve applied.

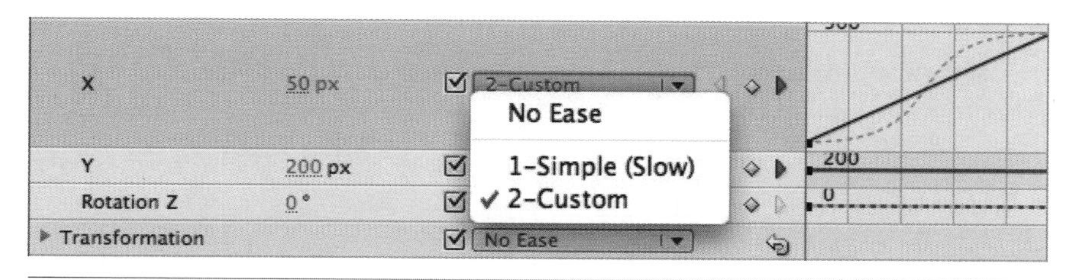

Figure 5-40. Applying a custom easing curve to a property in the Motion Editor

One of the really cool things about this implementation of custom easing curves is that the easing curve doesn't have to end at 100%. This was a limitation of the prior custom easing editor, which you'll learn about later on, but is not a factor in the new easing model. By creating an easing curve that does not finish at 100%, you can end your tween and leave properties at any desired values.

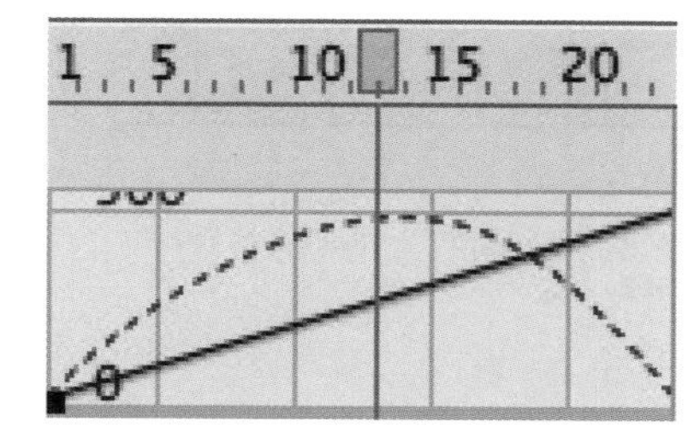

Figure 5-41. Applying a custom easing curve to a property in the Motion Editor

For example, Figure 5-41 shows an *x* property curve from 0 to 300. Without easing applied, the movie clip from this example will move from an *x* of 0 to an *x* of 300 in 24 frames. However, when the easing is applied, the movie clip will move from 0 at frame 1 to approximately 300 at frame 12 (because the easing curve reaches approximately 100% at that point) and then back to 0 at frame 24 (because the easing curve returns to 0% at the end of the tween).

Combining different properties with different easing curves can result in some interesting effects. A quick way to glance over sample settings is to apply the motion presets to a temporary file and look at the Motion Editor every time you try a new preset. After a handful of peeks at the inner workings of the presets, you'll be able to read the Motion Editor pretty easily.

Although it's possible to draw elaborate easing curves that factor in the locations and values of every property keyframe, this can be quite involved for complex animations. In some cases, when per-keyframe easing is required, the limitations of one ease per property tween in the Motion Editor will be a major annoyance. In those circumstances, it may be easier to use the classic motion tween format. The portfolio project demonstrates this scenario. Most animations will use the new motion tween format, but the rotating viewing wheel will use classic motion tweens so keyframe-specific easing can be employed.

Creating a Classic Tween

The classic tween format is the motion tween version used in prior Flash releases. Its creation and manipulation is similar to that of the shape tween discussed earlier in this chapter. However, just like the new motion tween, it animates symbols instead of morphing shapes.

1. Create a new file using File→New. This file will not be a part of the ongoing portfolio project.
2. Create a keyframe in frame 5 (F6).
3. Draw a small shape in frame 5, near the left edge of the Stage, and convert it to a movie clip (F8).
4. Create a keyframe (F6) in Frame 15 and drag the movie clip to the right edge of the Stage.
5. Click anywhere in the first frame span and use the Insert→Classic Tween menu command, or right-click (Windows) or Control-click (Mac) the span or movie clip and select Create Classic Tween from the pop-up menu (Figure 5-42).

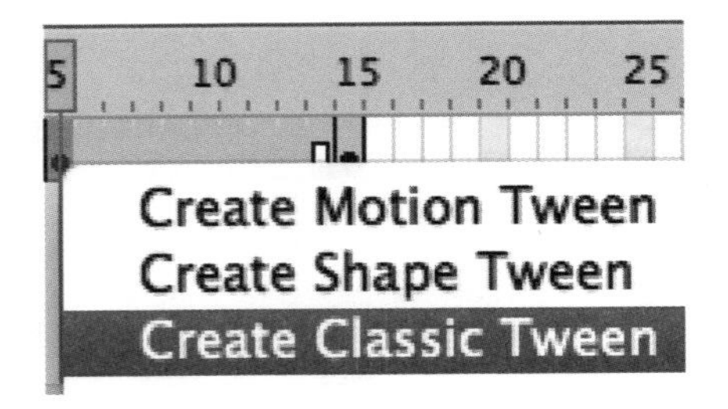

***Figure 5-42.** Creating a classic tween using the context-sensitive mouse menu*

The Timeline will look very similar to the shape tween Timeline, but the span will be purplish-blue instead of green (Figure 5-43). If you scrub through the Timeline, you'll see the movie clip move across the Stage. Compare your file to *classic_tween.fla* from the companion source files, then experiment further by changing other properties in each keyframe. For example, apply an alpha effect from the Color Effects *Style* menu in the Properties panel. Most importantly, however, apply easing to the tween.

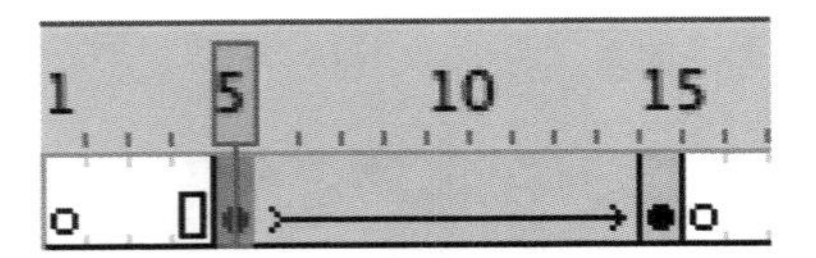

***Figure 5-43.** A purplish-blue tint and arrow in the Timeline indicate a classic tween*

Adding Easing with the Properties Panel

Adding a stock easing effect to a classic motion tween is very much like adjusting the very properties you tweened. Click in the first keyframe of the tween and, in the Properties panel, adjust the *Ease* value in the Tweening section (Figure 5-44). The value ranges from –100 (maximum ease in), to 0 (no easing), to 100 (maximum ease out). For more information about the difference between easing in and out, including sample motion paths, see the "Easing" section, earlier in this chapter. Sample files called *classic_tween_ease_in.fla* and *classic_tween_ease_out.fla* have been provided for you to compare against your own efforts.

***Figure 5-44.** Applying easing to a classic tween in the Properties panel*

Once you've had some experience with the differences between easing in and out, try to add another classic motion tween sequence, perhaps by practicing the Copy and Paste Frames processes discussed earlier in this chapter. Save a copy of your basic classic tween file first, as you'll want to work with the simple version when learning about motion paths near the end of this chapter. After saving a copy, apply different easing effects to each keyframe, and you'll be prepared for your work on the portfolio project later on. Compare your tests to the *classic_tween_ease_in_out.fla* source file for an example of multiple keyframe easing.

Custom Easing

Although applying stock easing effects to classic motion tweens is as straightforward as changing property values, sometimes a simple ease isn't enough. To achieve other easing effects, such as an effect that simulates bouncing, you need to draw a custom easing curve.

Although not as powerful as the new custom easing curves, you may prefer to use classic easing because of its per-keyframe application benefit. To create a custom easing curve, click the pencil icon to the right of the easing value, shown in Figure 5-44. This will open the Custom Ease In / Ease Out editor shown in Figure 5-45.

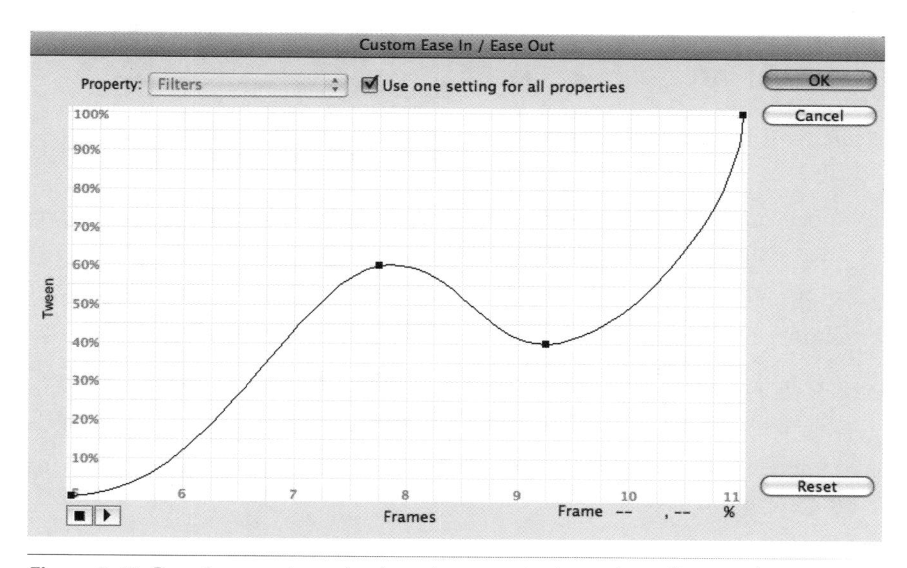

Figure 5-45. *Drawing a custom classic easing curve in the easing editor via the Properties panel*

Editing the custom classic easing curve is similar to creating an easing preset in the Motion Editor panel, including the values on the x- and y-axes of the graph, and the Bézier curve editing features. However, there is one limitation with the legacy format. Classic easing curves must end at 100% complete. That is, just like the Motion Editor easing curves, the first keyframe is in the lower-left corner of the graph at 0% complete and frame 1. You can add additional property keyframes, but the last keyframe must be anchored to the upper-right corner of the graph, at the end of the tween and 100% complete.

Another point worthy of discussion is that you can apply different easing curves to each property in the classic easing editor, just as you can in the Motion Editor. If you disable the *Use one setting for all properties* feature at the top of the editor, the adjacent menu offers the chance to create a unique curve for position, rotation, scale, color, and filters. Switching among these properties in the menu will also switch among any associated custom curves created.

Classic Motion Guide

The final significant difference between classic tweens and motion tweens applies to the use of a motion guide. It is not possible to simply paste a motion guide into a classic tween. Instead, you must paste it into a dedicated motion guide layer type, shown in Figure 5-46.

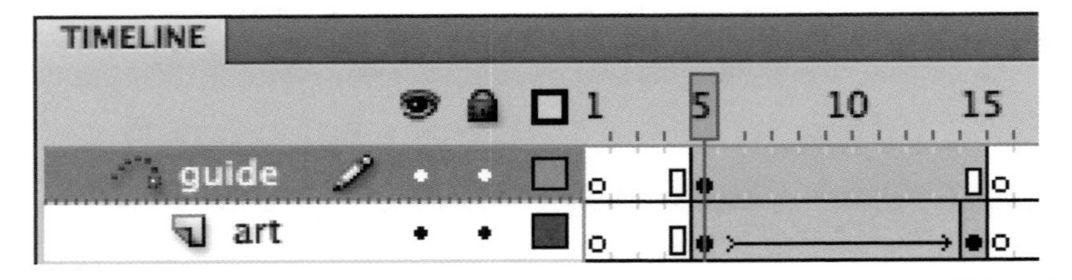

Figure 5-46. Transforming a layer with a stroke into a classic motion guide layer

1. Open the first classic tween exercise you created or, if you've edited that file further, open the *classic_tween.fla* source file.
2. Right-click (Windows) or Control-click (Mac) the classic tween layer and choose Add Classic Motion Guide from the pop-up menu.
3. Create keyframes (F6) in frames 5 and 16 to confine the motion guide to the span of the entire tween, keyframe to keyframe.
4. Draw a stroke with the Pencil tool in that guide layer to serve as a path for the animation to follow.
5. Click the first keyframe of the classic tween and, in the Properties panel, make sure Snap is enabled in the Tweening section.
6. Drag the movie clip instance on the Stage to the start of the path you drew, until it snaps into place.
7. Select the last keyframe and drag the movie clip in that frame until it snaps to the end of the motion guide to complete the process.

If you scrub through your Timeline, you'll see that the movie clip now follows the path you drew. Compare your work to the *classic_tween_motion_guide.fla* source file.

As the final experiment of the chapter, before working on the portfolio project, set up your tween so the animated object orients itself along the direction of the motion guide. This is ideal for elements that need to appear to be pointing in the direction in which they are traveling.

For example, think of a plane flying across the Stage and performing a 360-degree loop as it moves from left to right. Without orienting the object to the path, it would begin to fly right, appear to float upward parallel to the ground as it entered the loop, then fly backward at the top of the loop, and finally fall out of the loop and continue to fly to the right. On the other hand, if you oriented the plane to your path, the plane would appear more correctly to fly straight up, point left as it flew upside down, dive downward, and then swoop out of the loop and on its way.

To achieve the best results when trying this feature, edit your movie clip so it has a clear, correct orientation. For example, draw an arrow that will appear to point in a specific direction. If you prefer, continue with the *classic_tween_motion_guide.fla* source file. Then click in the first keyframe of the tween and, in the Tweening section of the Properties panel, enable the *Orient to path* option. Your movie clip should now follow your path with the correct orientation throughout.

You can also apply the *Orient to path* option to the Flash CS4 Professional version of motion tweens. After the motion tween is complete, select it in the Timeline and enable the feature in the Rotation section of the Properties panel.

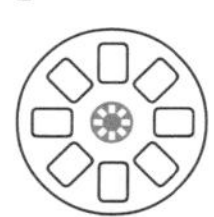

Project Progress

In this chapter, you'll build the project's skeletal structure and animate its interface. Before jumping into the step-by-step file construction, it's helpful to have a big-picture understanding of how the portfolio works. As you probably remember from the previous chapter, the design and navigation metaphor for the portfolio is a paper craft viewing wheel. A wheel rotates around its center spindle and eventually stops to reveal the contents of a specific screen.

Figure 5-47 shows a preview of part of the Timeline you'll build, focusing on a sample section screen of the portfolio. You will insert numbers into the text that follows, showing where each event occurs in the Timeline. By visualizing the navigation this way before you build it, the descriptions and the steps may be more obvious. It may also help to see the portfolio in action at the companion website, *http://www.LearningFlashCS4.com*, before proceeding.

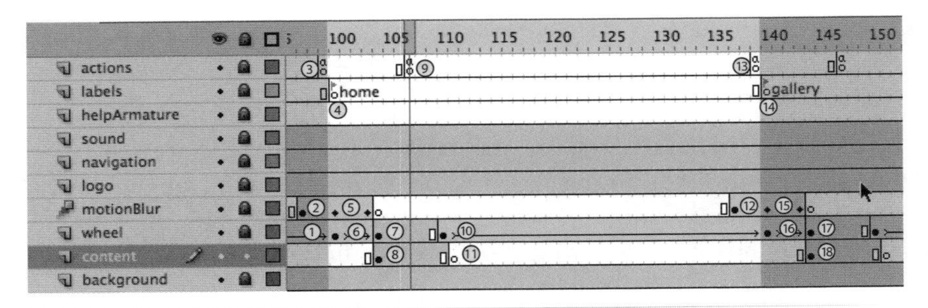

Figure 5-47. A step-by-step illustration of navigating between sections of the portfolio project

At first launch, the viewing wheel fades in over a black background and scales to full size. A paper background and logo then fade in, the wheel begins to spin, and the navigation and sound controls appear.

When the viewing wheel reaches full speed (1), a still image created from a motion blur of the wheel rotating fades in on top of the wheel for a partial second (2). The first section visited (*Home*) is stored into a variable by a frame script (3), and the playhead is sent to that frame label (4). The motion blur fades away (5), and the wheel slows down as it rotates into final position (6). Finally, the wheel stops rotating (7), the section content is revealed (8), and a frame script stops the playhead in the section screen (9).

This process is repeated each time the user switches portfolio sections. The user clicks a navigation bar button and stores the desired destination in the variable and starts the playhead moving. The wheel rotates (10), the previous section content disappears (11), and the motion blur fades in (12). When the playhead reaches the last frame in the section (13), the motion blur fades up enough to cover the navigation jump, and a script sends the playhead to the new desired section frame.

After the playhead jumps to the new section (14), the motion blur fades out (15), the wheel slows and stops spinning (16 and 17), and the new section's contents appear (18). By splitting the motion blur—fading up over the last few frames of the prior section, and down over the first few frames of the next section—you can hide any sudden change in the wheel's rotation during the transition between sections.

Inner Section Timeline Structure

With the portfolio navigation fresh in your mind, it's a good idea to start building your main file with the repeating structure for each portfolio section.

Adding layers and frame labels

The first task is to add a few new layers and divide the Timeline into sections:

1. Open your running portfolio project file or *portfolio_03.fla*.
2. Select the top layer in the Timeline panel and add three layers to the top using the New Layer button at the bottom-left of the panel (or by using Insert→Timeline→Layer). Name the new layers, from top to bottom, `actions`, `labels`, and `sound`, and lock them. Your Timeline layers should now resemble Figure 5-48.
3. Select the *background* layer, add a new layer above it, and name it `content`.
4. Scroll the Timeline panel until frame 260 is visible. With one process, click in frame 260 of the top layer and drag down to frame 260 of the bottom layer. This will select frame 260 in all layers. If you have any trouble doing this, look back over Figure 5-7. After selecting frame 260 in all layers, press F5 to add frames. Your file should span from frame 1 to frame 260 in all layers.
5. In the *labels* layer, click frame 100 and add a keyframe (F6). With the frame still selected, give it the label name `home` in the Properties panel. Repeat this process in frames 140, 180, and 220, applying label names `gallery`, `lab`, and `help`, respectively.

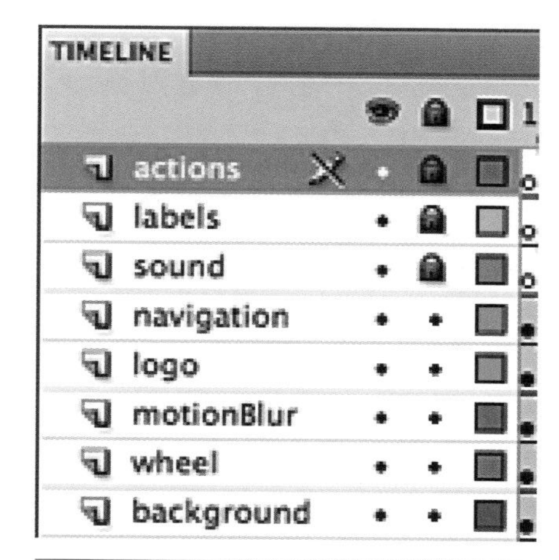

Figure 5-48. The layers of your main portfolio file

Inserting frame scripts

With the new layers added and the sections marked off with keyframes and frame labels, you'll insert the frame scripts required for navigation. You'll write the button scripts for the navigation bar later on.

6. In the *actions* layer, add keyframes that will stop the playhead in each portfolio section. These frames will stop the playhead 7 frames after every frame label, allowing enough time for the motion blur to fade out and the content to fade in. The frame numbers are 107, 147, 187, and 227. In each frame, add the following ActionScript:

```
stop();
```

7. In the *actions* layer, add keyframes that will serve as the end of each animated transition. When the playhead reaches these frames, a script will take the playhead to the frame label stored in a variable, providing a seamless navigation between sections. The end of each animated transition appears before every label (99, 139, 179, and 219) and before the end of the timeline (259). Add the following ActionScript to each of these frames except frame 99 (you will write a special script for frame 99 later in this chapter).

```
gotoAndPlay(nextSection);
```

Tweening the motion blur

The next task is to create the motion blur overlays for each section. You'll create the first fade in/fadeout sequence, then copy those frames for use in other sections.

8. If it's hidden (because of work you did in Chapter 4), show the *motionBlur* layer. In this layer, click in frame 97 and press F6 to create a keyframe. Repeat this process in frame 104. Click frame 1 of the layer to select its content and delete it, emptying the first section of the layer. Repeat this process in frame 104, deleting the content in frame 104 after the keyframe is created. Scroll through the Timeline and confirm that the only content in the *motionBlur* layer is between frames 97 and 103.

9. Right-click (Windows) or Control-click (Mac) the *motionBlur* between frames 97 and 103 and select Create Motion Tween from the pop-up menu. The layer frame span will turn blue, indicating that it has become a motion tween, and the first frame will automatically become the span's first keyframe.

10. Next, you need to create two property keyframes. Move the playhead to frame 100 and click the *motionBlur* movie clip on the Stage. In the Properties panel, in the Color Effect section, choose the *Alpha* Style and set its value to **100**. Repeat this process in frames 97 and 103, setting the *Alpha* of the movie clip to **0** in both frames. If you scrub through these frames, you should see the movie clip start with an alpha of 0, fade up to 100, and then fade back down again to 0.

Now that you have a completed motion blur tween, you need to copy these frames and paste them into the other sections. To prepare for this change, you'll need to create the 7-frame span required for the blur effect in the three other sections of the Timeline.

11. Add keyframe pairs to the *motionBlur* layer in frames 137 and 144 (for the Gallery screen), 177 and 184 (for the Lab screen), and 217 and 224 (for the Help screen). This will create the 7-frame span required for the motion blur effect in each section.

12. To copy the necessary frames, right-click (Windows) or Control-click (Mac) on the first motion blur tween, anywhere in the *motionBlur* layer between frames 97 and 103, and choose Copy Frames from the pop-up menu. This is the first of three important points of this task. Do not choose Copy out of habit. The latter option will only copy the contents of the current frame while the former will copy the selected frames in the Timeline.

13. To paste the tween, you must select *all the frames* between the keyframes you created to hold the corresponding blur effect in each of the other sections. This is the second important point in this task. If you simply click once in the frame span, additional unwanted frames will be added to the

layer when you paste. If you select the frames first, they will be replaced by the copied tween.

14. To select all the frames, click in the first frame of the span and, without releasing the mouse, drag to the last frame of the span. In the *motionBlur* layer, click in frame 137 and drag to frame 143, selecting all seven frames (Figure 5-49). Next, right-click or Control-click in the span and choose Paste Frames from the pop-up menu. This is the final important point in this task. Be sure to select Paste Frames instead of Paste. The result is that the motion blur tween will be copied from the Home section and pasted into the Gallery section.

15. Having duplicated the motion blur tween once, do so again two more times. Click and drag to select frames 177 through 183, right-click or Control-click the span and select Paste Frames, and then repeat the process for frames 217 through 223.

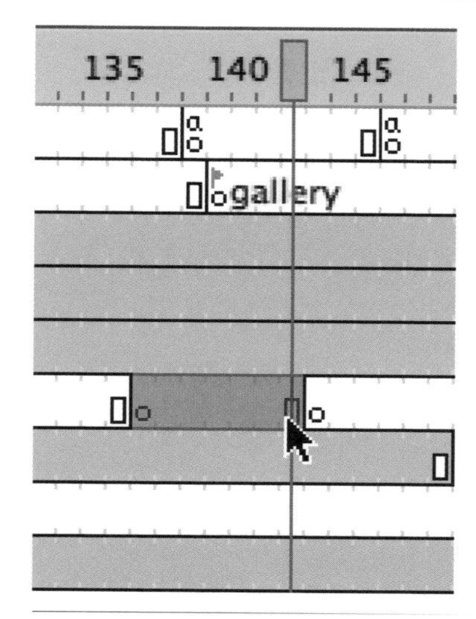

Figure 5-49. *Selecting a frame span before pasting frames*

Finally, you need to create *half* of the motion blur effect that fades up the blurred overlay at the end of the Help section. Remember, during navigation, the script at the end of the section will send the playback head to the second half of the fade in another section.

You will use the same Copy Frames and Paste Frames technique you used previously, but you need to select only the first half of another section's fade. You will need to hold a modifier key down and drag to select only the first half of the tween, but a helpful feature can make this tricky: you may remember from earlier in the chapter that you can grab the first or last frame of a tween span in the Timeline panel and drag it left or right to change the length of the span. This convenience can get in the way when you try to hold the Ctrl (Windows) or Command (Mac) key down to select only a portion of the frames in a span. When making your selection, if you start dragging the first frame in the span from left to right, Flash will think you are trying to shorten the span. So, you need to start the selection process from the middle property keyframe and drag left until you reach the first frame. You will be selecting the same four frames, but you'll be dragging from right to left instead of left to right.

16. Create a keyframe in frame 257 in the *motionBlur* layer. Similar to the process described in step 11, select the first half of the motion blur tween in the Help section by holding down the Ctrl key (Windows) or Command key (Mac), clicking in frame 220, and dragging to frame 217 (Figure 5-50). Right-click (Windows) or Control-click (Mac) the selection and choose Copy Frames from the pop-up menu. As described in step 12, select frames 257 through 260, right-click or Control-click the selection, and choose Paste Frames.

17. Save your work. When you're done, scrub through the Timeline panel and make sure the motion blur fades up and down near each frame label, and up at the end of the Timeline.

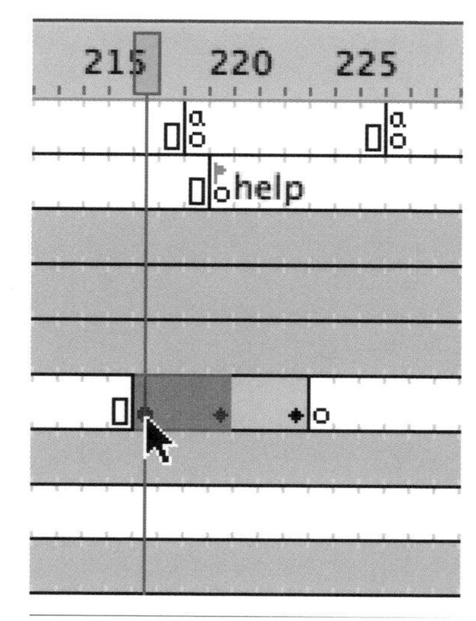

Figure 5-50. *Selecting part of a motion tween before copying frames*

Creating content spans and adding content

The last major task in building the Timeline structure that is specific to each section is to create frame spans in a new layer that will hold portfolio content. You will also add content created in prior chapters to two sections, and add the sound controller you created in Chapter 3 to the main interface.

18. Select the *content* layer in the Timeline panel by right-clicking (Windows) or Control-clicking (Mac) the layer's name and choosing Insert Layer from the pop-up menu.
19. Create keyframe pairs in the layer at frames 104 and 111 (Home), 144 and 151 (Gallery), 184 and 191 (Lab), and 224 and 231 (Help). This creates the necessary 7-frame spans for each section that will hold content.
20. Open the *home_page.fla* you created in Chapter 2. On the stage, select the instance of the *HomePage* movie clip (the only asset in that FLA's *content* layer) and copy it.
21. Switch back to your portfolio FLA and use Paste in Place (Shift+Ctrl+V in Windows or Shift-Command-V on Mac) to paste the movie clip into frame 104 of the *content* layer.
22. Repeat steps 18 and 19 for the Lab page content. Open the *lab_page.fla* created in Chapter 4, copy the *LabPage* movie clip from the Stage, and use Paste in Place to paste the movie clip into frame 184 of the portfolio FLA's *content* layer.
23. Finally, add the sound controller you created in Chapter 3. Open *sound_control.fla* and copy the *soundControl* movie clip. Switch back to your portfolio FLA and paste the movie clip into the *sound* layer. Using the Properties panel, set the movie clip's *x* and *y* locations to **590** and **420**, respectively.

You have completed the changes to the main portfolio document that are specific to each section. You can compare your file to *portfolio_06_01.fla* and continue on with that file, if you prefer. Next, you'll animate the layers that apply to all of the sections. Now is probably a good time to save your work and take a break!

Project-Wide Timeline Structure

Beyond screen-specific assets, the project includes interface elements used throughout the portfolio. In this Project Progress segment, you'll define frame spans for these elements and animate them with two kinds of tweens.

Defining the length of each layer frame span

Upon a user's first visit to the portfolio, the project assembles its interface one element at a time, introducing the viewer to functionality as it builds. To facilitate this, the content will not all appear in the same frame.

1. In the *sound* layer, add a keyframe (F6) in frame 90. Click in frame 1 of the layer and delete the content. This should remove the assets in the layer between frames 1 and 89.
2. In the *navigation* layer, add a keyframe (F6) in frame 90 and delete the content in frame 1.
3. In the *logo* layer, add a keyframe in frame 55 and delete the content in frame 1.
4. In the *wheel* layer, practice moving a keyframe. Click in frame 1 to select only the keyframe. Click and drag it to frame 2 to move the keyframe, leaving frame 1 empty.
5. In the *background* layer, add a keyframe in frame 50 and delete the content in frame 1.

User interface motion tweens

With the frame spans defined, you can now create new motion tweens for the *sound*, *navigation*, *logo*, and *background* layers. You'll address the *wheel* layer later on.

The interface elements will reveal themselves one by one, either by fading in or by moving into view from offstage. Because you imported many of the interface elements using the PSD File Importer, they are already in place. So, you can simplify the tween creation process by first adding property keyframes to the frames in which you want each tween to finish. You can then manipulate the starting position or opacity in the tween's first frame to accomplish the reveal.

6. Convert the *sound*, *navigation*, *logo*, and *background* layers into motion tweens. Right-click (Windows) or Control-click (Mac) in frame 120 of each layer (to be sure there is content in the frame) and select Create Motion Tween from the pop-up menu each time.
7. In the *sound* layer, click in frame 95. Open the Motion Editor panel and, in the Keyframe column of the Basic Motion section, click the diamond to add a keyframe for the *x* and *y* properties.
8. Repeat step 6 for the *navigation* layer, adding property keyframes for *x* and *y* in frame 95.
9. In the *logo* layer, click in frame 63. In the Motion Editor panel's Color Effects section, add Alpha from the plus (+) menu (Figure 5-51) so it appears as an editable property in the section. Set the value to 100%, (if it is not already).
10. Repeat step 9 for the *background* layer, adding a property keyframe for *Alpha* in frame 58.

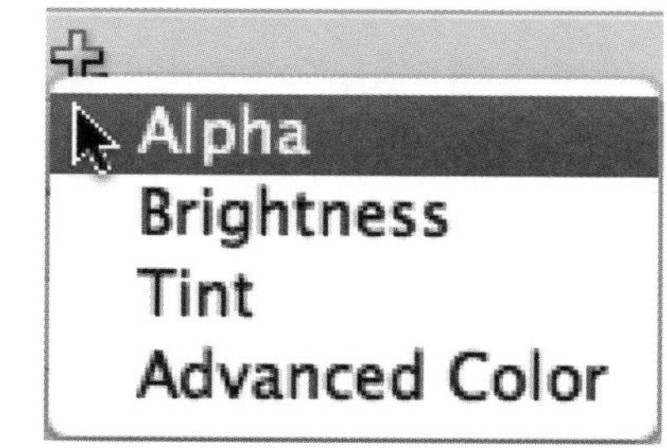

Figure 5-51. Adding an Alpha color effect in the Motion Editor panel

Now all you have to do is change the initial states of each of these tweens to assemble the interface.

11. In the *sound* layer, click in frame 90. Select the sound controller on the Stage with the Selection tool and, using the Properties panel, set its *x* and *y* properties (in the Position and Size section of the panel) to **790** and **420**, respectively.

12. Repeat step 11 for the *navigation* layer. In frame 90, click the navigation button bar and set its *x* and *y* properties to **207** and **575**, respectively.

13. In the *logo* layer, click in frame 55. Select the logo on the Stage and, using the Properties panel's Color Effects section, set the Style to *Alpha* and its value to **0**.

14. Repeat step 13 for the *background* layer. In frame 50, select the background image and set its *Alpha* property to **0**.

NOTE

Because you staggered the animated effects, such as the alpha and motion tweens used to build the portfolio interface, the animations won't all occur in the same frames. This can lighten the load on your computer's processor and reduce the number of times the animations perform sluggishly.

If you scrub the Timeline between frames 45 and 95, you should see the background fade in, followed by the logo fading in. Next, the navigation button bar should animate into view from below the stage and the sound controller should animate in from the right of the stage. After you've previewed your work, save your file.

Viewer wheel classic motion tween intro animation

The animating viewer wheel is a particularly good candidate for classic motion tweens because it's easier to add easing effects on a per-keyframe basis to classic tweens. As discussed previously, you can only apply easing settings in the new motion tween model to an entire tween span. This makes creating complex, frame-specific easing graphs more difficult and less attractive than using Flash's traditional tweening approach. This is especially true if the custom easing settings are not easily reused, as reuse is one of the compelling reasons for creating easing presets in the Motion Editor panel.

All of the steps in this section apply to the *wheel* layer. In Chapter 4, you finished the "Project Progress" section with the wheel in place at its final size. Therefore, to set up the intro animation, it will be easiest to work backward—staring at its end frame and working back to frame 2. Once that is complete, you can go on to create the similar animations required for each section.

15. Create keyframes (F6) at frames 45 and 21. Frame 45 is the last state of the intro animation and doesn't need to be changed. In frame 21, select the wheel on the stage and, using the Properties panel, set the *x* and *y* properties to **350** and **250**, respectively. Using the Transform panel, set both the *x* and *y* scales (at the top of the panel) to **20** percent. This will reduce and center the wheel on stage.

16. Right-click (Windows) or Control-click (Mac) in frame 21 and chose Create Classic Tween from the pop-up menu. The tween will turn blue, but will have a slight purple tinge and an arrow running through the span. If you scrub the Timeline panel through these frames, you'll see the wheel enlarge from 20% to 140% at its final size.

17. While still in frame 21, select the wheel on stage and copy it. In frame 2, select the large final size and position and delete it. Then paste the smaller wheel in place, using Edit→Paste in Place. Right-click (Windows) or Control-click (Mac) in frame 2 and choose Create Classic Tween from the pop-up menu. This first tween in the layer will fade in the wheel at its starting size.

18. Create a keyframe in frame 20 to serve as the final keyframe of this initial tween, which requires no change. In frame 2, select the wheel on the Stage and, using the Properties panel, set the Color Effect style to *Alpha* and give it a value of **0**. Scrubbing the Timeline panel among frames 2 through 20 will show the wheel fading in from 0% to 100%.

Viewer wheel classic motion tween section animations

With one exception, which we'll look at in a moment, the wheel section animations are consistent. At the end of each section (including the intro), the wheel makes one full clockwise spin. At the beginning of each section, the wheel spins one more quarter turn as it slows to a stop. You'll go through the first set of tweens in detail, then use the same process for the remaining three sections.

As in the previous section, all of the following tasks apply to the *wheel* layer. You will create several classic motion tweens so, for brevity, the instructions for converting a normal frame span into a tween will not be reviewed. In all cases, you can click on the first frame of the span and use the Insert→Classic Tween menu command, or right-click (Windows) or Control-click (Mac) the first frame of the span and choose Create Classic Tween from the pop-up menu.

19. Add keyframes to frames 65, 100, and 104. This span includes the spin up from the intro and the spin down at the start of the Home section.

20. At frame 65, create a classic tween. In the Tweening section of the Properties inspector, select CW (clockwise) from the *Rotate* menu and enter **1** to the right of the *Rotate* menu for one rotation.

21. At frame 100, create another classic tween. This time, however, you want to rotate the wheel manually. Select the keyframe in frame 104 and click the wheel on the Stage. In the Transform panel, enter **90** degrees in the *Rotate* value. The classic tween will use this transformation in the second keyframe and rotate the wheel one more quarter turn until it stops.

22. All that remains for this initial set of tweens is to add easing to the first half. Click in frame 65 and enter **–100** for the Ease value.

This last step adds a maximum Ease In to the first segment of the two-part tween, so the wheel slowly spins up to speed. The second part of the tween consists of only three frames so the wheel stops quickly. This span length is usually too short for easing to be noticeable, but, more importantly, the wheel slowdown occurs while the motion blur is fading out, further obscuring any visible easing. So, no easing is applied to the second part of the tween.

Now you'll repeat this process for the three remaining sections. If any of these steps seem abbreviated, simply review steps 19 through 22 to see how you should apply the values.

23. Add keyframes to frames 110, 140, and 144 to spin up leaving Home and spin down entering Gallery. Create classic tweens at frames 110 and 140. Select the tween at 110 and set *Rotate* to **1** clockwise rotation, with an Ease of **-100**. At frame 144, click the wheel on the Stage and enter a rotation value of **180** in the Transform panel. Note that this value has changed from 90 in the previous section. This is because the wheel needs to rotate another quarter turn.

24. Add keyframes to frames 150, 180, and 184 to spin up leaving Gallery and spin down entering Lab. Create classic tweens at frames 150 and 180. Select the tween at 150 and set *Rotate* to **1** clockwise rotation, with an Ease of **-100**. In frame 184, set the rotation of the wheel to **-90** in the Transform panel.

If you're wondering why the value is –90, you're not alone. In the previous two similar efforts, the addition of a quarter turn each time resulted in values of 90 and 180. Why does another quarter turn result in a value of –90 instead of 270? This is a side effect of Flash optimizing the way it handles angles. Flash will always take the shortest approach to rotation by converting large angles to their negative equivalents. For example, instead of rotating clockwise three quarters of a circle to 270 degrees, it's more efficient to rotate one quarter counter clockwise to –90 degrees and end up at the same location.

1. Add keyframes to frames 190, 220, and 224 to spin up leaving Lab and spin down entering Help. Create classic tweens at frames 190 and 220. Select the tween at 190 and set *Rotate* to **1** clockwise rotation, with an Ease of **-100**. In frame 224, set the rotation of the wheel to **0** in the Transform panel. Here, again, the negative angles are at work. The prior value of –90, plus one more quarter turn, yields 0.

2. The last tween covers only leaving the Help section, so there is only one segment. Add keyframes to frames 230 and 260. Create a classic tween at frames 230. Select the tween at 230 and set *Rotate* to **1** clockwise rotation, with an Ease of **-100**.

You're now done with the Timeline tweens. Save your work and scrub through each section in the Timeline. If the rotation of the wheel doesn't appear correct in any of the sections, review your settings. If you're still having trouble, check the companion website at *http://www.LearningFlashCS4.com* for additional information and continue on with the provided source file, *portfolio_06_02.fla*. All that remains is to add a script to activate the navigation buttons, allowing you to test your file.

Button Script

Earlier, you added a stop action to the content frame of each portfolio section and a script at the end of each section that tells the playhead to jump to a frame specified in the **nextSection** variable. Now you must add a script that will enable the buttons to populate that destination variable and play through the current section on to the next.

You will learn more about ActionScript, including more detail about this very script, in the next chapter. However, this code is included here because it's required to successfully test your tweens. Scrubbing through the Timeline is adequate, but it can't effectively show easing and doesn't give you an accurate picture of how your tweens look. By testing your file, you can see if any tweaks are required and adjust any tween values as needed.

Rather than just including this script as is, a brief explanation will follow. However, expect to learn and retain more about this code in the next chapter. For now, type this script into frame 99 of the *actions* layer. It's fine if you want to wait to look over the script after reading Chapter 6, but even a vague understanding of how this code works may make it easier to grasp when covering more detail later:

```
1   var nextSection:String;
2
3   navigation.home.addEventListener(MouseEvent.CLICK, onNavigate);
4   navigation.gallery.addEventListener(MouseEvent.CLICK, onNavigate);
5   navigation.lab.addEventListener(MouseEvent.CLICK, onNavigate);
6   navigation.help.addEventListener(MouseEvent.CLICK, onNavigate);
7
8   function onNavigate(evt:MouseEvent):void {
9       nextSection = evt.target.name;
10      play();
11  }
```

Here's a cursory look at how this script functions, with a more detailed explanation to follow in Chapter 6. Line 1 creates the variable that will hold the desired portfolio destination each time the user clicks a button.

Lines 3 through 6 assign specific code (lines 8 through 11) to each button. The buttons are named *home*, *gallery*, *lab*, and *help*, and all reside inside the navigation movie clip. The code is executed when the user clicks on the button. The *click* event includes both actions of pressing the mouse button down (mouse down) and releasing it (mouse up).

Finally, after each user interaction, the script can tell which button was chosen because it was the *target* of the user's mouse click. The name of that button is put into the destination variable and the playhead is set in motion.

When the playhead reaches the end of the section, the frame script you created earlier sends the playhead to the frame label that has the same name as the clicked button.

Testing Your Work

You've put a lot of effort into the project this chapter, so it's time to test your work. You should now be able to click any of the four navigation buttons and start the wheel rotating. After the motion blur fades in and out, the wheel should stop rotating on the desired frame.

So far, you've only created content for the Home and Lab sections, and haven't yet powered the Lab section with ActionScript. However, if you can successfully revisit the Home and Lab screens, your script is likely working. A further indication of success, when visiting the empty Gallery and Help screens, is that the rotation of the wheel changes with each section. Between the four sections, and including the art visible between frames as the wheel rotates, you should see the entire wheel.

Again, if you want to, compare your work to the online portfolio at the companion website, or to the *portfolio_06_02.fla* source file.

The Project Continues...

The next chapter will serve primarily as a primer for the ActionScript language, so no new work will be added to the portfolio project. However, the chapter will not be void of project progress. You will review the ActionScript you've learned so far and revisit the button navigation script you just added, as well as the script used in the Home screen assets you created in Chapters 1 and 2.

CHAPTER 6

ACTIONSCRIPT BASICS

Introduction

ActionScript is the internal programming language that Flash designers and developers use to add interactivity to projects. Sometimes a linear progression through the Timeline with animations that never vary is not enough. ActionScript can add variety, randomness, and user input and control to the mix.

Introduced in Flash 2, interactive control of Flash has been around for a long time. Flash 4 included support for written scripts. Flash 5, unveiled in 2000, contained the first reasonably full-featured version of a scripting language. This language was called ActionScript and was retroactively named ActionScript 1.0 (AS1) later on.

Since that time, there have been two major architectural changes to the language. Flash MX 2004 (actually released in September of 2003) included ActionScript 2.0 (AS2), a more robust iteration of ActionScript and the first to introduce formal object-oriented programming capabilities to Flash. Later, in 2007, Flash CS3 rebuilt ActionScript from scratch when it let ActionScript 3.0 (AS3) out of the cage.

Rather than enhancing the codebase of AS1 and AS2, and continuing with any baggage or flaws ActionScript adopted through its early existence, the code was reinvented for AS3. The prior code base was just too entrenched to accommodate sweeping improvements without breaking backward compatibility.

Instead, an entirely new codebase was developed and added to Flash Player alongside the legacy player code. The split codebases mean that AS3 can't intermingle with older versions of ActionScript in a single FLA, but it also means that current Flash Players still support projects created in any version of the language. Projects that use AS1 or AS2 will play back on virtually every Flash-enabled computer, while files that rely on ActionScript 3.0 require a contemporary player (version 9 or later) to function.

Although AS3 is a restart of sorts, it still shares some characteristics with AS1 and AS2, and even with other languages based on *ECMAScript*, the standard from which ActionScript evolved. ECMAScript actually began its life as

IN THIS CHAPTER

JavaScript so, if you have any experience coding web pages with JavaScript, you have a leg up when it comes to learning AS3.

In many ways, however, AS3 is entirely different. Here a few examples of changes and improvements that have reshaped AS3 into the fastest and most powerful version of ActionScript in any Flash release to date:

- **Consistent syntax** makes AS3 easier to pick up as you go along. Many rambling issues from prior versions of the language have been clarified and conventions followed more reliably. For example, in prior versions of ActionScript, properties (ways of describing objects, such as width and height) were sometimes preceded by underscores, and sometimes not.

  Once you gain a little bit of experience writing basic scripts, you'll find yourself hunting for specific, task-related solutions. At that point, you'll find the language consistent enough that you can actually start guessing at syntax and finding yourself right a lot of the time. This consistency also makes using help and other resources easier.

- **More detailed error reporting** makes finding problems in AS3 programs significantly easier. No longer will your code silently fail at runtime, leaving you to wonder where your troubles lay. Instead, AS3 not only improves runtime error reporting, it also introduces error and warning reports when your file is compiled to SWF for distribution. By learning about these errors at authoring time, you can usually catch and fix your bugs prior to introducing your application into the wild.

- **Stronger data typing** is the greatest aid to better error reporting. By declaring the type of your data as a number, for instance, you'll be notified if you accidentally use a text string instead. The consistent use of strong data typing makes AS3 more verbose than prior versions of the language but, again, this enables better error reporting and more reliably sound projects.

- **New display architecture** features consolidate and simplify the many ways that prior versions of ActionScript controlled visual assets. Using the new *display list*, it's now much simpler to control the visual stacking order of visual assets and to manipulate the familial relationships—parent, child, and sibling—between these assets.

- **New event architecture** features unify a method of processing events. All events are now handled with *event listeners*, which listen for specific events and react when those events occur. For example, a listener assigned to listen for a mouse event will ignore keyboard events. This makes event handling more efficient.

- **Improved handling of external assets** simplifies the task of loading data at runtime. Consistent approaches to loading assets can be applied to images, text, sound, video, and even other SWF files. It's much easier to work with XML in AS3, and you can even work with raw data, such as the bytes that make up a JPEG or sound.

- **Better sound management** provides greater control over sound playback in 32 discrete sound channels. You can also poll data from sounds during playback to visualize sounds with waveforms and similar graphics.

There are several more compelling attributes of AS3 that are likely of interest to intermediate to advanced coders; however, these fall outside the scope of this chapter. Some of these topics are discussed in the companion volume, *Learning ActionScript 3.0: A Beginner's Guide* (O'Reilly), described in the next section.

How Much ActionScript Is Covered in This Book?

AS3 is a robust language with significant breadth and depth. It also has a learning curve steeper than those associated with other versions of ActionScript. In this author's humble opinion, it's unrealistic to cover both the Flash authoring application and the ActionScript language simultaneously with any degree of effectiveness.

As such, this book is one of a pair that has been conceived to focus on these two areas of interest. Each volume is organized to make it easy to acquire and digest the material most appropriate for your needs. The book you're reading now focuses primarily on the Flash authoring application, while its companion volume, *Learning ActionScript 3.0*, focuses almost exclusively on ActionScript.

However, despite dividing this material into two separate volumes, ActionScript can't be ignored in the pages herein. Unless you intend to use Flash strictly for linear animations, you'll need a minimum amount of ActionScript to expand your Flash skills. You certainly need ActionScript to add interactivity, and you need at least one line of code just to stop a Timeline animation from looping forever.

Furthermore, this book is project-based, teaching topics both with isolated examples and the ongoing development of a single portfolio. The isolated examples require no long-term investment to grasp the ideas behind them, and the project allows you to apply your newly learned skills to an example real-world application.

Therefore, this chapter aims to provide an ActionScript primer, of sorts. It consists of material condensed from several chapters of *Learning ActionScript 3.0*, as well as content directly related to the portfolio project. This primer will cover the basic skills required to support each topic and project task through the remainder of this book. Additional ActionScript will be introduced periodically to cover chapter-specific content.

Try to remember that this material is here to get you started. Don't expect to learn the language from a primer, and don't worry if it seems like a lot to take in upon first reading. Feel free to use this chapter as a small reference, reading it in segments and coming back to it when you need additional help with language fundamentals.

You can then decide whether you'd like to explore ActionScript further, and whether to acquire this book's companion volume. *Learning ActionScript 3.0* covers a great deal more coding concepts, has considerably more detail, and is supported by its own companion website.

Introducing ActionScript Interface Elements

Although much of the Flash interface was covered in Chapter 1, ActionScript-related interface elements have been reserved for discussion in this chapter. At the entry level, we'll focus on two primary panels for ActionScript development: the Actions panel and the Output panel.

Actions Panel

The Actions panel is where you'll be writing your scripts (Figure 6-1). While it's also possible to write scripts in external files (discussed at length in the companion volume *Learning ActionScript 3.0*), this book concentrates on originating your scripts in the Flash Timeline, so the Actions panel will be your home.

Here's an overview of tools found within the Actions panel. In some cases, their functionality may not be self-explanatory, but relevant topics will be explained later in the chapter:

Panel panes

The main pane on the right is the Script pane where you write your scripts. The pane in the lower-left corner is the Script Navigator, used to select any script written in your FLA. The pane in the upper-left corner is the Actions Toolbox. From here, you can drag or double-click to add ActionScript to your script in progress. The left two panes can be minimized to provide more room for the Script pane.

Add a new item to the script

This menu provides access to the content in the Actions Toolbox for use when that pane is minimized.

Find

This button opens a find and replace dialog for editing your script.

Insert a target path

This button opens a graphical browser for selecting objects in your FLA such as movie clips. It will insert a path to that object in your script.

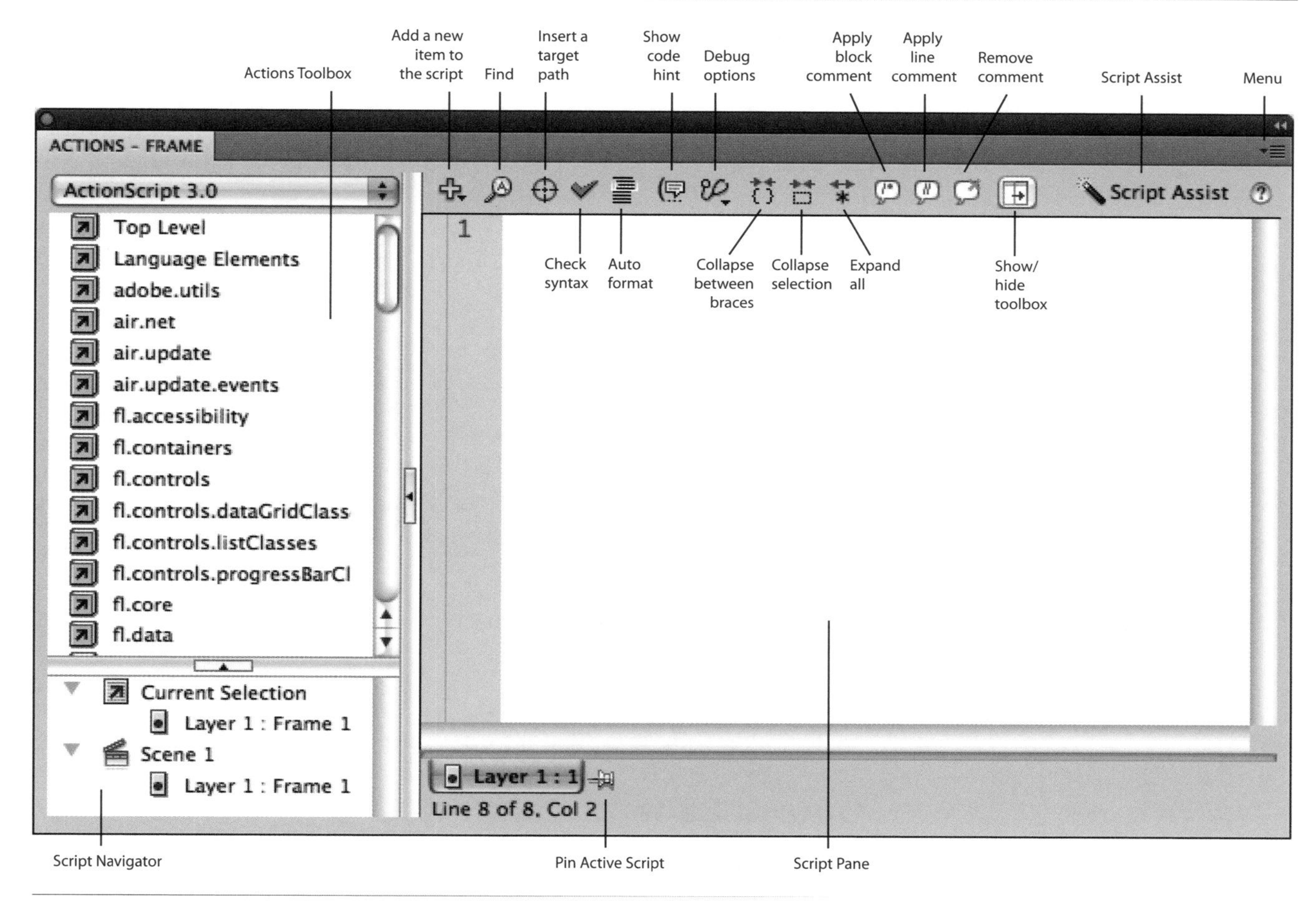

Figure 6-1. The Actions panel

Check syntax

This button checks your scripts for errors.

Auto format

This button formats error-free scripts, or alerts you to problems in your code.

Show code hint

When possible, this button shows you a floating syntax tip for the ActionScript surrounding the cursor while editing your script.

Debug options

Here you can insert, remove, or enable/disable break points for use in the ActionScript debugger. This is a tool for intermediate users and beyond and requires a bit of comfort and experience to use.

Code Collapse Controls

Using these three tools, you can temporarily hide code by collapsing multiple lines into one closed marker. You can collapse between braces ({}), collapse selected code, or expand all previously collapsed segments.

Comment Controls

These three tools help to add comments, or disable/enable code in blocks of multiple lines, or in a single line, as well as uncomment selected commented code.

Show/hide toolbox

This button expands and minimizes the Actions Toolbox.

Pin Active Script

Each time you change scripts, the previous script is replaced in the Script pane. This button will force script to stay in tabs at the bottom of the panel, allowing you to switch between them easily.

Script Assist

This button launches a rigid interface-driven script authoring system that is not discussed in this book.

Menu

This button opens the Actions panel menu, discussed later in this chapter.

Output Panel

The Output panel is a very simple, but very useful, text output panel that can be used to monitor your own understanding of a script (Figure 6-2). Its sole purpose is to display text generated by a script at authortime. You can't enter a script in the Output panel, and the panel doesn't exist at runtime.

You will use the Output panel only as a means of getting quick feedback from an example, or as a testing and debugging technique when writing scripts. You'll likely find the **`trace`** command very helpful, both in this book and in your own scripts, to send information to the panel.

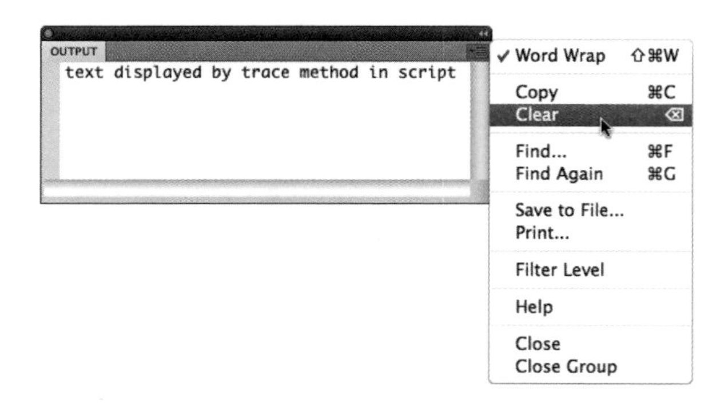

Figure 6-2. The Output panel

Basic Script Grammar

Because AS3 is a vast language, it can sometimes be a bit difficult to explain in a strictly linear fashion. For example, to understand how to effectively manipulate visual assets, you must learn about *properties*, *methods*, *events*, and *event listeners*. These are the basic building blocks of most scripted tasks that allow you to get and set characteristics of, issue instructions to, and react to input from many Flash objects.

To go into a reasonable depth, particularly with real-world samples, you typically must discuss two or more of these topics at the same time. For example, to create an interactive exercise that allows you to experiment with properties or methods, you need events. Similarly, to understand events, you usually need to set properties or call methods.

Therefore, this section will introduce you to some basic terms and an example use or two of these topics. Later, you'll follow up with more examples and additional discussions. For all three of these introductory passages, code samples will be provided for a movie clip that has been given an instance name of *mc*.

Introducing Properties

Properties are somewhat akin to adjectives in that they describe the object being modified or queried. For example, you can check or set the width of a movie clip. Most properties are read-write, meaning that you can both get and set their values. Some properties, however, are read-only, which means you can ask for, but not change, their values.

Here are examples of setting the *width*, *height*, and *rotation* of a movie clip. You can see these properties in use in the *properties.fla* source file.

```
mc.width = 100;
mc.height = 50;
mc.rotation = 90;
```

Introducing Methods

Methods are a bit like verbs. They are used to tell objects to do something, such as play and stop. In some cases, methods can simplify the setting of properties. You might use a method called **setSize()**, for example, to simultaneously set the width and height of something. Other methods are more unique, such as **navigateToURL()**, which will instruct a browser to display a web page.

Here are examples of telling a movie clip to play, stop, and go to the next frame. You can see these in use in the *methods.fla* source file.

```
mc.play();
mc.stop();
mc.nextFrame();
```

Introducing Events and Event Listeners

Events are the catalysts that trigger the actions you write, such as setting properties and calling methods. For instance, a user might click the mouse button, which results in a mouse event. That event then causes a function to execute, performing the desired actions. *Event handlers* are the ActionScript middlemen that trap the events and actually call the functions. ActionScript 3.0 unifies event handling into a consistent system of *event listeners*, which are set up to listen for the occurrence of a specific event and react accordingly.

Here is an example of an event listener, designed to listen for a mouse up event. You can see this in use in the *events.fla* source file:

```
mc.addEventListener(MouseEvent.MOUSE_UP, doIt);
function doIt(evt:MouseEvent):void {
    trace("do it");
}
```

You'll learn more about some important concepts in action here, such as how the block of code called a *function* works, what's inside the parentheses of the function name, and what **void** means, but here is the basic idea.

An event listener is listening for a specific mouse event—a mouse up event only, or when the mouse button is released after being pressed. The listener is attached to the movie clip, so when the movie clip is clicked, a mouse up event will be detected. When that happens, the function called **doIt()** is executed, and the letters "do it" are displayed in the Flash Output panel.

This flow of information—from the click of the mouse, to the *trapping* (receipt) of the event, to the execution of the function and the eventual output of the text message—is the cycle of event processing. Additional details will be discussed later in this chapter, but this is the mechanism by which interactivity is controlled.

Basic Syntax Issues

The infrastructure beneath a programming language is often overlooked, but understanding what may seem like smaller topics will make it easier for you to adapt to ActionScript and form good habits.

Dot Syntax

If you look over the code examples earlier in the "Basic Script Grammar" section, you'll see that a dot (.) separates the movie clip instance name from its properties and methods. This is sometimes referred to as *dot syntax* or *dot notation*. Essentially, this system strings together a series of items, from highest to lowest in the object hierarchy, including only items relevant to the task at hand. In this case, the first relevant item is the movie clip instance and the last is the property or method. Considering another example, in which you want to check the width of a movie clip that is inside another movie clip, the first item will be the parent, or container movie clip, then comes the nested movie clip, and then comes its width:

```
mc1.mc2.width;
```

This dot syntax will be used in virtually every example for the rest of this book, and it will soon become as familiar as your own language.

Case Sensitivity

Simply put, ActionScript is case-sensitive. For any word that is already part of the ActionScript lexicon, you must replicate that case exactly. For example, neither "True" nor "TRUE" will work when you're trying to represent the *Boolean* (true/false) value of **`true`**. Another example is keyboard input. When you want to verify a user's key input, "Claire" and "CLAIRE" are not the same.

For words you make up, such as *variable* names (a variable is a container for storing data, discussed later in the chapter), you must be consistent. If you name a variable myMovieClip, you can't refer to it later as MyMovieClip.

Although you can use any case you like for your own names, a few conventions exist to make code more readable—particularly among multiple programmers working on the same project, where standardized practices really help. A few examples of conventions that are widely adopted include:

Camel Case

This is a naming convention in which variable names are lowercase, except for the first letter of compound words. For example, "movie" is lowercase, but "myMovieClip" capitalizes the second and third words. Camel case is typically the default naming convention, and the next two examples are exceptions to the rule.

Classes

Classes are often used to create instances of objects, such as movie clips, for a net result similar to dragging a movie clip from the Library to the Stage. Classes are, essentially, collections of related code responsible for the creation and behavior of objects. You'll learn more about classes throughout this chapter and use them throughout this book. The first letter of a class name is capitalized to set it apart from other possible variable or instance names. For example, instead of using "movieClip" as the name of the class responsible for movie clip behavior, the actual class is called *MovieClip*.

Constants

Constants are variables that don't change their values. Constants are typically written in all uppercase. Constants that represent keys on the keyboard, for example, include *SPACE* and *TAB*. In ActionScript 3.0, constants are usually organized into classes and are referenced through the class name. The aforementioned keyboard constants, for instance, are part of the *Keyboard* class and are referred to by the usage *Keyboard.SPACE* and *Keyboard.TAB*.

Execution Order

In general, ActionScript executes in a top-to-bottom, left-to-right order. That is, each line executes one after another, working from left to right. While this is typically a reliable rule of thumb, several things can change this order in subtle ways. For example, *subroutines* of one type or another can be called in the middle of a script. This causes the execution order of the original script to pause while the remote routine is executed. When the subroutine has completed, the execution of the original script continues where it left off. Execution flow will be explained in context in all scripts in this book.

Use of the Semicolon(;)

The official use of the semicolon in most ECMAScript languages is simply to allow execution of more than one instruction on a single line. This is rare in the average script, and we will look at this technique when discussing loops.

However, the semicolon is also used to indicate the end of a line. Typically, this is not required, but there are cases in AS3 that rely on the semicolon to indicate the end of a line. These are cases in which a single line would be too difficult to read, so the line is broken up by carriage returns, even though it's essentially a single execution.

A good example of this is writing XML in a human-readable form in a script. The ActionScript compiler is smart enough to understand that line breaks in XML are just for readability, and it looks for the semicolon to indicate the end of a line.

For this reason, and because forging this habit makes it easier to transition into learning other languages where the semicolon is required, place a semicolon at the end of every line.

Evaluating an Expression

It's helpful to note that you are usually not solving an equation when you see an expression with like values on the left and right of an equals sign. For example, if you see something like `x = x + 1`, it's unlikely that you will be solving for the value of `x`. Instead, this line is assigning a new value to `x` by adding `1` to its previous value.

Absolute Versus Relative Addresses

Much like a computer operating system's directory or the file structure of a website, ActionScript refers to the location of its objects in a hierarchical fashion. You can reference an object, such as a movie clip, using an absolute or relative path. Absolute paths can be easy because you most likely know the exact path to any object starting from the main Timeline. However, they are quite rigid and will break if you change the nested relationship of any of the referenced objects. Relative paths can be a bit harder to call to mind at any given moment, but they are more flexible. Working from a movie clip and going up one level to its parent and down one level to a sibling will work from anywhere, be it in the main Timeline, another movie clip, or nested even deeper, because the various stages aren't named.

Tables 6-1 and 6-2 draw parallels to the operating system and website analogies. Table 6-1 references the **root**, which in this case is another way to refer to the main Timeline. The companion website has more information about how to use **root**.

Table 6-1. *Absolute paths from main Timeline to nested movie clip*

ActionScript	Windows OS	Mac OS	Website
`root.mc1.mc2`	*c:\folder1\folder2*	*Macintosh/ folder1/ folder2*	*http://www.domain.com/dir/dir*

Table 6-2. *Relative paths from a third movie clip, up to the root, and down to the child of a sibling*

ActionScript	Windows OS	Mac OS	Website
`this.parent.mc1.mc2`	*../folder1/folder2*	*..\folder1\folder2*	*../dir/dir*

Comments

Comments are lines of text within your scripts that are *not* executed and are invaluable programmer's tools. The obvious purpose of a comment is to add a brief descriptive note that will explain the purpose, expected outcome, or possibly a caveat of a segment of your script. Comments are really important when working with other programmers, but they're also very helpful when archiving your own code. It's not uncommon to revisit code long after writing it and have to figure out what it does for your project. A single-line comment begins with two slashes:

```
//set width to size of left column
mc.width = 100;
```

There's another, possibly lesser-known purpose of comments, however. It's quite convenient to use comments to temporarily disable code. For example, you may want to quickly disable ActionScript sound playback or change navigation options. You can also use comments to try two different approaches to a programming task without deleting and rewriting. By completing two alternate versions of a code segment, you can test either at any time by commenting one out and the other back in. Symmetrical slash-asterisk pairs surround multiline comments:

```
/*
mc.width = 100;
mc.height = 50;
*/
mc.width = 50;
mc.height = 100;
```

This is obviously a very simple example just to demonstrate the comment toggling process. In this case, it may be as easy to switch the values each time you test, but it's also easy to become distracted and lose track of the values you tested. As a proof of concept, this example switches a movie clip between horizontal and vertical sizes. To try this code the other way, you could uncomment the first block and comment the second block:

```
mc.width = 100;
mc.height = 50;
/*
mc.width = 50;
mc.height = 100;
*/
```

Checking and Formatting Your Scripts

As you write your scripts, it's helpful to check your progress. As Adobe's John Dowdell likes to say, "test early, test often." There are a few tools built into Flash's interface that can help you quickly examine the health of your scripts:

Checking syntax

A quick and easy way to see if your script is in good shape is to click the Check Syntax button at the top of the Actions panel (see Figure 6-1). If your script is OK, a dialog will tell you as much. If there are problems, the compiler will alert you via the Compiler Errors panel.

Errors and warnings

When the compiler detects errors or warnings, it will add them to the Compiler Errors panel (Figure 6-3) so you can try to find the problem and correct it. Error reports include the error, a brief description of the problem, and even a line number where the error is thought to have occurred. If you double-click the error, the Flash interface will switch to the Actions panel and scroll to the errant line number.

NOTE

If you ever want to learn more about the warnings or errors that may appear during development or runtime, here are a few resources to explore:

http://help.adobe.com/en_US/AS3LCR/Flash_10.0/compilerWarnings.html

http://help.adobe.com/en_US/AS3LCR/Flash_10.0/compilerErrors.html

http://help.adobe.com/en_US/AS3LCR/Flash_10.0/runtimeErrors.html

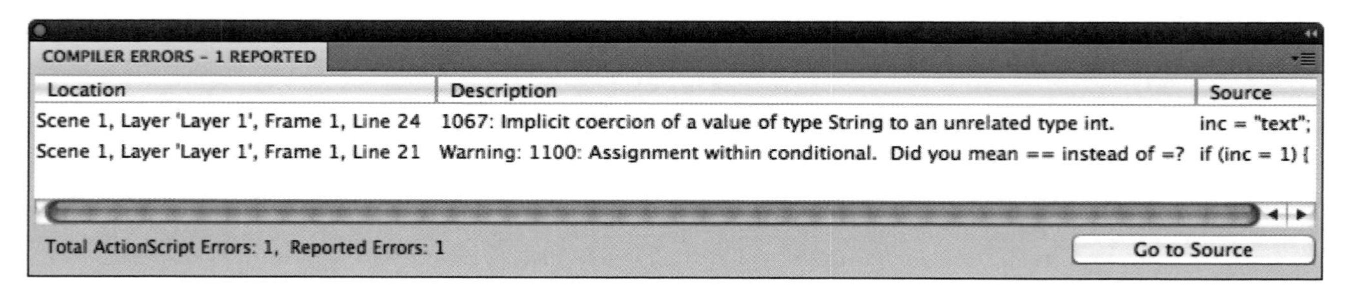

Figure 6-3. The Compiler Errors panel

Formatting scripts

As insignificant as it sounds, formatting a script can help you find problems. Formatting can help because Flash will check the integrity of your scripts before proceeding. If errors exist, formatting will be abandoned. In this way, you're giving your code a once-overy ever time you format it. In addition, formatting a script will indent it properly, place braces and spaces where specified, and more. These assists help you to spot lines of code that may be out of place.

To format a script, click the Auto Format button at the top of the Actions panel (see Figure 6-1). Flash's Preferences (Flash→Preferences on Mac, Edit→Preferences on Windows) allow you to customize (to some extent) how the script is formatted (Figure 6-4). Additional basic options, such as word wrapping, are available from the Actions panel's menu (Figure 6-5), accessible from the upper-right corner of the panel.

Reload Code Hints
Pin Script ⌘=
Close Script ⌘–
Close All Scripts ⇧⌘–
Go to Line... ⌘,
Find and Replace... ⌘F
Find Again ⌘G
Auto Format ⇧⌘F
Check Syntax ⌘T
Show Code Hint ^␣
Import Script... ⇧⌘I
Export Script... ⇧⌘P
Print...
Script Assist ⇧⌘E
Esc Shortcut Keys
Hidden Characters ⇧⌘8
✓ Line Numbers ⇧⌘L
✓ Word Wrap ⇧⌘W
Preferences... ⌘U
Help
Close
Close Group

Figure 6-5. The Actions panel menu options

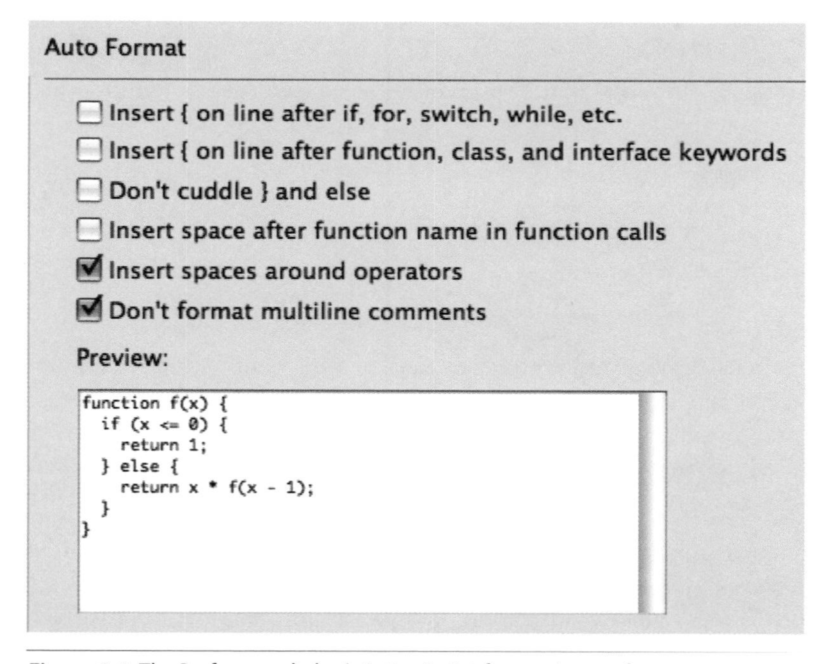

Figure 6-4. The Preference dialog's ActionScript formatting preferences

Code collapse

To hone in on a particular segment of a longer script, improving focus and clarity, you can use the code collapse feature (Figure 6-6). Highlight the lines of code you want to hide temporarily and click on the open arrow next to the first or last selected line number. To show the code again, click on the closed arrow.

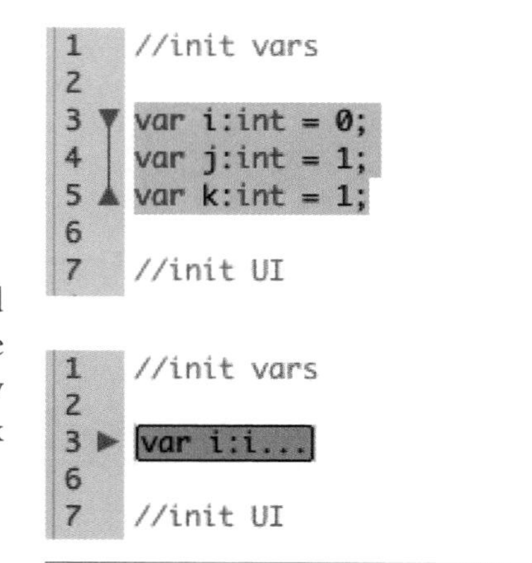

Figure 6-6. Before and after code collapse in the Actions panel

Variables and Data Types

Variables are best described as containers into which you place information for later recall. Imagine if you were unable to store any information for later use. You would not be able to compare values against previously described information (such as usernames or passwords), your scripts would suffer performance lags due to repeated unnecessary calculations, and you wouldn't be able to carry any prior experiences through to the next possible implementation of a task. In general, you wouldn't be able to do anything that required data that your application had to "remember."

NOTE

You'll learn on the next page that variables used for the first time must be declared with the **var** *keyword.*

Variables make all this and more possible, and are relatively straightforward. In basic terms, you need only create a variable with a unique name so it can be referenced separately from other variables and the ActionScript language itself, and then populate it with a value. Ignoring usage for a moment, a simple example is remembering the number 1 with the following:

```
myVariable = 1;
```

There are just a few rules and best practices to consider when naming variables. They must:

- Be one word
- Include only alphanumeric characters, dollar signs ($), or underscores (_)
- Not start with a number
- Not already be a keyword or reserved word in ActionScript

To help you catch bugs and unexpected uses of data, ActionScript can monitor what's placed into variables. You can tell the ActionScript compiler that you want a certain type of data to be stored in a variable, and when your file is compiled into a SWF, you will be warned if the variable contains a different kind of data.

This compile-time error checking can prevent problems from sneaking into your projects and is one of the best things about AS3. For example, if you try to perform a mathematical operation on a passage of text, Flash will issue a warning so you can correct the problem before you distribute your work to a client or the public.

To make this work, you must indicate what you intend to store in each variable—this is called *declaring* the variable. To declare a variable, precede its first use with the **var** keyword and use the syntax **<variable name>:<data type>** to specify the type of data to be stored. For instance, the previous example of remembering the number 1 should be written this way:

```
var myVariable:Number = 1;
```

There are several native data types including, but not limited to, those listed in Table 6-3.

Table 6-3. Variable types

Data type	Example	Description
`Number`	`4.5`	Any number, including floating point values (decimals)
`int`	`-5`	Any integer or whole number
`uint`	`1`	Unsigned integer or any nonnegative whole number
`String`	`"hello"`	Text or a string of characters
`Boolean`	`true`	True or false
`Array`	`[2, 9, 17]`	More than one value in a single variable
`Object`	`myObject`	The basic structure of every ActionScript entity, but also a custom form that can store multiple values as an alternative to **`Array`**

There are also dozens of additional data types that describe which kind of object is used. For example, the following line of code uses the **`MovieClip`** class (the built-in code that makes a movie clip behave the way it does) to create an empty movie clip at runtime:

```
var myMC:MovieClip = new MovieClip();
```

It's impractical to list every possible data type here, but this book will reference data types frequently, and using them will soon become second nature to you.

Casting Data Types

In some cases, you will need to change one data type to another. This is called *casting*. A good example of the need for casting data types is when using a text field to capture numeric input from the user. When a user types into a text field, the data captured is, logically, text. However, when the intended use of that information is mathematical, you must convert the input from text data to numeric data.

There are two ways to cast from one data type to another. The first is by using the **`as`** operator. Applying the **`as`** operator to existing data, followed by the desired data type, will make the conversion.

```
var num:Number = userAnswer.text as Number;
```

The second method is to use the desired data type class as a method and place the original data inside the parentheses. The following code, for example, converts a number to an integer:

```
var num2:int = int(num);
```

Operators

An important basic idea in any programming language is that of *operators*. Operators are symbols that represent action taken upon objects, such as setting, comparing, or testing values. Some operators will seem like common sense, such as arithmetic operators: addition (+), subtraction (–), multiplication (*), division (/), and more. Others, however, may not be as obvious.

For example, did you know that, in addition to adding two numbers, the plus operator (+) can join two strings of text? A single equals sign (x = 1) is an assignment operator, because it is used to assign a value to something, but did you know there is also a double equals sign (==) operator? The latter is used to *compare* two values to determine whether they are equal.

There are also a few categories of operators that you may never have seen before:

Shortcut arithmetic operators

Shortcut arithmetic operators combine two tasks into one operator. The following are standard operators:

`x++` means `x = x + 1` (add 1 to current value)

`x--` means `x = x - 1` (subtract 1 from current value)

`x += n` means `x = x + n` (add value on right of equals sign)

`x -= n` means `x = x - n` (subtract value on right of equals sign)

`x *= n` means `x = x * n` (multiply by value on right of equals sign)

`x /= n` means `x = x / n` (divide by value on right of equals sign)

Comparison operators

Comparison operators are typically used in conditional (`if`) statements for comparing values. The following examples include equal to, not equal to, greater than, greater than or equal to, less than, and less than or equal to.

```
if (x == 1) { }
if (x != 1) { }
if (x > 1) { }
if (x >= 1) { }
if (x < 1) { }
if (x <= 1) { }
```

Logical operators

Logical operators are also used in conditional statements. They group tests together using *and* (&&), or *or* (||) to create a new test. The combined test relies on either one or the other original tests passing (*or*) or both tests passing (*and*). Another logical operator tests for falsehood using *not* (!). You'll learn about conditionals later, but it's good to be able to recognize these operators:

```
if (x == 1 && y == 2) { }
if (x == 1 || y == 2) { }
if (!myClip.visible) { }
```

Scope and this

Scope is the realm or space within which an object lives. As you learn more about ActionScript, particularly when you start using writing classes, you'll find scope a more central issue. Scope is still notable when you're just starting out, however, when it comes to use of **this**.

The keyword **this** is essentially shorthand for, "whichever object or scope you're working with now." For example, think of a movie clip inside Flash's main timeline. Both of the movie clips have a unique scope, so a variable or function defined inside the movie clip will not exist in the main timeline, and vice versa.

It's easiest to understand the usage of **this** in context, but here are a couple of examples to get you started. If you want to refer to the width of a movie clip, from within that same movie clip, you write:

```
this.width;
```

The **this** identifier refers to the movie clip in which the script was written. Similarly, if you want to check the width of the movie clip's parent, **this** is still the basis of the code:

```
this.parent.width;
```

In both cases, **this** is a reference point from which you start your path. It's fairly common to drop the keyword when going down the chain of objects from the current scope (as in the first example), but it's usually used, or even required, when going up to a higher scope (as in the second example). This is because Flash must understand which ancestor is needed when traversing through the hierarchy. Imagine a family reunion in which several extended family members, including cousins and multiple generations, are present, and you are looking for your mother, father, or grandparent. If you just said "parent," any number of parents might answer. If you instead said "my parent" or "my mother's parent," that would be specific enough to get you headed in the right direction.

Functions

Functions are an indispensable part of programming in that they wrap code into blocks that can be executed only when needed. They also allow you to reuse and edit code blocks efficiently, without having to copy, paste, and edit repeatedly. Without functions, all code would be executed in a linear progression from start to finish, and edits would require changes to every single occurrence of any repeated code.

Creating a basic function requires little more than surrounding the code you wish to trigger with a wrapper that allows you to give the block a name. Triggering that function later requires only that you call the function by name. The following syntax shows a function that traces a string to the Output panel. The function is defined and then, to illustrate the process, immediately called.

In a real-world scenario, the function is usually called at some later time or from some other place, such as when the user clicks a button with the mouse. The output is depicted in the comment that follows the function call:

```
function showMsg() {
    trace("hello");
}
showMsg();
//hello
```

If efficiently reusing code and executing code only when needed were the only advantages of functions, you'd already have a useful improvement over linear execution of a script. This allows you to group your code into subroutines that you can trigger at any time and in any order. However, you can do much more with functions to gain even greater power.

Arguments

Assume you need to vary the purpose of the previous function slightly. Let's say you need to trace 10 different messages. To do that using only what you've learned so far, you'd have to create 10 functions and vary the string that is sent to the Output panel in each function.

However, you can accomplish this more easily using *arguments*. Arguments are like variables that have life only within their own functions. By adding an argument to the parenthesis next to the function name, you can pass a value into that argument when you call the function. In the following case, the argument is called **`msg`** and is expected to contain a data type of **`String`**:

```
function showMsg(msg:String) {
    trace(msg);
}
showMsg("goodbye");
//goodbye
```

By using **`msg`** in the body of the function, it takes on the value that is sent into the argument. In this example, the function no longer traces "hello" every time it's called. Instead, it traces the text sent into its argument when the function is called. When using arguments, it's ideal to type the data coming in so Flash knows how to react and can issue any warnings or errors.

Return Values

It's also possible to return a value from a function, increasing its usefulness. Returning a value to the script from which it was called means a function can vary its input *and* its output. The following examples convert temperature values from Celsius to Fahrenheit and Fahrenheit to Celsius. In both cases, a value is sent into the function and a calculated value is returned to the script. The first example immediately traces the result sent back from the function, while the second example stores the value returned from the function in a variable. This mimics real-life usage in that you can immediately act upon the returned value *or* store and process the data at a later time.

```
function celToFar(cel:Number):Number {
    return (9 / 5) * cel + 32;
}
trace(celToFar(20));
//68

function farToCel(far:Number):Number {
    return (5 / 9) * (far - 32);
}
var celDeg:Number = farToCel(68);
trace(celDeg);
//20
```

When returning a value from a function, you should also declare the data type of the return value. As with applying other data types, use a colon followed by the type specific to that function, but place the data type between the argument's close parenthesis and the opening function brace. Once you get used to this practice, it's best to specify **void** as a return type when your function does not return a value. This will cause an alert if you attempt to return a value after originally planning not to do so.

Conditionals

You will often need to make a decision in your script, choosing to do one thing under one circumstance and another thing under a different circumstance. These situations are usually addressed by *conditionals*. Put simply, a conditional is a test that determines whether a particular condition is met. If the condition is met, the test evaluates to **true**, and specific code is executed accordingly. If the condition is not met, either no further action is taken or an alternate set of code is executed.

if

The most common form conditional is the **if** statement. The structure of the statement's basic structure is the **if** keyword, followed by the conditional test in parentheses and the code to be executed (if the statement evaluates to **true**) in braces. The first three lines in the following example establish a set of facts. The **if** statement evaluates the given facts (this initial set of facts will be used for this and subsequent examples in this section).

```
var a:Number = 1;
var b:String = "hello";
var c:Boolean = false;

if (a == 1) {
    trace("option a");
}
```

To evaluate the truth of the test inside the parentheses, conditionals often make use of *comparison* and *logical operators*. A comparison operator compares two values, such as equals (==), less than (<), and greater than or equal to (>=), to name a few.

NOTE

The test in the preceding example uses a double equals sign. This is a comparison operator that asks, "Is this equal to?" This distinction is very important because the accidental use of a single equals sign will cause unexpected results. A single equals sign is an assignment operator that assigns the value on the right side of the equation to the object on the left side of the equation. Because this assignment occurs, the test will always evaluate to **true**.

A logical operator evaluates the logic of an expression. Included in this category are *and* (**&&**), *or* (**||**), and *not* (**!**). These allow you to ask if "this *and* that" are true, or if "this *or* that" is true, or if "this" is *not* true.

For example, the following code would return **false** because *both* conditions must be true. The value of a is 1, but the value of b is "hello." Because the second test fails, the combined test fails. As a result, nothing would appear in the Output panel.

```
if (a == 1 && b == "goodbye") {
    trace("options a and b");
}
```

In the next example, the test would evaluate to **true**, because *one* of the two conditions (the first) is true. As a result, "option a or b" would appear in the Output panel.

```
if (a == 1 || b == "goodbye") {
    trace("option a or b");
}
```

Finally, the following would also evaluate to **true** because the *not* operator correctly determines that **c** is not true (remember that every **if** statement, at its core, is testing for truth).

```
if (!c) {
    trace("not option c");
}
```

You can also use the *not* operator in a comparison. When combined with a single equals sign, the pair means "not equal to." Therefore, the following will fail because **a** does equal 1, and nothing will appear in the Output panel:

```
if (a != 1) {
    trace("a does not equal 1");
}
```

Additional power can be added to the **if** statement by adding an unconditional alternative (true no matter what). In this case, an alternative set of code is executed no matter what the value being tested is, simply because the test did not pass. With the following new code added to the previous example, the second (else) trace will occur:

```
if (a != 1) {
    trace("a does not equal 1");
} else {
    trace("a does equal 1");
}
```

Finally, you can make the statement even more robust by adding a conditional alternative (or an additional test) to the structure. In this example, the second trace will occur:

```
if (a == 2) {
    trace("a does not equal 1");
} else if (a == 1) {
    trace("a does equal 1");
}
```

The **if** statement requires one **if**. You can use only one optional **else**, but you can use any number of optional additional **else if** tests. In all cases, however, only one result can come from the structure. Consider the following example, in which all three results could potentially execute—the first two because they are true, and the last because it's an unconditional alternative:

```
if (a == 1) {
    trace("option a");
} else if (b == "hello") {
    trace("option b");
} else {
    trace("option other");
}
```

In this case, only "option a" would appear in the Output panel, because the first truth would exit the **if** structure. If you needed more than one execution to occur, you would need to use two or more conditionals. The following structure, for example, executes the first trace in each **if**, by design:

```
if (a == 1) {
    trace("option a");
}
if (b == "hello") {
    trace("option b");
} else {
    trace("option other");
}
```

switch

An **if** statement can be as simple or as complex as needed. However, long **if** structures can be difficult to read and are sometimes better expressed using the **switch** statement. The **switch** statement also has a unique feature that lets you control which, if *any*, instructions are executed, even when a test evaluates to **false**.

Imagine an **if** statement asking if a variable is 1, **else if** it's 2, **else if** it's 3, and so on. A test like that can become difficult to read quickly. An alternate structure appears as follows:

```
switch (a) {
    case 1 :
        trace("one");
        break;
    case 2 :
        trace("two");
        break;
    case 3 :
        trace("three");
        break;
    default :
        trace("other");
        break;
}
```

In this case, "one" would appear in the Output panel. The **switch** line contains the object or expression you want to test. Each case line offers a possible value. Following the colon are the instructions to execute upon a successful test, and each **break** line prevents any following instructions from executing. When not used, the next instructions in line will execute, even if that test is false.

For example, if **a** equals 1, the following will place both "one" and "two" in the Output panel, even though **a** does not equal 2:

```
switch (a) {
    case 1 :
        trace("one");
    case 2 :
        trace("two");
        break;
}
```

This **break** feature does not exist with the **if** statement and, if used with care, makes **switch** an efficient alternative to a more complex series of multiple **if** statements. **Switch** statements must have one **switch** and one **case**, an optional unconditional alternative in the form of **default**, and an optional **break** for each **case** and **default**. The final **break** is not needed, but you may prefer to include it for consistency.

Loops

It's quite common to execute many repetitive instructions in your scripts. However, including them line by line, one copy after another, is inefficient and difficult to edit and maintain. Wrapping repetitive tasks in an efficient structure is the role of *loops*. A programming loop is probably just what you think it is: you use it to go through the structure and then loop back to the start and do it again. There are a few kinds of loops, and the type you choose to use can help determine how many times your instructions are executed.

for Loop

The first type of loop structure is the **for** loop. This loop executes its contents a finite number of times. For example, you may wish to create a grid of 25 movie clips or check to see which of 5 radio buttons a user has selected.

For our purposes, suppose you want to trace content to the Output panel three times. To loop through a process effectively, you must first start with an initial value, such as 0, so you know you have not yet traced anything to the Output panel. The next step is to test to see if you have exceeded your limit. The first time through, 0 does not exceed the limit of three times. The next step is to trace the content once, and the final step is to increment the initial value, registering that you've traced the desired content once. The process then starts over until, ultimately, you will exceed the limit of the loop. The syntax for a basic **for** loop is as follows:

```
for (var i:int = 0; i < 3; i++) {
    trace("hello");
}
```

> **NOTE**
>
> *Note the use of the semicolon to execute more than one step in a single line.*

You may notice the declaration and typing of the counter, **i**. This is a common technique because the **i** variable is often used only for counting and, therefore, is created on the spot and not used again. If you have already declared and typed the counter, you can omit this step. Next is the loop test. In this case, the counter variable must have a value that is less than 3. Finally, the double plus sign (++) is equivalent to **i = i + 1**, or "add 1 to the current value of **i**."

The result is three occurrences of the word "hello" in the Output panel. It's also possible to count down by reversing the values in steps 1 and 2, and then decrementing the counter:

```
for (var i:int = 3; i > 0; i--) {
    trace("hello");
}
```

In other words, as long as the value of **i** is greater than 0, execute the loop and subtract one from the counter each time. This is less common, and works in this case because the loop only traces a string. However, if you need to use the actual value of **i** inside the loop, that need may dictate whether you count up or down. For example, if you created 10 movie clips and called them **mc0**, **mc1**, **mc2**, and so on, it may be clearer to count up.

while Loop

The other loop you'll likely use is the **while** loop. Instead of executing its contents a finite number of times, it executes as long as something remains true.

For example, look at a very simple case of choosing a random number. ActionScript generates a random number using a method of the *Math* class called **random()**. This method chooses a random number between 0 and 1. So, say you want to choose a random number greater than or equal to 0.5. For the sake of discussion, you have a 50% chance of choosing a desired number each time, so you may end up with the wrong choice several times in a row. To be sure you get a qualifying number, you can add this to your script:

```
var num:Number = 0;
while (num < 0.5) {
    num = Math.random();
}
```

Starting with a default value of 0, **num** will be less than 0.5 the first time into the loop. A random number is then put into the **num** variable, and if it's less than 0.5, the loop will execute again. This will go on until a random number that is greater than 0.5 is chosen, thus exiting the loop.

> **WARNING**
>
> *Use **while** loops with caution until you are comfortable with them. It's very easy to accidentally write an infinite loop with no exit, which will crash your machine. Do not try this code yourself, but here is a significantly simplified example of an infinite loop:*
>
> ```
> var flag:Boolean = true;
> while (flag) {
>     trace ("I am an infinite
>     loop");
> }
> ```
>
> *In this case, the flag variable remains true and, thus, the loop can never fail.*

A Loop Caveat

It's very important to understand that although they are compact and convenient, loop structures are not always the best method to achieve an outcome. This is because loops are very processor-intensive. Once a loop begins its process, nothing else will execute until the loop has been exited. For this reason, it may be wise to avoid `for` and `while` loops when interim visual updates are required.

In other words, when a loop serves as an initialization for a process that is updated only once upon its completion, such as creating the aforementioned grid of 25 movie clips, you are less likely to have a problem. The script enters the loop, 25 clips are created, the loop is completed, a frame update can then occur, and you see all 25 clips.

However, if you want each of the 25 clips to appear one by one, those interim visual updates of the playhead cannot occur while the processor is consumed by the loop. In this situation, it's more desirable to create a loop using methods that do not interfere with the normal playhead updates. A *frame loop* is just such a method. A frame loop is simply a repeating frame event, executing an instruction each time the playhead is updated. The events occur concurrently with any other events in the ordinary functioning of the file, so visual updates, for example, can continue while the frame loop is executing.

Frame loops will be explained in greater detail later in this chapter. For now, the important thing is to remember that frame loops offer a possible alternative to `for` and `while` loops.

Arrays

Basic variables can contain only one value. If you set a variable to 1 and then later set that same variable to 2, the value will be reassigned to 2.

However, there are times when you need one variable to contain more than one value. Think of a set of groceries, including 50 items. A variable approach to this task would be to define 50 variables and populate each with a grocery item. That's the equivalent of 50 pieces of paper, each containing one grocery item. That's not a shopping list you are likely to use. It's unwieldy, can only be created at authortime, and you'd have to recall and manage all variable names every time you wanted to access the grocery items.

An array equivalent, however, resembles the way we handle this situation in real life. You can write a list of 50 groceries on one piece of paper. You can add to the list while at the store, cross each item off as it's acquired, and you only have to manage one piece of paper.

Creating an array is quite easy. You can prepopulate an array by setting a variable (typed as an **Array**) to a comma-separated list of items, surrounded by brackets. You can also create an empty array by using the *Array* class. Both techniques are illustrated here:

```
var myArray:Array = [1, 2, 3];
var yourArray:Array = new Array();
```

In both cases, you can add to or remove from the array at runtime. For example, you can add a value to an array using the **push()** method, which pushes the value into the array at the end. You can remove an item from the end of an array using the **pop()** method.

```
var myArray:Array = new Array();
myArray.push(1);
trace(myArray);
// 1 appears in the Output panel
myArray.push(2);
// the array now has two items: 1, 2
trace(myArray.pop());
// the pop() method removes the last item, displaying its value of 2
trace(myArray);
// the lone remaining item in the array, 1, is displayed
```

There are a dozen or so other array methods, allowing you to add to or remove from the front of an array, sort its contents, check for the position of a found item within the array, compare each value against a control value, and more.

You can also add to or retrieve values from locations within the array by using brackets and including the index, or position, of the array you need. Keep in mind that ActionScript uses zero-based arrays, which means that the first value is at position 0, the second is at position 1, the next at position 2, and so on. So, to retrieve the existing fifth value from an array, you must request the item at position 4:

```
var myArray:Array = ["a", "b", "c", "d", "e"];
trace(myArray[4]);
//"e" appears in the Output panel
```

Multidimensional Arrays

Arrays can even contain other arrays. The result is called a *multidimensional array*. Arrays within arrays can resemble database structures or tables. Here are two examples of creating multidimensional arrays:

```
var multiArray:Array = new Array();
multiArray.push([1,2,3]);
multiArray.push([4,5,6]);

var multiArray2:Array = new Array();
var array1:Array = [1,2,3];
var array2:Array = [4,5,6];
multiArray2.push(array1);
multiArray2.push(array2);
```

The example in the first three lines pushes two arrays that are created on the fly into a multidimensional array. The second example pushes two existing arrays, which have already been stored in their own variables, into a multi-dimensional array. In both cases, the resulting array looks like this:

```
// [[1,2,3],[4,5,6]]
```

As stated earlier, this is nothing more than an array of arrays. As such, accessing values from the multidimensional array is similar to accessing values from one-dimensional arrays. First, you follow the array name with brackets and an index that identifies which array you want to query further. Because that, too, is an array, you follow it with another set of brackets and index. The following example contains the syntax required to pull 4 out of the first position of the second array in either multidimensional array in the example (the first example, *multiArray*, is used here):

```
trace(multiArray[1][0]);
// 4
```

Associative Arrays and Objects

Associative arrays store a *pair* of items—the value and an associated property name, or *key*, to describe that value. For example, a student might be represented this way:

```
var student1:Array = new Array();
student1["name"] = "Jodi";
student1["email"] = "jodi@maildomain.com";
student1["phone"] = "212-555-1212";
```

You can access the value the same way you set the value:

```
trace(student1["name"]);
// Jodi
```

Although the values can be any data type, each property in an associative array must be a string.

Another way to accomplish this task is to use an *object*. Creating a custom object allows you to get and set your property values using the familiar dot syntax employed throughout ActionScript. There are two ways to define objects. The first is to write them out explicitly, as shown in the following example. The syntax looks a bit like an array, with two differences. First, the entity is wrapped with braces, not brackets. Second, instead of single values, commas separate property/value pairs. To access or populate the values, simply reference the property using dot syntax:

```
var student2:Object = {name:"Sally", email:"sally@maildomain.com",
    phone:"212-555-1212"};
trace(student2.name);
// Sally
```

You can create an array of objects this way. You will see this approach in Chapter 9 when you work with components. To access an object's property, use the same syntax, but first determine which item in the array you want to manipulate. Pull that object out of the array using brackets, then access its property using dot syntax as previously described:

```
var studentGroup:Aray = new Array();
studentGroup.push({name:"Jodi", email:"jodi@maildomain.com",
    phone:"212-555-1212"});
studentGroup.push({name:"Sally", email:"sally@maildomain.com",
    phone:"212-555-1212"});
studentGroup.push({name:"Claire", email:"claire@maildomain.com",
    phone:"212-555-1212"});
trace(studentGroup[0].name);
// Jodi
```

Finally, the clearest way to create and populate an object is to create a new instance of the object and add its properties as needed:

```
var student3:Object = new Object();
student3.name = "Claire";
student3.email = "claire@maildomain.com";
student3.phone = "212-555-1212";
trace(student3.name);
// Claire
```

Properties

If you think of properties as ways of describing an object, using them becomes second nature. Asking where a movie clip is, for example, or setting the size of a movie clip are both descriptive steps and both use properties.

When referencing a property, you must begin with an instance, because you must decide which element to query or change. If you consider a test file with only one movie clip, instantiated as "box," all that remains is referencing the property and either getting or setting its value.

Table 6-4 contains syntax for making five changes to movie clip properties. Later, when you learn more about events, you'll change these properties interactively.

***Table 6-4.** Movie clip properties*

Description	Property	Syntax for setting value	Units and/or range
Location	*x*, *y*	`box.x = 100;` `box.y = 100;`	pixels
Scale (1)	*scaleX*, *scaleY*	`box.scaleX = 0.5` `box.scaleY = 0.5`	percent / 0–1
Scale (2)	*width*, *height*	`box.width = 72;` `box.height = 72;`	pixels
Rotation	*rotation*	`box.rotation = 45;`	degrees / 0–360
Transparency	*alpha*	`box.alpha = 0.5`	percent / 0–1
Visibility	*visible*	`box.visible = false;`	Boolean

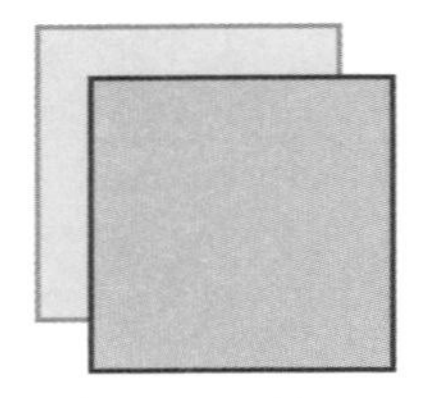
box.x += 10;
box.y += 10;

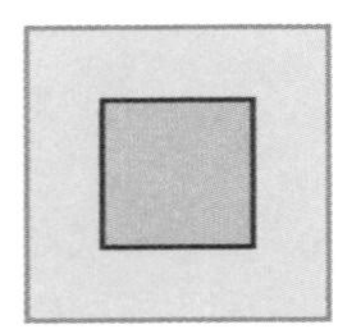
box.scaleX = 0.5;
box.scaleY = 0.5;

box.rotation = 20;

box.alpha = 0.5;

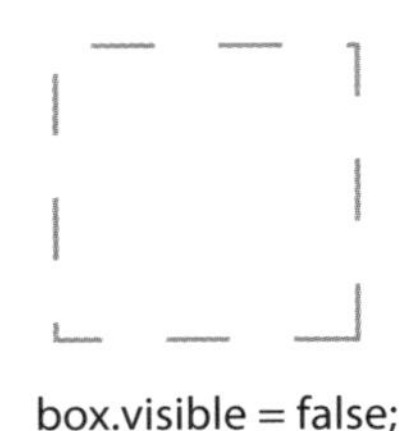
box.visible = false;

Figure 6-7. Changes to five movie clip properties

Figure 6-7 shows the visual change made by each property included in Table 6-4. The light-colored square is the original state and the darker color represents the square after a property change (the **alpha** property shows only the final state). The dashed stroke for the visible property indicates that the box is not visible.

A few changes have simplified and unified the way properties are referenced in AS3. First, the properties do not begin with an underscore. Rather than varying property syntax, some with and some without leading underscores, no properties begin with the underscore character.

Second, some value ranges that used to be 0–100 are now 0–1. Examples include **scaleX**, **scaleY**, and **alpha**. Instead of using 50 to set a 50% value, specify 0.5.

Finally, the first scaling method uses the properties **scaleX** and **scaleY** rather than **_xscale** and **_yscale**, which are their AS1/AS2 equivalents. Typically, AS3 properties will cite the x and y versions of a property as a suffix to make the code easier to read.

Table 6-4 shows syntax for *setting* a property. Querying the value of a property, also known as *getting* the property, is just as easy. For example, to trace the box's alpha value or store it in a variable, you can write either of the following:

```
trace(box.alpha);
var bAlpha:Number = box.alpha;
```

You can also set the properties of a class instance. For example, the following code creates an instance of the *Point* class, and then sets the x and y values of that point:

```
var myPoint:Point = new Point();
myPoint.x = 20;
myPoint.y = 20;
```

This code is usually more readable (or at least virtually self-commenting) than trying to make all possible property assignments when the instance is created. For example, in Chapter 7, you'll apply filters, such as a drop shadow filter, to movie clips with ActionScript. You can customize a filter in one line using this syntax:

```
var ds:DropShadowFilter = new DropShadowFilter(5, 45, 0x000000, 0.5, 5,
    5, 0.5, 1, false, false, false);
```

However, you may not remember all of the properties that are being set in that syntax, or you may not remember the order in which they are set. If readability and clarity are more important than brevity, you can accomplish the same task using this:

```
var ds:DropShadowFilter = new DropShadowFilter();
ds.distance = 5;
ds.angle = 45;
ds.color = 0x000000;
ds.alpha = 0.5;
ds.blurX = 5;
ds.blurY = 5;
ds.strength = 1;
ds.quality = 1;
ds.inner = false;
ds.knockout = false;
ds.hideObject = false;
```

Methods

Methods, the verbs of the ActionScript language, instruct their respective objects to act. Like properties, methods appear consistently in the dot syntax that is the foundation of ActionScript, following the object calling the method. For example, if the movie clip "box" in the main Timeline issues the **stop()** method, the syntax would appear like this:

```
box.stop();
```

Methods also sometimes require values to be passed to the object being manipulated. For example, although the **stop()** method will stop a movie clip from playing the visible frame, another method will stop playback after first going to a specific frame. The following example tells the movie clip "box" to go to frame 3 and stop:

```
box.gotoAndStop(3);
```

Events

Events make the Flash world go 'round. They are responsible for setting your scripts in motion, causing them to execute. A button can be triggered by a mouse event, and text fields react to keyboard events. Even calling your own custom functions is a means of issuing a custom event.

Events come in many varieties. In addition to the obvious events like mouse and keyboard input, most ActionScript classes have their own events. For example, events are fired when watching a video, working with text, and resizing the Stage. To take advantage of these events to drive your application, you need to be able to detect their occurrences.

In previous versions of ActionScript, there were a variety of ways to trap events. In AS3, trapping events is simplified by relying on one approach for all event handling, the use of *event listeners*.

Event Listeners

The concept of event listeners is pretty simple. Imagine that you are in a lecture hall that holds 100 people. Only one person in the audience has been given instructions about how to respond when the lecturer asks a specific question. In this case, one person has been told to listen for a specific event and to act on the provided instructions when this event occurs.

Now imagine that many more responses are required. For example, when the lecturer takes the stage, someone must dim the lights. When the lecturer clicks a hand-held beeping device, an audio/visual technician must advance to the next slide in the presentation. When each video ends, the lecturer must react by introducing the next exhibit in the lecture. Finally, when an audience member raises a hand, an usher must bring a microphone to assist the audience member in asking his or her question.

These are all reactions to specific events that are occurring throughout the lecture. Some are planned and directed to a specific recipient, such as the beeping that triggers the technician to advance to the next video in the series. Others are unplanned, such as when, or even if, an audience member has a question. Yet each appropriate party in the mix has been told which event to listen for and how to react when that event occurs.

Creating an event listener, in its most basic form, is also fairly straightforward. The first step is to identify the host of the listener. That is, which object should be told to listen for a specific event. One easy-to-understand example is instructing a button to listen for mouse events that might trigger its scripted behavior.

Once you have identified an element that should listen for an event, the next step is choosing an event appropriate for that element. For example, it makes sense for a button to listen for a mouse event, but it makes less sense for the same button to listen for the end of a video or the resizing of the Stage. It would be more appropriate for the video player to listen for the end of the video and for the Stage to listen for any resize event. Each respective element could then act or instruct others to act when that event occurs, which leads to the third main step in setting up a listener.

To identify the instructions that must be executed when an event occurs, you simply write a function, then tell the event listener to call that function when the event is heard. That function uses an argument to receive information about the event that called it, allowing the function to use key bits of data during its execution.

To tie it all together, the **`addEventListener()`** method puts the listener into service and assigns the function to be executed when the event is heard.

Suppose you want a button called *rotate_right_btn* to listen for a mouse up event and call the function **onRotateRight()** when the event is heard. The code to accomplish this looks like the following script:

```
rotate_right_btn.addEventListener(MouseEvent.MOUSE_UP, onRotateRight);
function onRotateRight(evt:MouseEvent):void {
    box.rotation += 20;
}
```

In the first line, **addEventListener()** is attached to the button. The method requires two mandatory parameters. The first is the event for which you want to listen. In AS3, similar events are grouped together into classes to make it easier for you to check against their data types. Checking to make sure the incoming event is of type **MouseEvent** prevents a **KeyboardEvent** from triggering the listener function.

The *MouseEvent* class contains constants that refer to mouse events like mouse up and mouse down. This example uses the **MOUSE_UP** constant to reference the mouse up event.

The second parameter is the function that should be called when the event is received. In this example, it is a reference to the **onRotateRight()** function, which begins on the second line. The function used in an event listener is just like any other function, with one exception: the argument in a listener function is *not optional*.

In the following code, for example, the argument is named **evt** and receives information about the element that triggered the event. You can use information from this argument in the function, which you'll do in a moment. The argument should be typed to the expected data to improve error checking. In this case, because the function listens for a *MouseEvent*, that is the data type is used for the argument.

To illustrate this, look at the impact of another mouse event, with more than one listener is in play.

```
1   myMovieClip.addEventListener(MouseEvent.MOUSE_DOWN, onStartDrag);
2   myMovieClip.addEventListener(MouseEvent.MOUSE_UP, onStopDrag);
3   function onStartDrag(evt:MouseEvent):void {
4       evt.target.startDrag();
5   }
6   function onStopDrag(evt:MouseEvent):void {
7       evt.target.stopDrag();
8   }
```

In this code, two event listeners are assigned to a movie clip. One listens for a **mouse down** event and another listens for **mouse up**. They each invoke different functions. In both functions, however, the **target** property of the event, which is retrieved from the function argument, identifies which element received the mouse event. This allows the instruction in line 3 to start dragging the selected movie clip, and allows the instruction in line 6 to stop dragging the selected movie clip without specifying the movie clip using its instance name in either case.

This generic approach is very useful because it makes the function much more flexible. It means the function can act upon any appropriate item that is clicked and passed into its argument. In other words, the same function could start and stop dragging any movie clip to which the same listener was added. In the companion source files, the *start_stop_drag.fla* file shows this by adding the following lines to the previous example:

```
9   myMovieClip2.addEventListener(MouseEvent.MOUSE_DOWN, onStartDrag);
10  myMovieClip2.addEventListener(MouseEvent.MOUSE_UP, onStopDrag);
```

You can drag and drop each movie clip simply by adding another movie clip to the exercise and specifying the same listeners.

Using Mouse Events to Control Properties and Methods

Now you can use syntax from the "Properties," "Methods," and "Events" sections of this chapter to set up interactive control over a movie clip. In the companion source code, you'll find a file called *props_methods_events.fla*. It contains nothing more than the example movie clip "box," and two buttons in the Library that will be used repeatedly to change the five properties discussed earlier. The movie clip contains numbers to show which of its frames is visible at any time, and the instance names of each button will reflect its purpose. Included are *move_up_btn*, *scale_down_btn*, *rotate_right_btn*, *fade_up_btn*, and *toggle_visibility_btn*, among others. The start of the main chapter project consists of several buttons that will modify properties of the center movie clip. Figure 6-8 shows the layout of the file.

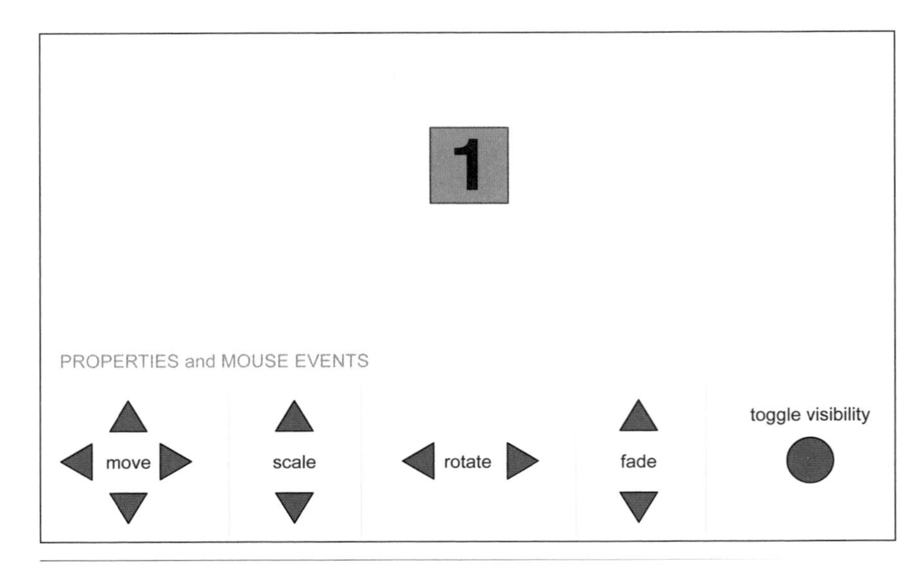

Figure 6-8. Layout of the props_methods_events.fla file

Starting with movement, you need to define one or more functions to update the location of the movie clip. In this case, you'll define a separate function for each type of movement.

```
1   function onMoveLeft(evt:MouseEvent):void {
2       box.x -= 20;
3   }
4   function onMoveRight(evt:MouseEvent):void {
5       box.x += 20;
6   }
7   function onMoveUp(evt:MouseEvent):void {
8       box.y -= 20;
9   }
10  function onMoveDown(evt:MouseEvent):void {
11      box.y += 20;
12  }
```

You can see that the structure of the functions is consistent. Once you have defined the functions, all you have to do is add the listeners to the appropriate buttons:

```
13  move_left_btn.addEventListener(MouseEvent.MOUSE_UP, onMoveLeft);
14  move_right_btn.addEventListener(MouseEvent.MOUSE_UP, onMoveRight);
15  move_up_btn.addEventListener(MouseEvent.MOUSE_UP, onMoveUp);
16  move_down_btn.addEventListener(MouseEvent.MOUSE_UP, onMoveDown);
```

This simple process is then repeated for each of the buttons on stage. The remainder of the script collects the aforementioned properties and event listeners in the same pattern used with the four movement listeners to complete the demo.

```
17  scale_up_btn.addEventListener(MouseEvent.MOUSE_UP, onScaleUp);
18  scale_down_btn.addEventListener(MouseEvent.MOUSE_UP, onScaleDown);
19
20  rotate_left_btn.addEventListener(MouseEvent.MOUSE_UP, onRotateLeft);
21  rotate_right_btn.addEventListener(MouseEvent.MOUSE_UP,
    onRotateRight);
22
23  fade_in_btn.addEventListener(MouseEvent.MOUSE_UP, onFadeIn);
24  fade_out_btn.addEventListener(MouseEvent.MOUSE_UP, onFadeOut);
25
26  toggle_visible_btn.addEventListener(MouseEvent.MOUSE_UP,
    onToggleVisible);
27
28  function onScaleUp(evt:MouseEvent):void {
29      box.scaleX += 0.2;
30      box.scaleY += 0.2;
31  }
32  function onScaleDown(evt:MouseEvent):void {
33      box.scaleX -= 0.2;
34      box.scaleY -= 0.2;
35  }
36
37  function onRotateLeft(evt:MouseEvent):void {
38      box.rotation -= 20;
39
40  }
41  function onRotateRight(evt:MouseEvent):void {
42      box.rotation += 20;
```

```
43  }
44
45  function onFadeIn(evt:MouseEvent):void {
46      box.alpha += 0.2;
47  }
48  function onFadeOut(evt:MouseEvent):void {
49      box.alpha -= 0.2;
50  }
51
52  function onToggleVisible(evt:MouseEvent):void {
53      box.visible = !box.visible;
54  }
```

Finally, you can add an example of a mouse event triggering a method. The final 10 lines of this script will toggle the movie clip between playing and stopped states. Because the **stop()** and **play()** methods are not Boolean opposites, you can't use a single *not* (!) operator to switch the states. Instead, you need to use a conditional statement that tests the value of a Boolean variable. Each time the play state is reversed, the Boolean variable can be toggled using the *not* (!) operator to prepare for the next test.

```
55  var isPlaying:Boolean = true;
56  box.addEventListener(MouseEvent.CLICK, onTogglePlay);
57  function onTogglePlay(evt:MouseEvent):void {
58      if (isPlaying) {
59          box.stop();
60      } else {
61          box.play();
62      }
63      isPlaying = !isPlaying;
64  }
```

Line 55 creates the variable **isPlaying** and sets it to **true**. The default state of the movie clip is playing. Lines 56 through 64 contain the event listener. The listener triggers the **onTogglePlay()** function when a user clicks the box movie clip.

Line 58 tests to see if **isPlaying** is **true**. If so, line 59 stops the movie clip. Otherwise, line 61 plays the movie clip. In either case, the value of **isPlaying** is reversed (if it was **false**, it's set to **true**, and vice versa).

Frame Events

Frame events are not triggered by user input the way mouse events are. Instead, they occur without intervention as the Flash file plays. Each time the playhead enters a frame that contains a frame script, that script is executed. This means that frame scripts execute only once for the life of the frame, making them excellent for seldom-executed tasks such as initializations. For a frame script to execute more than once, the playhead must leave the frame and return, either because of an ActionScript navigation instruction or a playback loop that returns the playhead to frame 1 when it reaches the end of the Timeline.

However, using an event listener, you can listen for a recurring enter frame event in a movie clip that is independent from the playhead. This type of event can trigger even when the playhead is stopped; this is a great way to run continuous animation. An enter frame event is fired at the same pace as the document frame rate. For example, if the default frame rate is 12 frames per second, the default enter frame frequency is 12 times per second.

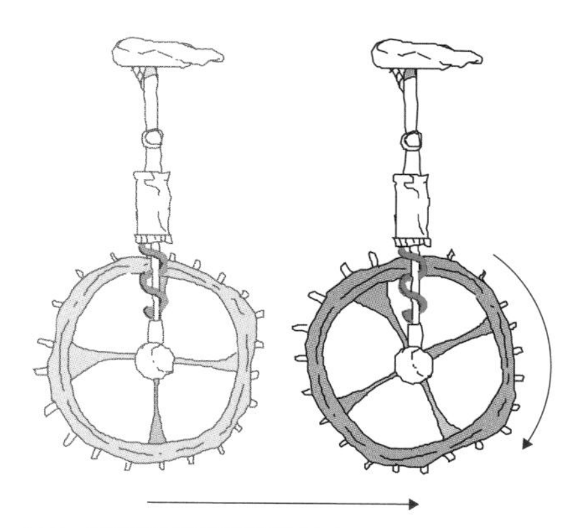

***Figure 6-9.** Visual depiction of the unicycle movements*

The *frame_events.fla* file in the companion source code demonstrates this event by updating the position of a unicycle every enter frame. It places the unicycle at the horizontal location of the mouse and rotates the wheel child movie clip. Figure 6-9 shows the effect. As the user moves the mouse to the right on the Stage, the unicycle will move to the right, and the wheel will rotate clockwise.

Here is the code for this example. The first line adds an enter frame event listener to the main Timeline. The listener function then sets the unicycle's *x* location to the mouse's *x* location using the *mouseX* property. It also sets the *rotation* property of the *wheel* clip (nested inside *cycle*) to the same value.

```
stage.addEventListener(Event.ENTER_FRAME, onFrameLoop);
function onFrameLoop(evt:Event):void {
     cycle.x = mouseX;
     cycle.wheel.rotation = mouseX;
}
```

Removing Event Listeners

While event listeners make most event handling easy to add and maintain, leaving them in place when unneeded can wreak havoc. From a logic standpoint, consider what could happen if you kept an unwanted listener in operation. Imagine a week-long promotion for radio station 101 FM, which rewards the 101st customer to enter a store each day of that week. The manager of the store is set up to listen for "customer enter" events and when customer 101 enters the store, oodles of prizes and cash are bestowed upon the lucky winner. Now imagine if you left that listener in place after the promo week was over. Oodles of prizes and cash would continue to be awarded at great, unexpected expense.

NOTE

This example also demonstrates a scripting shortcut aided by ActionScript. When specifying a rotation greater than 360 degrees, ActionScript will understand and use the correct value; that is, 360 degrees is one full rotation around a circle, returning to degree 0 (720 degrees is twice around the circle and also equates to 0). Similarly, 370 degrees is equivalent to 10 degrees, as it's 10 degrees past degree 0, and so on. This allows you to set the rotation of the wheel movie clip to the x coordinate of the mouse without worrying about moving past the 360-pixel point on the Stage.

Unwanted events are not the only problem, however. Every listener occupies a small amount of memory. Injudiciously creating many event listeners without cleaning up after yourself can result in dwindling memory. Therefore, it's a good idea to remove listeners when you know they will no longer be needed.

To do so, simply use the **`removeEventListener()`** method. By specifying the owner of the relevant event and the listener function that is triggered, you can remove that listener so it no longer reacts to future events. The **`removeEventListener()`** method requires two parameters: the event and function specified when the listener was created. Specifying the event *and* function is important, as you may have multiple listeners set up for the same event.

The following example rotates the hand of a watch 2 degrees every time an enter frame event is fired. When the hand rotates to 90 degrees, the listener is removed and the rotation stops. (See the *remove_listener.fla* source file.)

```
this.addEventListener(Event.ENTER_FRAME, onLoop);
function onLoop(evt:Event):void {
    watch.hand.rotation += 2;
    if (watch.hand.rotation >= 90) {
        this.removeEventListener(Event.ENTER_FRAME, onLoop);
    }
}
```

The Display List

One of the most dramatic changes introduced by AS3, particularly for designers accustomed to prior versions of the language, is the way in which code is used to access visual elements. AS3 brings with it an entirely new way of handling visual assets, called the *display list*. It's literally a hierarchical list of everything you can see, including movie clips, buttons, text fields, and more.

There's a lot of power and flexibility available to a scripter who uses the display list to its full potential. This book's companion volume includes deeper coverage of the display list, but here you'll focus on the tasks you'll find yourself repeating over and over: adding assets to the list, referencing assets in the list by position or name, and removing assets from the list.

Adding and Removing Children

If you have experience with prior versions of ActionScript, you'll probably be excited to hear that the many varied ways of adding an asset to the visual world of Flash have been unified and simplified. If you're new to ActionScript, the news is even better. You only have to learn one approach to getting an asset into the public eye.

Objects that you can add to the display list are called, appropriately, *display objects*. For all display objects, you'll use a variant of `new MovieClip()`—the `new` keyword followed by the object class for which you want to create an instance. Other examples include `new TextField()`, `new Video()`, `new Bitmap()`. Even adding an existing movie clip from the Library is consistent with this syntax, as you'll see in a moment.

While there are many object types that you can add to the display list, you'll focus on movie clips initially. Just like any other class, the `new` keyword creates an instance of the display object and you can use the object class (`MovieClip`) as a data type:

```
var mc:MovieClip = new MovieClip();
```

But that's only the first half of the story. This code creates an object (in this case, a movie clip), but it only places that object in memory. It hasn't yet become a part of the list of assets the user can see. The second half of the story is adding the object to the display list.

Using addChild()

The primary way to add a movie clip to the display list is by using the **addChild()** method. To add the movie clip you created a moment ago to the main Timeline, you can place this statement after the prior instruction to create the movie clip:

```
addChild(mc);
```

Despite its simplicity, this code does imply a destination for the movie clip. By omitting an explicit destination, you cause the movie clip to be added to the scope in which the script was written—in this case, the main Timeline.

You can specify a particular location to which the movie clip will be added, but not all display objects will be happy to adopt a child. For example, neither the *Video* nor *Bitmap* display object types can have children. To include a child, the destination must be a *display object container*. Examples of display object containers include *Stage*, *MovieClip*, and *Sprite* (a one-frame movie clip with no Timeline), but for the purposes of this chapter, you'll continue to work with movie clips.

So, if you wanted to add the **mc** movie clip nested inside another movie clip, called **mc2**, you would provide a destination object for the **addChild** method to act upon:

```
mc2.addChild(mc);
```

You don't even have to specify a depth (visible stacking order), because the display list automatically handles that for you. In fact, you can even use the same code for changing the depths of existing display objects.

Adding Symbol Instances to the Display List from the Library

Thus far you've either referenced display objects using instance names that have been applied on the Stage (through the Properties panel) or limited the dynamic creation of display objects to empty movie clips.

However, you will likely find a need to dynamically create and use instances of movie clips that already exist in your library. This process is nearly identical to creating and using empty movie clips, but one additional step is required: you must first set up the symbol first by adding a *linkage class*. In its most basic use, this is nothing more than a unique name that allows you to create an instance of the symbol dynamically.

To see this process, look at the companion source file *addChild.fla*. In the Library, you will find a unicycle. Select the movie clip in your library, then click the Symbol Properties button (it looks like an "i" at the bottom of the Library) for access to all the symbol properties.

In the resulting dialog, shown in Figure 6-10, enable the Export for ActionScript option and add **Unicycle** to the *Class* field.

You will also likely notice that Flash adds the *MovieClip* class (in this case) to the *Base* class field for you. This makes it possible to automatically access the properties, methods, and events available to the *MovieClip* class. For example, you can automatically manipulate the x and y coordinates of your new custom movie clip.

Now that you've given your movie clip a class name, you can create an instance of that custom movie clip class the same way you created an instance of the generic movie clip class. Instead of using **MovieClip()**, however, you will use **Unicycle()** to create the movie clip. The same call of the **addChild()** method adds the newly created movie clip to the display list.

```
var cycle:MovieClip = new Unicycle();
addChild(cycle);
```

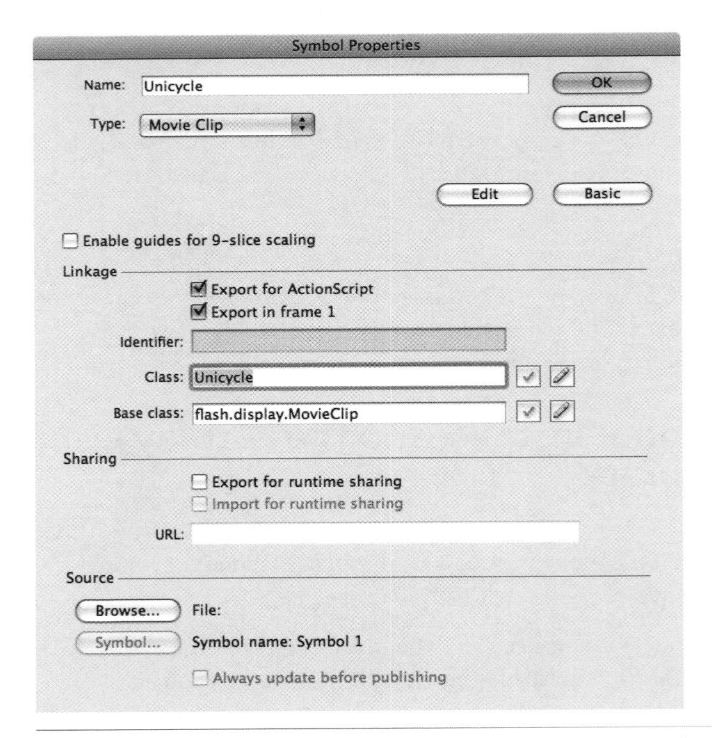

Figure 6-10. Entering a class name for a movie clip in the library Linkage settings

Using addChildAt()

The **addChild()** method adds the display object to the end of the display list, which always places the child on top of other display objects. In some cases, however, you may need to add a child at a specific position in the display list. For example, you may wish to insert an item into the middle of a vertical stack of display objects.

This example, found in the *addChildAt.fla* source file, adds a Library movie clip with the class name *Ball* to the start of the display list with every mouse click. The ultimate effect is that a new ball is added below the previous balls and positioned down and to the right 10 pixels every time the mouse is clicked.

```
1   var inc:uint = 0;
2
3   stage.addEventListener(MouseEvent.CLICK, onClick);
4
5   function onClick(evt:MouseEvent):void {
6       var ball:MovieClip = new Ball();
7       ball.x = ball.y = 100 + inc * 10;
8       addChildAt(ball, 0);
9       inc++;
10  }
```

Line 1 initializes a variable that will be incremented with each ball added. Line 3 adds an event listener to the Stage, listening for a mouse click, so that any mouse click will trigger the listener's function. The function in lines 5 through 10 performs four basic tasks. In line 6, a new Ball movie clip is created.

Line 7 manipulates the x and y coordinates in a single instruction, setting **x** equal to **y**, which is equal to the value of an expression. This is handy when both **x** and **y** values are the same. In this case, the expression sets the new ball to **x** and **y** of **100** and adds a 10-pixel offset for each ball added. For example, when the first ball is added, **inc** is 0, so the additional pixel offset is 0*10, or 0. **inc** is incremented at the end of the function, in line 9. The next mouse click that calls the function will update the offset to 1*10, or 10, pixels for the second ball; 2*10, or 20, pixels offset for the third ball; and so on. Most importantly, line 8 adds the ball to the display list, but always at position 0, making sure the newest ball is always on the bottom.

Choosing when to use **addChild()** and when to use **addChildAt()** depends entirely on your needs. If you only need to add the display object to the display list, or if you want the object to appear on top of all other objects, use **addChild()**. If you need to insert the object anywhere below the top of the visual stacking order, use **addChildAt()** and specify a depth.

Any time you do specify a depth, the new object will be sandwiched between the surrounding objects, rather than overwriting whatever is in the specified depth. The display list can have no gaps in it, so everything above an insertion level is automatically moved up a notch in the list. For example, assume a file started with objects in levels 0 through 9 by virtue of adding 10 objects to the display list. Then assume you need to insert a new display object into level 5. All objects in levels 5 through 9 will automatically move up to levels 6 through 10 to accommodate the insert.

Removing Objects from the Display List and from Memory

It's equally important to know how to remove objects from the display list. The process for removing objects is nearly identical to the process for adding objects to the display list. To remove a specific display object from the display list, use the **removeChild()** method:

```
removeChild(ball);
```

To remove a display object at a specific level, use the **removeChildAt()** method:

```
removeChildAt(0);
```

The following example is the reverse of the **addChildAt()** script discussed in the prior section. It starts by using a **for** loop to add 20 balls to the stage, positioning them with the same technique used in the prior script. It then uses an event listener to *remove* the children with each click.

```
1   for (var inc:uint = 0; inc < 20; inc++) {
2       var ball:MovieClip = new Ball();
3       ball.x = 100 + inc * 10;
4       ball.y = 100 + inc * 10;
5       addChild(ball);
6   }
7
8   stage.addEventListener(MouseEvent.CLICK, onClick);
9   function onClick(evt:MouseEvent):void {
10      removeChildAt(0);
11  }
```

Preventing out-of-bounds errors

This script will work correctly as long as there is something in the display list. If, after removing the last ball, you click the Stage again, there will be a warning that, "the supplied index is out of bounds." This makes sense, because you are trying to remove a child from position 0 of the display list, when there is nothing in the display list at all.

To avoid this problem, you can first check to see if there are any children in the display object container that you are trying to empty. Making sure that the number of children exceeds zero will prevent the aforementioned error from occurring. The following is an updated **onClick()** function, replacing lines 9-11 in the previous code, with the new conditional in bold:

```
9   function onClick(evt:MouseEvent):void {
10      if (numChildren > 0) {
11          removeChildAt(0);
12      }
13  }
```

Removing objects from memory

As discussed previously for event listeners, it's always a good idea to try to keep track of your objects and remove them from memory when you are sure you will no longer need them.

Keeping track of objects is particularly relevant when discussing the display list because it's easy to remove an object from the display list and forget to remove it from RAM. When this is the case, the object will not be displayed but will still linger in memory. The following script, a simplification of the previous example, will remove a movie clip from both the display list and RAM:

```
1   var ball:MovieClip = new Ball();
2   ball.x = 100;
3   ball.y = 100;
4   addChild(ball);
5
6   stage.addEventListener(MouseEvent.CLICK, onClick);
7
8   function onClick(evt:MouseEvent):void {
9       this.removeChild(ball);
10      //ball removed from display list but still exists
11      trace(ball);
12      ball = null;
13      //ball now entirely removed
14      trace(ball);
15
16      stage.removeEventListener(MouseEvent.CLICK, onClick);
17  }
```

Lines 1 through 5 create and position the ball, then add it to the display list. Line 6 adds a mouse click listener to the Stage. The first line of function content, line 9, removes the ball from the display list using the **`removeChild()`** method. Although it's no longer displayed, it's still around, as shown by line 11, which traces the object to the Output panel. Line 12, however, sets the object to **`null`**, removing it entirely from memory—again, shown by tracing the object to the output panel in Line 14.

As an added review of best practices, line 16 removes the event listener.

Finding Children by Position and by Name

Many of the example scripts in this chapter demonstrate working with children that have been previously stored in a variable. However, you will likely need to find children in the display list with little more to go on than their position or name.

Finding a child by position is consistent with adding or removing children at a specific location in the display list. Using the **getChildAt()** method, you can work with the first child of a container using this familiar syntax:

```
var dispObj:DisplayObject = getChildAt(0);
```

If you don't know the location of a child that you wish to manipulate, you can try to find it by name using its instance name, instead of using the display list index. Assuming a child has an instance name of **circle**, you can store a reference to that child using this syntax:

```
var dispObj:DisplayObject = getChildByName("circle");
```

Finally, if you need to know the location of a display object in the display list but have only its name, you can use the **getChildIndex()** method to accomplish your goal.

```
var dispObj:DisplayObject = getChildByName("circle");
var dispObjIndex:int = getChildIndex(dispObj);
```

Casting a Display Object

In the preceding discussion, you used **DisplayObject** as the data type when retrieving a reference to a display object, rather than another type, like **MovieClip**, for example. This is because you may not know if the child is a movie clip or another type of display object.

In fact, Flash may not even know the data type, such as when referencing a parent movie clip created using the Flash interface (rather than ActionScript). Without the data type information supplied in the ActionScript creation process, Flash sees only the parent Timeline as a display object container.

To tell Flash that the container in question is a movie clip, you can cast it as such; that is, you can change the data type of that object to **MovieClip**. For example, consider a movie clip created in the Flash Player interface that needs to tell its parent, the main timeline, to go to frame 20. A simple line of ActionScript is all that would ordinarily be required:

```
parent.gotoAndStop(20);
```

However, since Flash doesn't know that **gotoAndStop()** is a legal method of the display object container (the Stage, for example, can't go to frame 20, and neither can a sprite), you will get the following error:

```
Call to a possibly undefined method gotoAndStop through a reference
    with static type flash.display:DisplayObjectContainer.
```

To tell Flash the method is legal for the main timeline, you need to state that the parent is of a data type that supports the method. In this case, the main timeline is a movie clip, so you can say:

```
MovieClip(parent).gotoAndStop(20);
```

This will prevent the error from occurring, and the movie clip will be able to successfully send the main timeline to frame 20.

Timeline Control

One of the most basic ActionScript skills you need to embrace is navigating within your Flash movies. You will often use these skills to control the playback of the main Timeline or movie clips nested therein.

The first thing to learn is how to start and stop playback of the main Timeline or a movie clip, and then add an initial jump to another frame. Figure 6-11 shows *navigation_01.fla*, which contains four Timeline tweens of black circles. For added visual impact, the circles use the Invert blend mode to create an interesting optical illusion of rotating cylinders. You can start and stop playback at any point, as well as starting and stopping in a specific frame—frame one, in this example. Initially, you'll rely on frame numbers to specify where to start and stop.

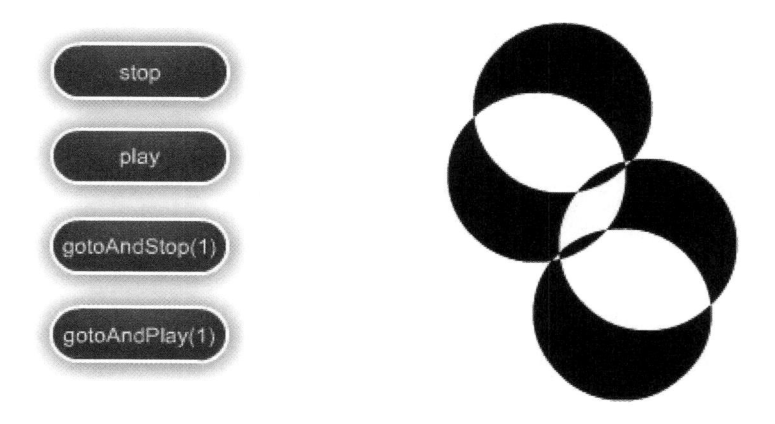

Figure 6-11. *navigation_01.fla demonstrates simple navigation*

You've already seen the **stop()** action at work in a frame script as a passive means of halting playback at the end of an animation or, perhaps, to support a menu screen or similar single frame. In the following code, look at invoking the **stop()** action via user input, such as clicking a button.

In the first frame of the *actions* layer, you'll find the following code:

```
1   stopBtn.addEventListener(MouseEvent.CLICK, onStopClick);
2
3   function onStopClick(evt:MouseEvent):void {
4       stop();
5   }
```

This code does not introduce anything new, other than the aforementioned use of **stop()** as a method triggered by user interaction. Line 1 is an event listener added to a button named *stopBtn*. It uses a mouse click to call **onStopClick**.

The effect of this setup is to add to the *stopBtn* functionality for stopping the main movie. All playback of the main Timeline will cease when the user clicks the button. Adding new lines to the script (shown here in bold) will allow you to restart playback. The code structure is similar to the previous example, but invokes the **play()** method on the *playBtn* instead. Using this pair of buttons, you can start and stop playback at any time without relocating the playback head in the process.

```
1   stopBtn.addEventListener(MouseEvent.CLICK, onStopClick);
2   playBtn.addEventListener(MouseEvent.CLICK, onPlayClick);
3
4   function onStopClick(evt:MouseEvent):void {
5       stop();
6   }
7   function onPlayClick(evt:MouseEvent):void {
8       play();
9   }
```

Using **stop()** and **play()** in this fashion is useful for controlling a linear animation, much the way a controller bar might control audio or video playback. However, it's less common in the case of menus or other navigation devices because typically you must jump to a specific point in your Timeline before stopping or playing.

For example, you might have generic sections that could apply to any project, such as home (start), about (info), and help. If restricted to the use of **stop()** and **play()**, you'd be forced to play through one section to get to another.

Adding again to the example script, the following content shown in bold adds a slight variation. The buttons in the new script function in similar ways, but instead of stopping in or playing from in the current frame, the new buttons go to a specified frame first. For example, if you had previously stopped playback in frame 20, triggering **play()** again would begin playback at frame 20. However, if you use **gotoAndPlay()** and specify frame 1 as a destination (shown in the script that follows), you will resume playback at frame 1 rather than at frame 20. There are no structural differences in this code, so simply add the content shown in bold to your ongoing script:

```
1   stopBtn.addEventListener(MouseEvent.CLICK, onStopClick);
2   playBtn.addEventListener(MouseEvent.CLICK, onPlayClick);
3   gotoPlayBtn.addEventListener(MouseEvent.CLICK, onGotoPlayClick);
4   gotoStopBtn.addEventListener(MouseEvent.CLICK, onGotoStopClick);
5
6   function onStopClick(evt:MouseEvent):void {
7       stop();
8   }
9   function onPlayClick(evt:MouseEvent):void {
10      play();
11  }
12  function onGotoPlayClick(evt:MouseEvent):void {
13      gotoAndPlay(1);
14  }
15  function onGotoStopClick(evt:MouseEvent):void {
16      gotoAndStop(1);
17  }
```

To add a nice level of diagnostic reporting to your playback, you can add two new properties to this script. Using the **trace()** method to send text to the Output panel, you can reference **totalFrames** to display the number of frames in your movie, and reference **currentFrame** to tell you which frame the playback head is displaying at the time the script is executed.

```
trace("This movie has " + totalFrames + " frames.");
trace(currentFrame);
```

The companion sample file, *navigator_02.fla*, demonstrates the use of these properties. It uses **totalFrames** at the start of playback, and uses **currentFrame** each time a button is clicked.

Frame Labels

There are specific advantages to using frame numbers with **goto** methods, including simplicity and use in numeric contexts (such as with a loop or other type of counter). However, frame numbers also have specific disadvantages. The most notable disadvantage is that edits that you make to your file subsequent to the composition of your script may result in a change to the frame sequence in your timeline.

For example, your help section may start at frame 100, but you may then insert or delete frames in a section of your timeline prior to that frame. This change may cause the help section to shift to a new frame, and your navigation script will no longer send the playback head to the help section.

One way around this problem is to use frame labels to mark the location of a specific segment of your timeline. As long as you shift content by inserting or deleting frames to all layers in your timeline, therefore maintaining sync among your layers, a frame label will move with your content.

For example, if your help section, previously at frame 100, is marked with a frame label called "help," adding 10 frames to all layers in your Timeline panel will not only shift the help content, but will also shift the frame label used to identify its location. So, although you will still be navigating to the "help" frame label after the addition of frames, you will correctly navigate to frame 110.

This is a useful feature when you are relying heavily on timeline tweens for file structure or transitions, or when you think you may be adding or deleting sections in your file. In fact, frame labels free you to simply rearrange your timeline if desired. The capability to go to a specific frame label, no matter where it is, means that you don't have to arrange your file linearly, and you are free to add last-minute changes to the end of your timeline without having to remember an odd sequence of frame numbers to jump to content.

The sample file, *frame_labels_01.fla*, demonstrates the use of frame labels instead of frame numbers when using a **goto** method. It also illustrates another important and useful concept, which is that you can use these methods to control the playback of movie clips as well as the main timeline.

Instead of controlling the playback of a linear animation, *frame_labels_01.fla* moves the playback head between the frames of a movie clip called *pages*. This is a common technique for swapping content in Flash because you can keep your main Timeline simple and just jump the movie clip from frame to frame to reveal each new screen. Figure 6-12 displays the "page1" frame of *frame_labels_01.fla*. The Timeline inset shows the frame labels.

Figure 6-12. The "page1" frame of frame_labels_01.fla

The initial setup of this example requires that you prevent the movie clip from playing on its own so that you can exert the desired control over its playback. There are several ways to do this. The first, and perhaps most obvious, approach is to put a `stop()` action in the first frame of the movie clip. You will see this technique used often.

The second is to add the `stop()` method to a main timeline script, but to target a movie clip instead of the main timeline. To do this, precede the method with the object you wish to stop, as shown in line 1 of the following script. In this case, you're stopping the movie clip called *pages*.

You'll look at a third method for stopping movie clips at the end of this chapter, but for now, let's focus on the simple changes this file introduces. In addition to stopping the *pages* movie clip in line 1, this script adds listeners to buttons *one*, *two*, and *three*, which cause the movie clip to change frames in lines 8, 11, and 14, respectively.

```
1   pages.stop();
2
3   one.addEventListener(MouseEvent.CLICK, onOneClick);
4   two.addEventListener(MouseEvent.CLICK, onTwoClick);
5   three.addEventListener(MouseEvent.CLICK, onThreeClick);
6
7   function onOneClick(evt:MouseEvent):void {
8       pages.gotoAndStop("page1");
9   }
10  function onTwoClick(evt:MouseEvent):void {
11      pages.gotoAndStop("page2");
12  }
13  function onThreeClick(evt:MouseEvent):void {
14      pages.gotoAndStop("page3");
15  }
```

The code is essentially the same as the ActionScript you've seen before. To test the effectiveness of using frame labels, simply add or delete frames across all layers before one of the existing frame labels. Despite changing the frame count, you will find that the navigation still works as desired.

Frame Rate

Also new to AS3 is the ability to dynamically change the frame rate at which your file plays at runtime. The default frame rate of a Flash CS4 movie is 24 frames per second. Previously, whichever frame rate you chose was the frame rate you were stuck with for the life of your SWF. It's now possible to update the speed at which your file plays by changing the **frameRate** property of the *stage*, as demonstrated in the sample file *frame_rate.fla*.

This simple script increments or decrements the frame rate by **5** frames per second with each click of a button. You may also notice another simple example of error checking, in the function used by the slower button, to prevent a frame rate of zero or below. Start the file and watch it run for a second or two at the default frame rate of 24 frames per second. Then, experiment with additional frame rates to see how they change the movie clip animation.

```
1   info.text = stage.frameRate;
2
3   faster.addEventListener(MouseEvent.CLICK, onFasterClick);
4   slower.addEventListener(MouseEvent.CLICK, onSlowerClick);
5
6   function onFasterClick(evt:MouseEvent):void {
7       stage.frameRate += 5;
8       info.text = stage.frameRate;
9   }
10  function onSlowerClick(evt:MouseEvent):void {
11      if (stage.frameRate > 5) {
12          stage.frameRate -= 5;
13      }
14      info.text = stage.frameRate;
15  }
```

This **frameRate** property requires little explanation, but its impact should not be underestimated. Other interactive environments have long been able to vary playback speed, and this is a welcome change to ActionScript for many enthusiastic developers, especially animators. Be it for a *Matrix* parody or a sports game, slow mo has never been easier.

Project Progress

You won't add anything new to the portfolio project in this chapter, but it'll help to review the scripts you wrote in previous chapters now that you have a little more experience with AS3.

Chapter 3: The Deco Tool

To begin, you wrote a script in Chapter 3 that animated the symbols created by the Deco tool (Figure 6-13).

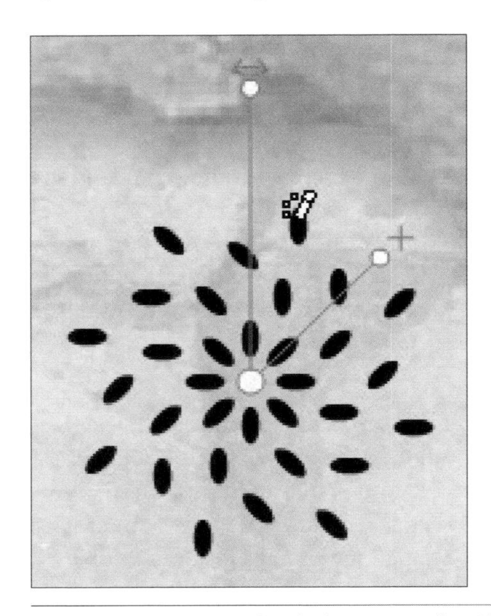

Figure 6-13. Art created with the Deco tool in Chapter 3

The Deco tool quickly and automatically adds movie clips to a parent movie clip container, arranging the children in circular patterns. However, you did not give each of the child ovals instance names, so you might wonder how they can be controlled with ActionScript.

The answer is by using display list methods and properties. The script you entered in Chapter 3 uses the `numChildren` property to determine how many ovals are in the parent movie clip, and then loops through that number, gaining access to each child using the `getChildAt()` method. The combination of these tools makes it possible for the script to manipulate each oval individually.

```
1   ovals.addEventListener(Event.ENTER_FRAME, onEnter);
2   function onEnter(evt:Event):void {
3       var numOvals:int = ovals.numChildren;
4       for (var i:int = 0; i < numOvals; i++) {
5           ovals.getChildAt(i).rotation += 10;
6       }
7       ovals.rotation = mouseX;
8   }
```

Line 1 adds an enter frame listener to the parent movie clip, which has an instance name of *ovals*, which was applied through the Properties panel. Lines 2 through 8 comprise the listener function, **`onEnter()`**. This function includes the mandatory event argument, typed to the *Event* class, and returns no value.

Line 3 checks the number of children in the parent movie clip and places that integer into the **`numOvals`** variable.

Lines 4 through 6 contain a **`for`** loop that loops through those children one at a time. Each time through the loop, the child at the next highest level in the display list is rotated 10 degrees. That is, the first time through the loop, the loop counter is 0. As such, the child in the *ovals* movie clip that is at level 0 is rotated. The next time through the loop, the *ovals* child at level 1 is rotated, and so on.

After the **`for`** loop fully executes, the parent movie clip (not the individual oval-shaped children) is also rotated, this time to the same value as the x position of the mouse. The entire process is repeated every time the enter frame event is fired or, by default, 24 times per second.

This means that you can move the mouse left and right to control the rotation of the container movie clip. All the while, the individual ovals inside the container continue to rotate, contributing to an interesting visual effect.

Chapter 5: The Portfolio Project Navigation

In Chapter 5, you added a script to control the navigation of the portfolio project. With every click of a button, this script populates a variable and sets the portfolio in motion. A corresponding script then sends the playhead to the frame chosen by the button click, and a **`stop()`** action halts the playhead on the desired screen.

```
1   var nextSection:String;
2
3   navigation.home.addEventListener(MouseEvent.CLICK, onNavigate);
4   navigation.gallery.addEventListener(MouseEvent.CLICK, onNavigate);
5   navigation.lab.addEventListener(MouseEvent.CLICK, onNavigate);
6   navigation.help.addEventListener(MouseEvent.CLICK, onNavigate);
7
8   function onNavigate(evt:MouseEvent):void {
9       nextSection = evt.target.name;
10      play();
11  }
```

Line 1 creates the needed variable and types its data as a string of text.

Lines 3 through 6 add event listeners to the navigation buttons that sit within the navigation bar container. Each button calls the same function when a mouse click is detected.

Lines 8 through 11 define the listener function, which expects a *MouseEvent* and returns no value.

Line 9 parses the incoming event information to get the name of the target object that received the click, and then puts that name into the **nextSection** variable. The function then sets the playhead in motion.

At the end of the section, a separate frame script tells the playhead to go to the desired frame and play through the section so the transitions can complete their visual updates.

```
gotoAndPlay(nextSection);
```

Finally, after the transitions are complete, the playhead is stopped in the middle of the content frame span by another independent script:

```
stop();
```

The Project Continues...

The next chapter will introduce filters and blend modes, both of which you will apply to portfolio assets.

CHAPTER 7

FILTERS AND BLEND MODES

IN THIS CHAPTER

Introduction

Although many designers and developers like to capitalize heavily on Flash's vector-based features, the application has a lot to offer from the world of pixels. Perhaps the most useful way in which vectors and pixels coexist in Flash is through the use of bitmap compositing features. In this context, *compositing* pertains to combining visual elements to give those elements a new appearance. This includes the use of transparency, blending (combining the pixel color values of two bitmaps), and filters (such as blur or drop shadow). Whether your asset is a bitmap or vector, Flash can use traditional bitmap compositing techniques to combine elements during authoring or at runtime.

Bitmap Caching

Flash can act upon vectors as if they were bitmaps by using a feature called *bitmap caching*. Bitmap caching is a process by which Flash Player temporarily takes a bitmap snapshot of your asset and uses that snapshot, rather than the original asset, for bitmap compositing. There are two important ideas going on here. First is the concept that you can apply an effect typically reserved for bitmaps, such as blurring, to vectors. Second is the fact that the effect is applied to a cached snapshot of the original asset, which means the asset remains in its original format. For example, if the original asset is a vector, it remains crisp at any resolution.

Bitmap caching not only makes certain features possible, it can also improve performance. This is because it's easier for your computer to manipulate bitmaps than to recalculate all of the math needed to display the points, lines, curves, fills, and so on that comprise a vector asset. In short, Flash takes a picture of your asset after every visual change, ensuring that its appearance will remain at maximum quality, and uses the bitmap for the heavy lifting.

How often you change the asset can have a big impact on performance. For example, if you just intend to move a movie clip around on the stage, a bitmap only needs to be cached once. However, if you materially change the movie clip through scaling, rotating, or changing its opacity (alpha), Flash will need to cache the bitmap version of the asset after every such change. If bitmap caching is used injudiciously, it can actually affect performance adversely. Limit the use of the feature when doing a lot of these kinds of transformations, and test your work with caching disabled, if your feature set permits. This will let you know if caching is helping or hurting performance.

Sometimes bitmap caching is applied automatically by Flash Player, as when using a filter effect on text elements. You'll learn how to use filters in just a moment. However, you can manually apply bitmap caching to buttons or movie clips. Because this step is used in later sections of this chapter, it's helpful to learn how to enable bitmap caching using both the interface and ActionScript. Enabling the feature without doing anything else won't show any change, but it's necessary groundwork for upcoming discussions.

The Properties Panel

NOTE

This chapter provides examples of bitmap compositing features created in the Flash interface (the primary focus) and with ActionScript to help you practice and expand the ActionScript skills you learned in Chapter 6. You'll find more such opportunities in subsequent chapters as space allows.

The *Cache as bitmap* option appears in the Display section of the Properties panel (Figure 7-1) when you select a button or movie clip. To get a feel for the feature, try this short activity:

1. Create a new file. Use the Rectangle tool to draw a rectangle anywhere on the stage.
2. Use the Selection tool to select the shape and convert it to a movie clip (Modify→Convert to Symbol). The name and registration point of the symbol are unimportant in this case, but make sure it's a movie clip.
3. In the Properties panel, look under Display and enable the *Cache as bitmap* option. Bitmap caching is now enabled. This file is not required for the final project.

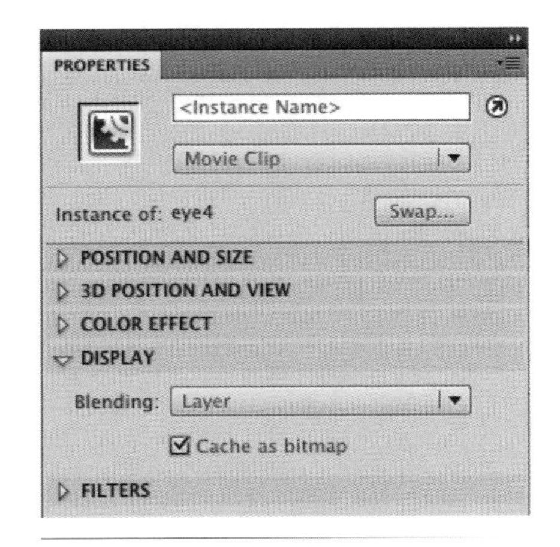

Figure 7-1. A movie clip's Cache as bitmap option in the Display section of the Properties panel

The ActionScript Method

An alternative to using the Flash interface to enable bitmap caching is to use ActionScript to do so at runtime.

1. To rely solely on ActionScript, select your movie clip on the Stage with the Selection tool and disable *Cache as bitmap* in the Properties panel. The ActionScript property will work regardless of the Properties panel values, but resetting them makes the tutorial clearer.
2. Using the Properties panel, give the movie clip an instance name of `myMovieClip`.
3. Create a new layer in the Timeline, name it `actions`, and lock it.

4. Select the frame in the new layer and open the Actions panel (Window→Actions). Enter the following script:

```
myMovieClip.cacheAsBitmap = true;
```

After following the Properties panel or ActionScript steps described, you can apply compositing features that require bitmap caching to *myMovieClip*.

Filters

Flash features seven filter effects that are akin to Adobe Photoshop's layer styles. You can apply one or more filters and one or more copies of each filter to compatible display objects (shapes and graphic symbols are not supported, for example). You can copy and paste, hide, reset, and delete them. You can also save a library of custom presets for later use.

Filter Inventory

Figure 7-2 shows a quick example of each basic filter configuration in use, with a control (unaffected object) in the lower-right corner of the figure. The source file *filters.fla* demonstrates each filter.

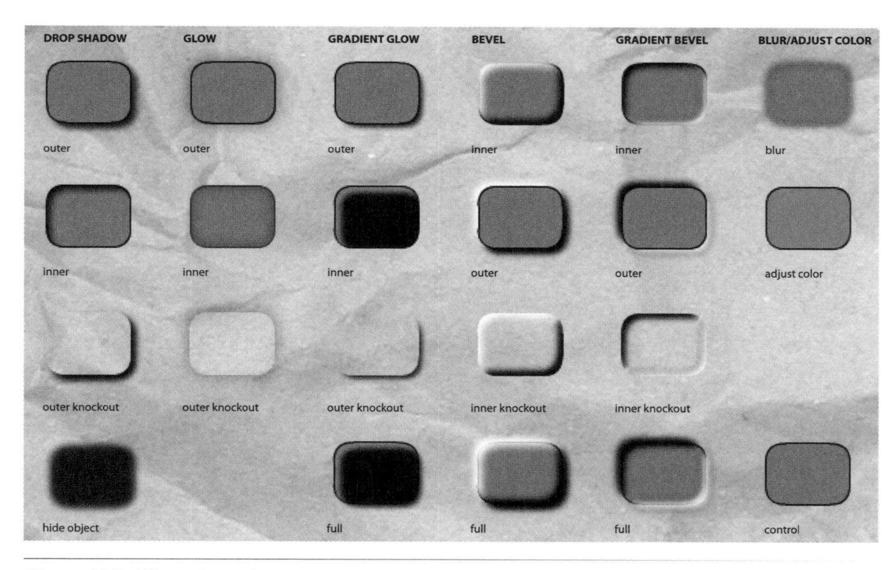

Figure 7-2. Filters in action

Drop shadow

Creates a drop shadow with the following configurable properties: *BlurX* (width), *BlurY* (height), *Strength* (darkness/opacity), *Quality* (degree of blurriness of the shadow), *Angle* (of virtual light source), *Distance* (from the object casting the shadow), *Knockout* (removes the area of the object from the shadow and shows only the remaining shadow), *Inner Shadow* (applies a shadow within object boundaries), *Hide Object* (hides the object and shows only the complete shadow), and *Color*.

Blur

Blurs the object with the following configurable properties: *BlurX* (width), *BlurY* (height), and *Quality* (degree of blurriness).

Glow

Creates a glow with the following configurable properties: *BlurX* (width), *BlurY* (height), *Strength* (sharpness of glow edge), *Quality* (degree of blurriness of glow), *Color*, *Knockout* (removes the area of the object from the glow and shows only the remaining glow), and *Inner Glow* (applies a glow within the object boundaries).

Bevel

Bevels, or edges, the object with the following configurable properties: *BlurX* (horizontal softness), *BlurY* (vertical softness), *Strength* (opacity), *Quality* (degree of blurriness of the bevel), *Shadow* (color of the shadow portions of the bevel), *Highlight* (color of the highlight portions of the bevel), *Angle* (of virtual light source), *Distance* (depth of the bevel), *Knockout* (removes the object and shows only the bevel), and *Type* (applies a bevel to the inner portion of the object, outer portion of the object, or the full object—inside and outside).

Gradient glow

Applies a glow to the object, functioning the same way as the Glow filter, but uses a gradient instead of a solid color. Adds the following configurable properties to those found in Glow: *Angle* (of gradient), *Distance* (of glow from the object), and *Type* (applies a bevel to the inner portion of the object, the outer portion of the object, or the full object—inside and outside). The Glow filter's *Color* property is replaced with *Gradient* (features a gradient editor similar to Color panel, but the first end color must be 100% transparent).

Gradient bevel

Applies a bevel to the object, functioning the same way as the Bevel filter, but uses a gradient instead of highlight and shadow colors. The Bevel filter's *Shadow* and *Highlight* properties are replaced with *Gradient* (features a gradient editor similar to Color panel, but the first end color must be 100% transparent).

Adjust color

Adjusts the color of the object with *Contrast*, *Brightness*, *Saturation* (how vibrant the color is), and *Hue* (color range) sliders.

Where applicable, high quality approximates a Gaussian blur on the effect, while low quality is better for performance.

The Properties Panel

Configuring a filter in the Properties panel is almost as straightforward as manipulating any other property. You just need to use a few added interface elements, which are found in a row in the lower-left corner of the Filters section of the panel (Figure 7-3).

The first button in the row (single document icon) adds a filter from a pop-up menu of filters that ship with Flash. Using this menu, you can also remove, enable, or disable all filters. The second button (multiple document icon) allows you to manage your own custom presets. You can name and save presets, and rename and delete them. The third button (clipboard icon) copies all or selected filters, and pastes filters from the clipboard. The fourth button (eye icon) enables or disables the filter. The fifth button (arrow) resets the filter to the preset values. The last button (trashcan icon) deletes a filter.

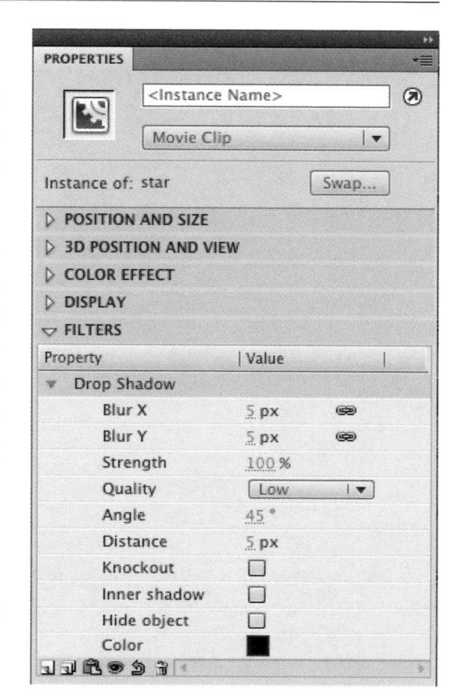

Figure 7-3. A Drop Shadow filter applied in the Properties panel

The ActionScript

Applying a filter in ActionScript adds flexibility and requires two steps. First create the filter, and then you must apply it to your display object.

Configuring a filter

It is possible to configure a filter when it is created, like the drop shadow created here:

```
var ds:DropShadowFilter = new DropShadowFilter(5, 45, 0x000000, 0.5, 5,
    5, 0.5, 1, false, false, false);
```

Because you must remember the order of all the properties to do this, however, it's not recommended. Applying the properties as separate lines of script is easier to read:

```
var ds:DropShadowFilter = new DropShadowFilter();
ds.distance = 5;
ds.angle = 45;
ds.color = 0x000000;
ds.alpha = 0.5;
ds.blurX = 5;
ds.blurY = 5;
ds.strength = 1;
ds.quality = 1;
ds.inner = false;
ds.knockout = false;
ds.hideObject = false;
```

All properties in these filters have default values, so you can create a filter with preset values with a single line of code. If you want to alter any of the preset values, you can adjust only those parameters, as shown with this bevel:

```
var bvl:BevelFilter = new BevelFilter();
bvl.distance = 10;
```

NOTE

The companion website includes information about sharing your custom filter presets with others.

Applying a filter

Applying a filter to a movie clip instance is as simple as adding the variable you used to create that filter to an array of all such filters applied to the object. The array is added to the aptly named *filters* property of the movie clip. For example, you can apply the drop shadow and bevel filters to a movie clip with an instance name of *myMovieClip* using the following ActionScript:

```
myMovieClip.filters = [ds, bvl];
```

You can apply a single filter in exactly the same way:

```
myMovieClip.filters = [ds];
```

It may seem strange at first to place a single value into an array, but it must be done this way to accommodate more than one filter. That is, without the array structure, you might only be able to apply a single filter to a movie clip. That would be too limiting.

NOTE

On the companion website, you'll find ActionScript techniques that replicate the functionality of the filter interface buttons in the Properties panel. Among these techniques are how to create, enable/disable, and reset your filters, as well as how to maintain a library of presets.

If, after applying a filter, you make a change to a filter property, you must reapply the filter to the display object. The following simplified example alters a filter property after application and then reapplies the filter to update the change:

```
myMovieClip.filters = [ds];
ds.distance = 0;
myMovieClip.filters = [ds];
```

Filters in Practice

This exercise demonstrates how to take advantage of the *hide object* feature of the Drop Shadow filter. Figure 7-4 shows the portfolio protagonist, Scaly, minding his own business. On the left is the original movie clip, selected to show that the transform point has been dragged down to near the bottom center of the clip.

Figure 7-4. Unaffected movie clip (left), normal drop shadow filter applied (middle), and filter applied to transformed duplicate movie clip with hide object enabled

The middle image in Figure 7-4 is the same movie clip with a drop shadow filter applied in a typical manner. Think of the shadow using an imaginary light source somewhere behind you above your right shoulder. This gives Scaley some depth, but it's not very realistic.

On the right, the same movie clip is used, but a second copy of the movie clip has been added to a layer underneath. That copy has the same filter applied, but the *hide object* feature has been enabled, and the second movie clip has been scaled and skewed to add perspective.

To try this exercise yourself, open the *shadow_hide_object_01.fla* from the companion source files. If you want to use your own assets and adapt the instructions as you go, please pay careful attention to the transform point of the movie clips. You will need to adjust your transform point and apply scale, skew, and possibly x- and y-coordinate values that work with your assets.

In the provided source file, you will see the setup depicted in Figure 7-4, but without the filters applied. The visible characters on the left and right are in locked layers. The character in the middle is in the *normalFilter* layer, and a duplicate of the character on the right is in the *perspectiveFilter* layer, both of which are unlocked.

1. Select the middle character and use the Properties panel to add a Drop Shadow filter. Change the *BlurX* and *BlurY* to **10**, the *Strength* to **30**, and the *Angle* to **110**. This completes the normal filter, which should resemble the shadow in the center of Figure 7-4.
2. Select the filter name in the Properties panel and copy it using the clipboard icon button in the lower left of the panel.
3. Select the duplicate character on the right (because the topmost layer is locked, clicking on the character selects the duplicate underneath). Using the same clipboard icon button, paste the filter onto the selected character. Both the middle and right characters should look the same now because you haven't begun to transform the shadow.
4. With the right duplicate movie clip selected, open the Transform panel (Figure 7-5). Click the link icon in the top row until you unlink the horizontal and vertical scale settings. Set the horizontal scale to **75** and the vertical scale to **50**.
5. In the panel's second row, select the *Skew* radio button and enter **-30** in the first value for horizontal skew. The duplicate movie clip should now be scaled and skewed the way the shadow appears in Figure 7-4.
6. All that remains is to hide the object. Select the duplicate movie clip and, in the Filters section of the Properties panel, click the *Hide object* option. Finit.

Your file should now look like Figure 7-4. The best part is that the shadow is really a duplicate movie clip instance, so if the movie clip is animated, the shadow will animate, too!

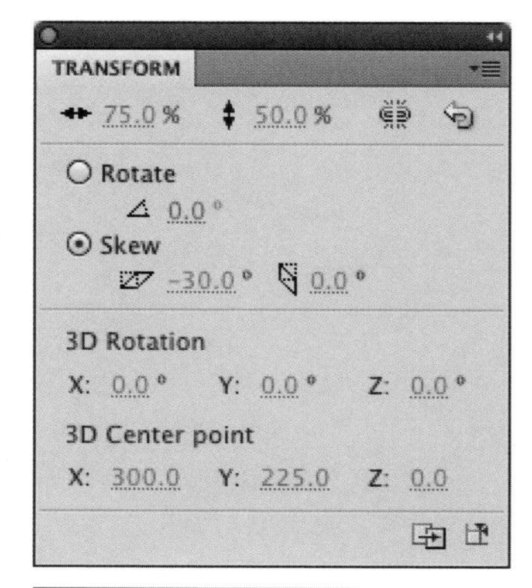

Figure 7-5. Transform settings for duplicate, shadow-only movie clip

Blend Modes

Flash includes a dozen blend modes that function similarly to their namesakes in Adobe Photoshop and Adobe Fireworks. Blend modes are used to blend one compatible display object (such as a movie clip or button) with another, based on the colors of both.

Blend Mode Inventory

Figure 7-6 shows an example of each visual blend mode, and the source file *blend_modes.fla* demonstrates them in use. The list that follows explains each blend mode and uses *target* to refer to the underlying art, or the art to which the *blending* artwork is applied. For the most part, the underlying art is the crumpled paper background from the portfolio project, and the blending art is the eyeball vignette seen in large part in the lower-right corner of the figure. The only visual change to this eye is the removal of the small star shape applied by the *Erase* blend mode.

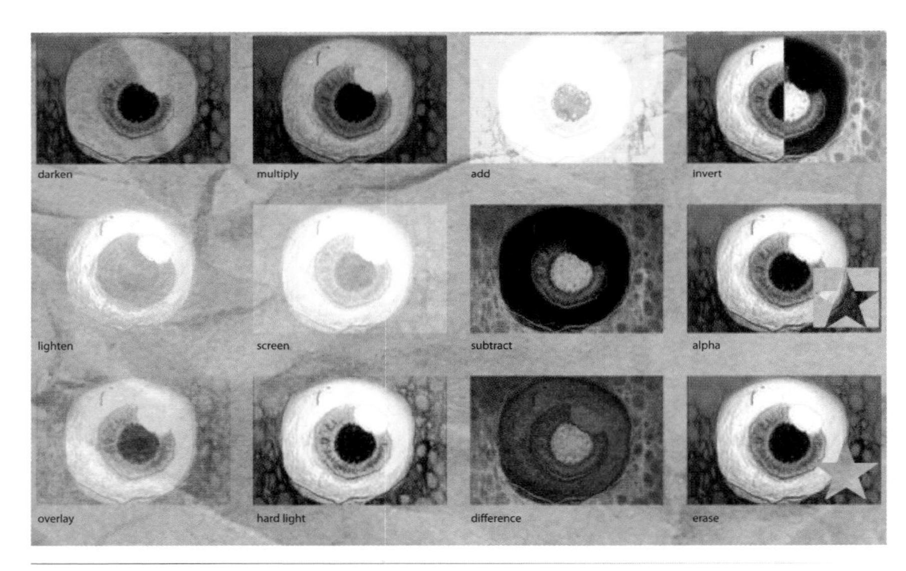

Figure 7-6. Blend modes in action

The three exceptions to this are *Invert*, *Alpha*, and *Erase*. Overlaying the eyeball with a black rectangle movie clip and applying the *Invert* blend mode to the rectangle produces a clearer demonstration of the effect. As for *Alpha* and *Erase*, these blend modes use alpha data and work only when the two assets being composited are in a container display object. This arrangement will be explained in detail, but the visual difference in Figure 7-6 between these two modes exists because a second movie clip containing a PNG of a star on a transparent background is placed atop the eye.

The most important thing to remember when looking over these blend modes is that trial and error with your own artwork is probably more helpful than trying to understand how each blend mode works in detail. Some are easy to understand, such as *Darken*, in which only blending pixels with colors that are darker than the underlying pixels will be visible, but others are a bit more complicated. You will likely become familiar with the blend modes you use most often, and you can always try a variety of modes to see how they look.

Darken

Darken applies only color from the blending object that is darker than the color in the target object.

Multiply

The *Multiply* blend mode multiplies the target color by the blending color, resulting in a more uniform application of darker colors.

Lighten

The opposite of *Darken*, this blend mode applies only blending color that is lighter than the target color.

Screen

For all intents and purposes, *Screen* is the opposite of *Multiply*. Screen multiplies the target color by the inverse of the blending color, resulting in a more uniform application of lighter colors.

Overlay

Using *Overlay*, blending colors darker than 50% gray are applied to the target color using the *Multiply* effect. Blending colors lighter than 50% are applied to the target color using the *Screen* effect.

Hard Light

Similar to *Overlay*, color application is determined by a mid-gray threshold. Blending colors darker than 50% gray are applied to the target color using the *Darken* effect, while blending colors darker than 50% gray are applied to the target color using the *Lighten* effect.

Add

Add sums the blending color values and target color values, resulting in a higher-contrast, bleaching effect.

Subtract

The opposite of *Add*, *Subtract* reduces the target color values by the values of the blending colors. This results in a higher-contrast, darkening effect.

Difference

Using *Difference*, blending colors darker than 50% gray are applied to the target color using the *Add* effect. Blending colors darker than 50% gray are applied to the target color using the *Subtract* effect, resulting in a look similar to a color negative.

Invert

The *Invert* blend mode doesn't rely on the color values of the blending object. Instead, target colors are inverted by any blending color. The degree to which the effect is applied is based on the transparency value of each pixel. For example, the effect shown in Figure 7-6 is achieved using an opaque black rectangle movie clip blended onto the eye movie clip. However, the same effect would occur if the rectangle were white.

Alpha

Using *Alpha*, transparent and semitransparent alpha values in the blending object remove all colors from both the blending object and target. In Figure 7-6, this blend mode uses the transparency around a star to knock out the color. *Alpha* requires a catalyst to work. The blending and target display objects must be in a parent container (such as another movie clip) and the *Layer* blend mode must be applied to the parent. *Alpha* will not work when applied only to a display object in the main Timeline.

Erase

Erase works like *Alpha*, including the need for a parent container and use of the *Layer* blend mode, but opaque and semiopaque alpha values in the blending object remove the colors from the blending object and target. Figure 7-6 shows that the opaque star knocks out the color, not the surrounding alpha data.

Layer

Layer is an important facilitator blend mode that *precomposes* the contents of a display object container to which the blend mode is applied. That is, the *Layer* blend mode composites the contents of the container first, and then treats the container as a single object. It does not have any visual effect on its own. *Layer* is required for the *Alpha* and *Erase* blend modes to work, but also improves the appearances of opacity changes to parent movie clips. See the "Blend Modes in Practice" section, later in this chapter, for more information.

Normal

This blend mode applies no change whatsoever. Use *Normal* to remove blend mode effects.

Blending: Think of Colors As Numbers

Typically, a little bit of trial and error will teach you just enough to use blend modes as much as you need to use them. If, however, you want to understand more about the way they work, this may help.

Some designers have difficulty understanding the inner workings of a few of the blend modes because they are thinking about mixing colors in the physical world. For example, how can adding 50% black to 50% black give you white? If you fall into this category, it helps to think of colors not as pigments, or degrees of transparency, but as numbers.

Think of the hexadecimal values, for instance, used to specify colors in ActionScript and in many of Flash's color interface controls. 0x000000 (#000000 in HTML) is black, and 0xFFFFFF (#FFFFFF in HTML) is white. If you add one mid-gray (0x7FFFFF) to another mid-gray (0x80000), you will get white (0xFFFFFF). This should make blend modes like *Add* and *Subtract* easier to understand so you can more easily predict the outcome when using them.

The Properties Panel

Applying a blend mode using the Properties panel is very simple. All you need to do is select a movie clip or button, look in the Display section of the Properties panel (Figure 7-1), and choose the desired blend mode from the Blending menu.

ActionScript

Using ActionScript is also straightforward. References to all blend modes are stored in a single **BlendMode** class. They are stored by name in *public static constants*. This means that their names don't change, and they are available to all scripts without requiring you to first create an instance of the class. For more information about instantiating classes, see Chapter 6.

Constants are also usually easy to remember and often eliminate case-sensitivity mistakes because they are comprised of all uppercase letters. To use a blend mode, use the following syntax:

```
myMovieClip.blendMode = BlendMode.DARKEN;
```

This example assumes that you are applying the blend mode to a movie clip with an instance name of *myMovieClip*. In the case of blend modes, all the constants are one word, so the hard light effect is applied this way:

```
myMovieClip.blendMode = BlendMode.HARDLIGHT;
```

Blend Modes in Practice

This section shows you how to use the Flash-specific *Layer* blend mode to solve a common problem. A sample source file, *layer_blend_mode.fla*, demonstrates the problem and solution.

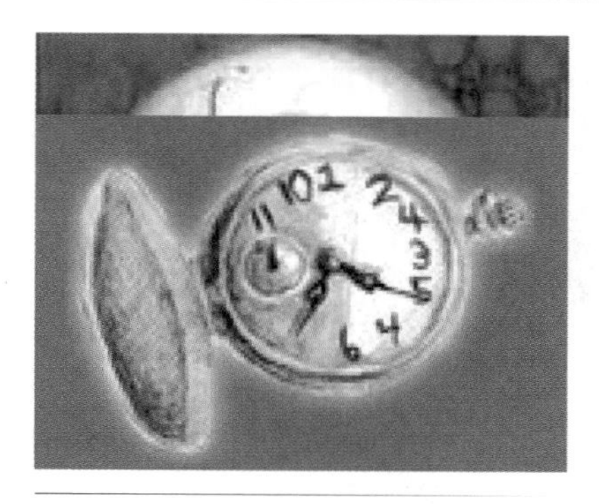

Figure 7-7. Movie clip prior to setting its Alpha property to 50%

Figure 7-7 shows a movie clip that contains two assets: an image of a pocket watch overlapping much of an image of an eyeball. If you set the *Alpha* value of that movie clip to 50%, the expected result is a movie clip with the same overall appearance, but faded to 50%.

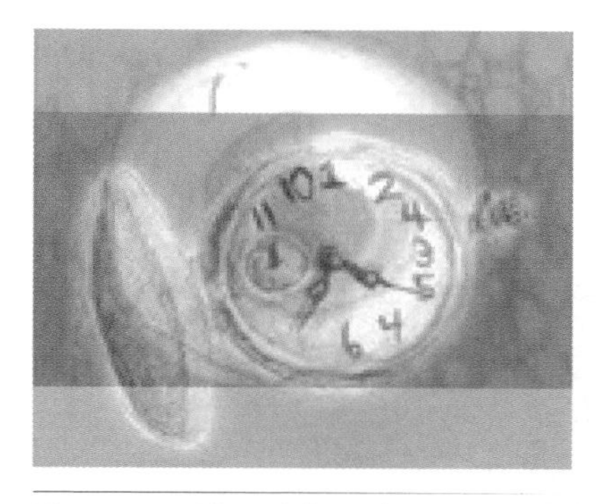

Figure 7-8. After Alpha value is applied to movie clip

Unfortunately, that's not what Flash will give you. Instead, Flash will apply the opacity change to each of the items inside the movie clip, meaning that both the eyeball and pocket watch will be independently faded to 50%. The result is shown in Figure 7-8.

If, however, you first apply the *Layer* blend mode to the movie clip, the blend mode will precompose the movie clip so it can be affected as a whole by the *Alpha* setting. Figure 7-9 shows the final result, after both the *Layer* blend mode and 50% *Alpha* setting have been applied.

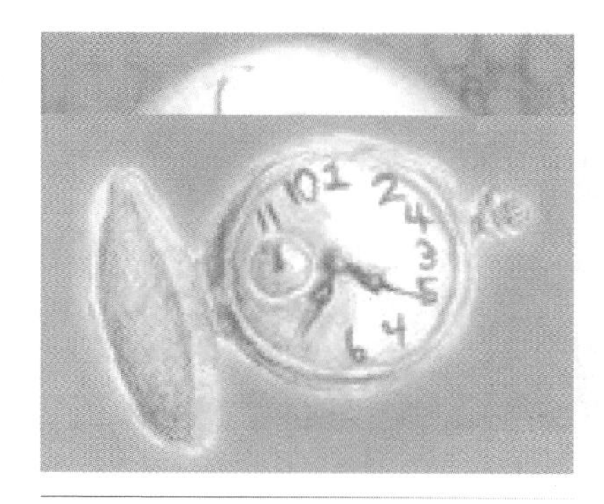

Figure 7-9. Faded movie clip after the Layer blend mode is applied

Alpha Masks

Building the sound control in Chapter 3 demonstrated that Timeline layer masks are good for showing only portions of a layer at runtime. Essentially, they are adept at cropping underlying content. However, as you'll soon see, Timeline layer masks can have a sharp edge only. Even if you use a gradient with alpha data in a mask layer, all nontransparent pixels are considered opaque in a mask layer and, therefore, soft-edge masks are impossible using this technique.

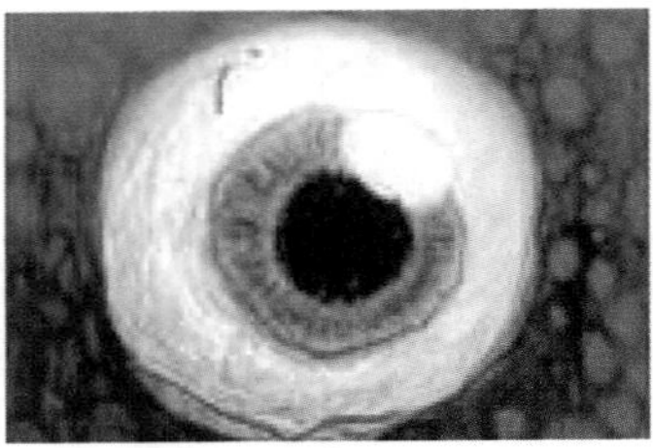

Unmasked Movie Clip

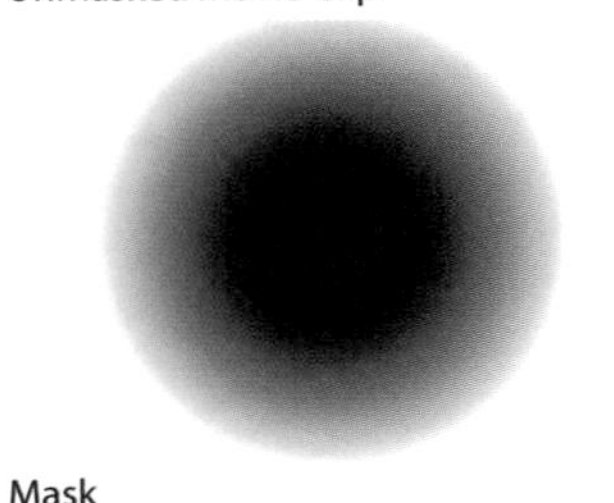

Mask

Figure 7-10. The parts used to create all masks in this chapter

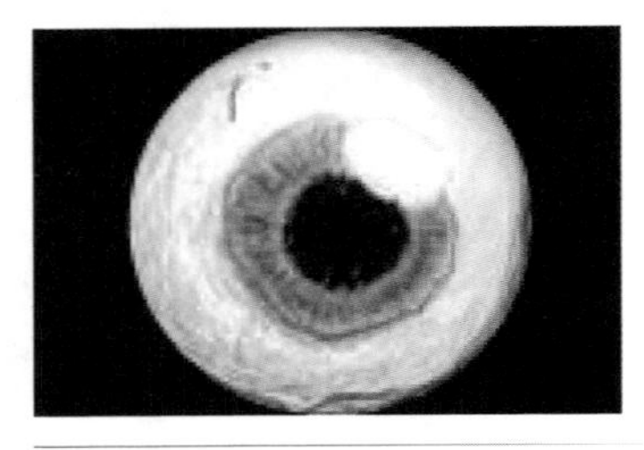

Figure 7-11. A mask created using a Timeline mask layer

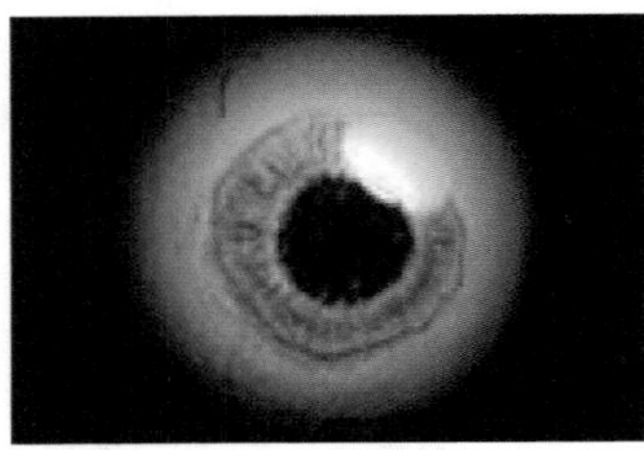

Alpha

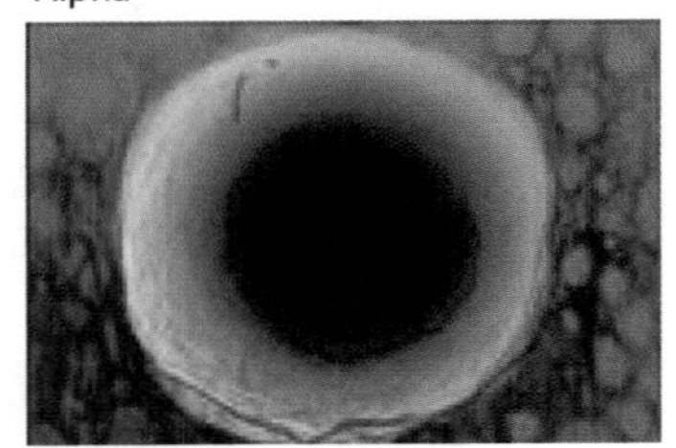

Erase

Figure 7-12. Using the Alpha and Erase blend modes, in a container to which the Layer blend mode is applied

Figure 7-10 shows the parts used for creating all the masks in this chapter. The unmasked movie clip contains a familiar orb, and the Mask movie clip is filled with a radial gradient that spans from opaque black to transparent black. The sample source files *masking_techniques_01.fla* and *masking_techniques_02.fla*, as well as the images that follow, use a black stage to make the masked eyeball easier to see.

The Timeline Limitation

In a Timeline mask layer, any nontransparent color will be included in the mask, so even gradients that contain *Alpha* values of less than 100% can't create soft edges. Figure 7-11 shows the result of a Timeline mask using the parts in Figure 7-10.

The Blend Mode Solutions

Using the *Alpha* blend mode, you can create the look of a soft-edge mask, shown in Figure 7-12. To follow along, open *masking_techniques_01.fla*:

1. Place the mask on top of the eyeball. The mask should partially cover the eyeball with black from the radial gradient.
2. Select only the mask and give it a blend mode of *Alpha*. The mask will disappear without affecting the eyeball, but that is expected behavior and is only temporary.
3. Select both movie clips and create a new movie clip (Modify→Convert to Bitmap). The mask will still be invisible and the eyeball will still appear unaffected.
4. Select the new movie clip and apply the *Layer* blend mode. The Alpha blend mode within the movie clip will take effect, and you should see the same thing shown at the top of Figure 7-12.

You can easily reverse the area of the eyeball movie clip that is masked by following the same process, but switching the internal blend mode in step 2 with *Erase*. Using the Erase blend mode gives the effect seen at the bottom of Figure 7-12.

The ActionScript Solution

Finally, you can create a soft-edge mask with ActionScript using just a few lines of code. Both the content display object (the eyeball movie clip, in this case) and the mask display object (the radial gradient movie clip) must have bitmap caching enabled. From there, the only thing that remains is to assign the mask movie clip as the value of the content movie clip's *mask* property.

```
eyeballMC.cacheAsBitmap = true;
maskMC.cacheAsBitmap = true;
eyeballMC.mask = maskMC;
```

To see the result, test your movie (Control→Test Movie). It should resemble Figure 7-13. Compare your work with the final version of the source file, *masking_techniques_02.fla*.

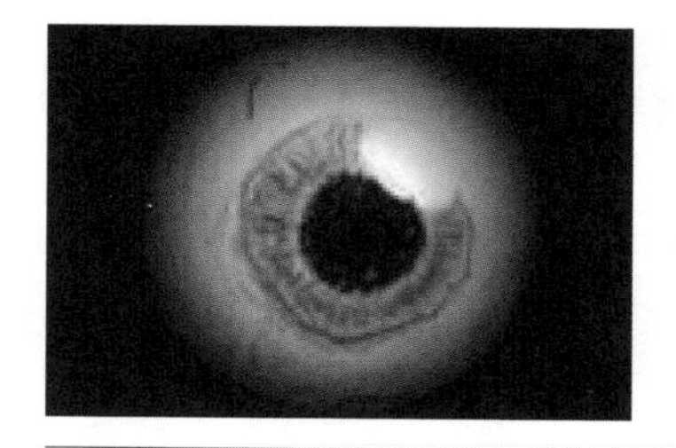

Figure 7-13. Masks created using the Timeline, blend modes, and ActionScript

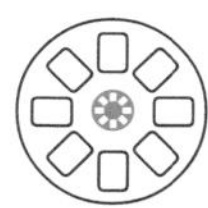

Project Progress

In this chapter, you will examine blend modes and add filter effects to the portfolio, including creating the Up, Over, and Down states of the navigation buttons.

Confirming the Imported Blend Mode

In Chapter 4, you imported the user interface shell from Adobe Illustrator. One of the features that remains editable after its move from Illustrator to Flash is support for blend modes. Happily, an *Overlay* blend mode is now applied to your project without you having to do much at all.

Figure 7-14 shows what your logo might have looked like without the blend mode applied (top) and what it looks like with the blend mode applied (bottom). In this section of Project Progress, you will confirm that the blend mode is correctly applied or apply it manually if a problem exists.

1. Open your project file. Based on previous chapters, it is likely called *portfolio.fla*.
2. Scroll your Timeline panel to frame 63 in the *logo* layer. This shows the interface after fading and zooming in, and is a likely place to find the logo.
3. Look in the upper-right corner of the Stage to locate the *logo* movie clip. If its colors are blending with the wheel below it, you can skip to the next Project Progress section. If the logo appears white, or if you want to see that the blend mode is still editable, continue.
4. Double-click the movie clip. You should switch into editing mode, editing the logo in place.
5. If necessary, lock the *black* and *red* layers, and unlock the *white* layer. This prevents you from selecting the wrong asset.
6. Click the logo to select the white layer and look in the Display section of the Properties panel. You should see that the *Overlay* blend mode has been applied during the AI import process. If it hasn't been applied, apply it now. If you made a change, save your work.

Figure 7-14. Logo before (top) and after (bottom) blend mode is applied

Filters

Next, apply Glow filters to the logo, sound control, and wheel. You'll also apply drop shadows to the navigation buttons to create visual changes in

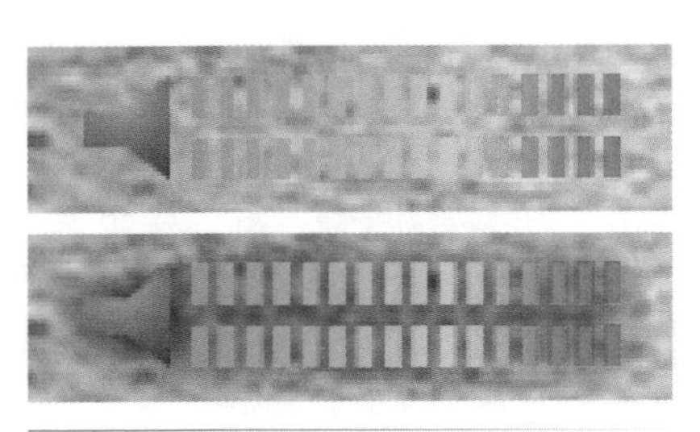

Figure 7-15. The logo and sound controls before and after a glow is applied

their *Up*, *Down*, and *Over* states. Along the way, you'll learn a little bit more about blend modes.

Make the UI glow

Adding a black Glow filter to the logo, sound control, and wheel will contribute to the illusion that they are free-floating. Figure 7-15 shows the logo and sound control elements before and after the Glow filter was applied. Figure 7-16 shows a detail of the same transformation to the wheel.

1. Return to editing the *logo* movie clip and lock all layers but *white*.
2. Select the white logo element and add a Glow filter. Use all the default values except for *BlurX* and *BlurY*. Change these values to `15`.
3. Leave editing mode (double-click a faded stage element or use the Edit→Edit Document menu command) and look at the shadow. Look closely and notice that the strength of the glow isn't very impressive and, what's more, it seems to vary. This is because a blend mode is applied to every part of an element, including any filters applied. This means that the filter is subject to the *Overlay* blend mode, too, and is blending into the underlying artwork. That's not what you want.
4. Double-click the movie clip again to edit it. In the Timeline, select the *FOLIO* layer and add a new layer. This creates a new layer above *FOLIO* and below *white*. Name this layer *glow*.
5. Select the white logo element and copy it. Then select the new *glow* layer and paste the element in place (Edit→Paste in Place). This pastes the copy in the same location as the original.
6. You now have two blend modes and two filter effects. You want one of each. Lock the *glow* layer so you can easily work with the *white* layer. Select the logo element and, in the Properties panel, select the Glow filter. Delete it using the trashcan icon button near the lower-left corner of the panel.
7. Next, lock the *white* layer and unlock the *glow* layer. Select this logo element and, in the Display section of the Properties panel, change the blend mode to `Normal`. Immediately, the glow around the logo becomes stronger. You now have one blend mode and one filter, but you're not done with the logo yet.
8. In the Filters section of the Properties panel, enable the *Knockout* feature. This leaves the glow but hides the duplicate logo element in the *glow* layer so the *Overlay* blend mode from the *white* layer will still work. Now all that's left is the text.
9. While still in the Filters section of the Properties panel, select the Glow filter and copy it using the clipboard icon button near the lower-left corner of the panel.

10. Lock the *glow* layer and unlock the *FOLIO* layer. Select the text element in this layer and, in the Properties panel, use the clipboard icon button again to paste the filter effect. The same glow from the previous steps is applied to the text.

Now that you're finished with the logo, it's time to move on to the sound control. You still have a Glow filter in the clipboard, so put it to work.

11. Leave editing mode and select the sound control. In the Filters section of the Properties panel, paste the Glow filter again. This adds the effect you used in the logo to the sound control. Both the logo and sound control should now look like the after stage of Figure 7-15.

NOTE

Remember, these filters are rendered live, not composited. So, as the sound control animates under ActionScript control (which you'll tackle in Chapter 9), the filter will update dynamically, too.

Making button states with drop shadows

The final exercise in this chapter is to add drop shadows to the navigation buttons to create button states. Start by creating the states for one button and then repeat the process carefully for the other three buttons. Figure 7-16 shows the *Up*, *Over*, and *Down* states, from top to bottom, of the Lab button.

You're using drop shadows rather than glows for a few reasons. First, it gives you experience with additional buttons. Second, the movement of the drop shadow when the button changes states is a little more obvious than adjustments to a glow filter would be, making the buttons seem a bit more responsive. Finally, the close proximity of the buttons to the background paper on which they rest makes a drop shadow more appropriate. A drop shadow was added to the button background in Photoshop and a shadow element was created during the PSD import.

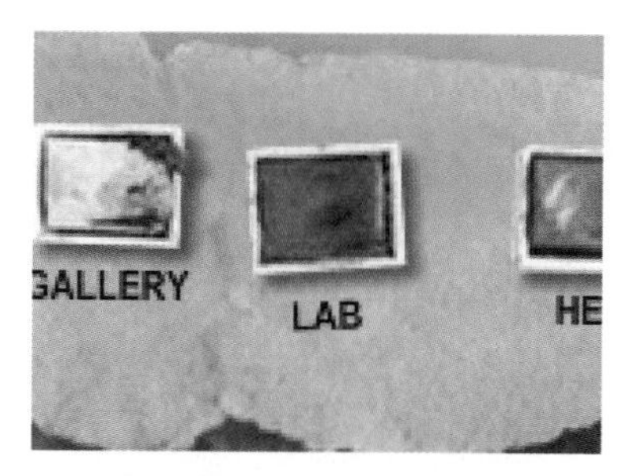

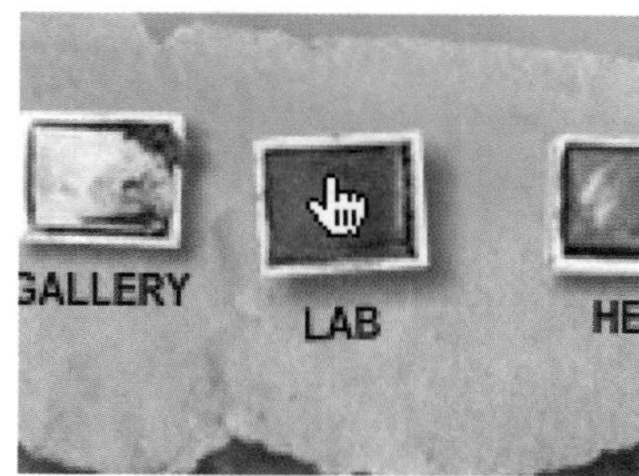

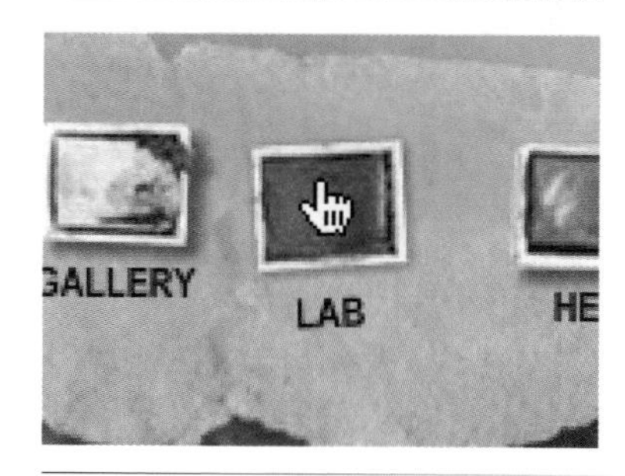

Figure 7-16. Interactive states of a button with drop shadows applied

WARNING

Unlike blend modes, Photoshop layer styles do not remain editable after a PSD import. The styles are faithfully reproduced and, depending on your import settings, you can create them as separate elements that you can later transform. However, you cannot edit them in the Properties panel.

1. Double-click the navigation movie clip to get access to the button art.
2. Select the blue movie clip and create a new button symbol (Edit→Convert to Symbol). Choose a center registration point again and name the button `labButton`.
3. Double-click the new button to edit it and select the movie clip in the Up state.

4. In the Filters section of the Properties panel, add a Drop Shadow filter. Use all the default values except for Strength (change to 50).
5. In the Timeline, click in the layer under the *Over* state and add a keyframe (Modify→Timeline→Convert to Keyframe, or F6).
6. Select the blue movie clip in this frame and change its filter values. Change the *BlurX*, *BlurY*, and *Distance* to 8. This makes the button appear to pop up when the mouse is placed over it.
7. Repeat the process described in step 5 to add a *Down* state to the button. Repeat the process described in step 6 and change the *Down* state filter values of *BlurX*, *BlurY*, and *Distance* to 5, 5, and 1, respectively. This makes the button appear to be pushed down to the background paper when clicked.
8. Save your work and test your movie (Control→Test Movie). The Lab button should have a slightly raised default state, the button should jump up a bit during mouse over, and the button should flatten to its background when clicked.
9. If the Lab button functions to your satisfaction, repeat this exercise with the three remaining buttons. If not, retrace your steps to see what went wrong or pick up with the provided source file, *portfolio_07_final.fla*.

The Project Continues...

In the next chapter, you'll rotate the logo in 3D and add easy parallax scrolling to the project.

CHAPTER 8

3D

Introduction

IN THIS CHAPTER

Three-dimensional manipulation of Flash assets is not new, but 3D has never been an integrated part of any version of the application until now.

Historically, 3D effects have been accessible to Flash developers in a variety of ways. The most basic approach to adding a 3D appearance to your projects has always been to import sequential images rendered by external 3D applications. Interactive 3D simulations became possible early on as coders began writing basic ActionScript solutions that approximated 3D with dynamic drawing techniques. Today, much more advanced 3D ActionScript packages are available from third-party developers. Several of these 3D engines contain an impressive array of 3D features, including the ability to load 3D models, control textures, lighting, and more. For your exploration, a short list of additional 3D resources is available in the upcoming "3D Outside the Box" sidebar.

Unfortunately, from a beginner's standpoint, these options are a step or two away, because their use requires intermediate to advanced ActionScript coding skills. Now the good news: for the first time, Flash CS4 Professional introduces simple 3D support available directly in the Flash interface, so even Flash neophytes can add a little 3D to their projects. When you need interactivity, you can use ActionScript, which includes the same 3D features, as well as a few more.

None of this means that Flash is now a full-fledged 3D animation or modeling application. Flash CS4's 3D capabilities are often referred to as *2.5D* because some true 3D features, such as support for models created in 3D modeling applications, are not supported. Essentially, Flash can distort movie clips in clever ways that simulate 3D. This chapter focuses on how to rotate, position, and animate movie clips within 3D space. You will also create a *parallax scrolling* effect for your portfolio project, in which objects at different 3D depths scroll at different speeds.

3D Outside the Box

Flash and ActionScript developers have been pushing the boundaries of 3D use in Flash for some time and have released several open source and commercial 3D-related products. This short list of additional resources focuses on a few of these efforts.

3D software dedicated to Flash development

Electric Rain's Swift 3D: *http://www.erain.com/*

For years, Swift3D has led the way in 3D graphics tools that specialize in Flash development. Swift3D can model 3D objects or import objects created in other 3D modeling applications, animate them, and then export compatible graphics for use in Flash. It can also export more complex 3D data specifically designed for use with ***Papervision3D***, an open source ActionScript package for rendering 3D environments.

Open source ActionScript 3D engines

Papervision3D: *http://blog.papervision3d.org/*

Sandy 3D: *http://www.flashsandy.org/*

Away 3D: *http://www.away3d.com/*

Feature-rich and ever expanding, these are free open source packages from some of the brightest ActionScript developers around. While the packages are similar in many respects, experienced 3D users will find feature differences, as well as pros and cons among the engines, so you should evaluate each to determine your personal preference. Papervision3D is the leader of the pack, with the largest feature set and best tie-ins with other 3D application developers.

Commercial ActionScript 3D engines

Alternativa Platform: *http://alternativaplatform.com/en/*

There's also a lot of interest surrounding this ActionScript package, which is free for noncommercial use but requires a license for commercial products.

COLLADA 3D format

COLLADA: *http://www.collada.org/*

Although it is a file format, for all intents and purposes, COLLADA warrants mention here because it is among the best tools for including 3D data in ActionScript 3D engines. When you're looking for 3D modeling applications, choosing one that can export COLLADA files, either directly or through a plug-in, will make it easier for you to create 3D environments in Flash. Blender, DAZ Studio, and Google SketchUp are examples of free software packages that support COLLADA. Commercial offerings include Autodesk 3D Studio Max, Maxon Cinema4D, NewTek Lightwave3D, Autodesk Maya, and Smith Micro Poser. Check the Products Directory on the COLLADA site for more information.

Moving Objects in 3D Space

The two basic ways to move through 3D space are translation and rotation. In simple terms, *translation* is movement along a specific axis and *rotation* is spinning around a specific axis. Just like the 2D axes you've worked with for several chapters, 3D space also contains an x-axis and a y-axis. Furthermore, each of these 3D axes points in the same direction as its 2D counterpart. Imagine a wall in front of you. The x-axis runs left and right. Adding to the *x* position value moves an object to the right, while subtracting from the *x* position value moves the object to the left. The y-axis runs up and down. Adding to the *y* position value moves an object down, while subtracting from the *y* position value moves the object up.

3D space, however, adds a third axis—the *z-axis*. This z-axis is used to simulate depth in a 3D world, and will be discussed in a bit more detail later in this chapter. For now, think about the idea that something can be close to you or far away from you. The z-axis runs perpendicular to your imaginary wall. Adding to the *z* position value pushes an object further away from you, into the wall. Subtracting from the *z* position value brings an object closer to you.

Switching your frame of reference to the Flash Stage, the x- and y-axes still run left-right and up-down, respectively, just like the similarly named 2D axes. The z-axis runs perpendicular to, or in and out of, the Stage. All three axes are shown in Figure 8-1.

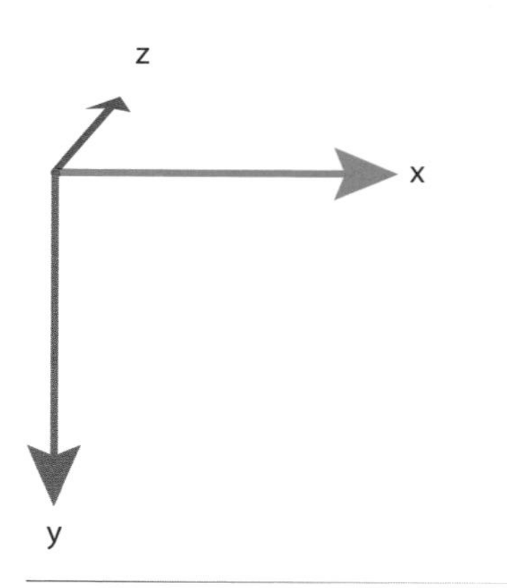

Figure 8-1. 3D axes, x, y, and z

Translating 3D Objects

Movement along a 3D axis is easily demonstrated using the *3D Translation* tool. This tool, along with the related *3D Rotation* tool, can be found near the top of the Tools panel, shown in Figure 8-2.

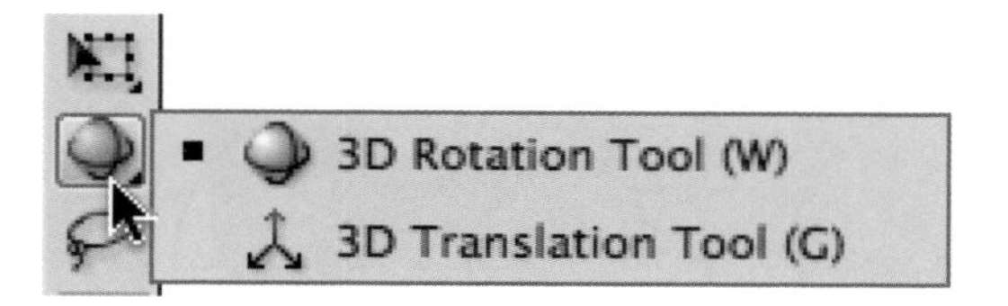

Figure 8-2. 3D tools

Using the Flash interface, you can manipulate only movie clips in 3D space. (ActionScript can also alter 3D properties of button symbol instances). To use the 3D Translation tool, click on any movie clip with the tool selected. The tool then displays 3D axes on the movie clip (Figure 8-3).

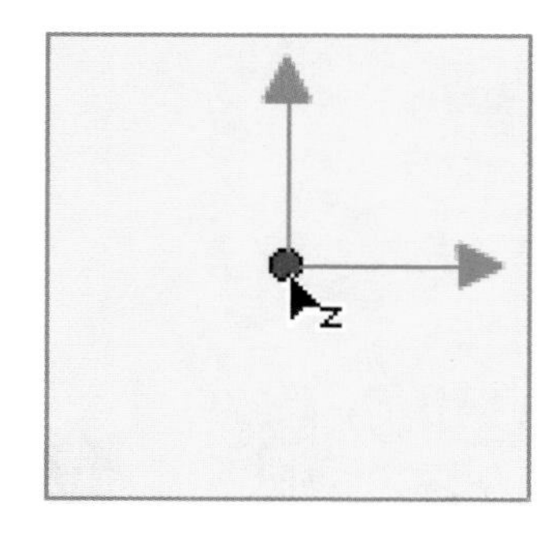

Figure 8-3. Moving an object in 3D space using the 3D Translation tool

Clicking and dragging any one of these axes constrains movement along that axis. Drag the red axis left or right to move the movie clip along the x-axis. Drag the green axis up and down to restrict the movie clip's movement to the y-axis. The blue dot at the center of the axes is equivalent to looking directly down the z-axis, as it is perpendicular to the stage. Click the blue dot and drag your mouse up and down to move the movie clip closer to or further away from your viewpoint. The movie clip will get bigger or smaller, accordingly.

It's also possible to translate movie clips in other ways. You can move in *x* and *y* directions just by dragging the movie clip with the Selection tool, or by adjusting the *x* and *y* properties of the movie clip in the 3D Position and View section of the Properties panel. To adjust the *z* value of a movie clip, change the *z* property in the Properties panel.

Rotating 3D Objects

In addition to translating movie clips along 3D axes, it is also possible to rotate movie clips around these same axes. Figure 8-4 shows black arrows that indicate rotation paths around each axis. The direction of these arrows shows how the movie clip rotates when positive values are used for the *x*, *y*, or *z* rotation properties. For example, the arrow around the z-axis indicates that positive values of the *z* rotation property spin the movie clip clockwise, while negative values spin the movie clip counterclockwise.

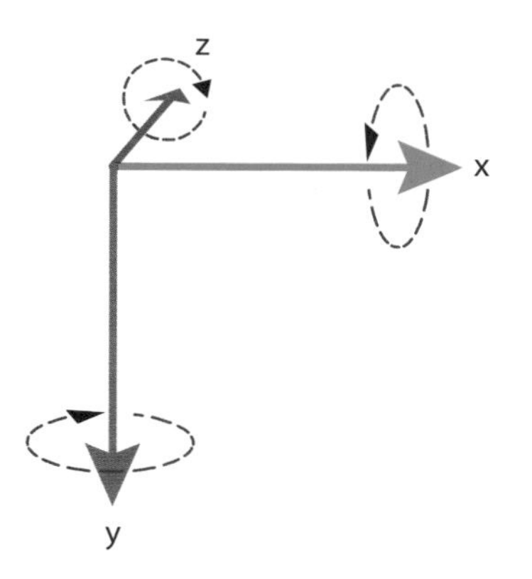

Figure 8-4. Rotation along the 3D axes, x, y, and z

You can use the Transform panel for precise numeric adjı rotation, but you can also use the 3D Rotation tool for a ı approach.

Using the 3D Rotation tool is somewhat similar to using th tool. When you click a movie clip, an interactive tool interfa center of the movie clip. A bit more information must be coı rotation along any of three axes, however, so the interface fo tool is a bit more detailed. Figure 8-5 shows four represent and Figure 8-6 shows the effect of each corresponding m movie clip.

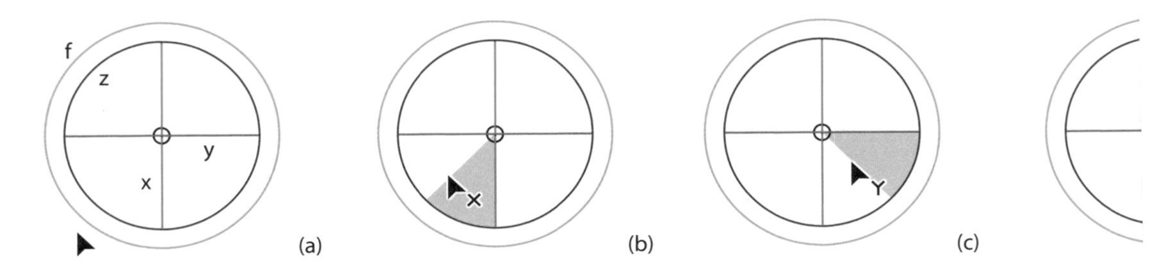

Figure 8-5. Feedback provided by the 3D Rotation tool when rotating space

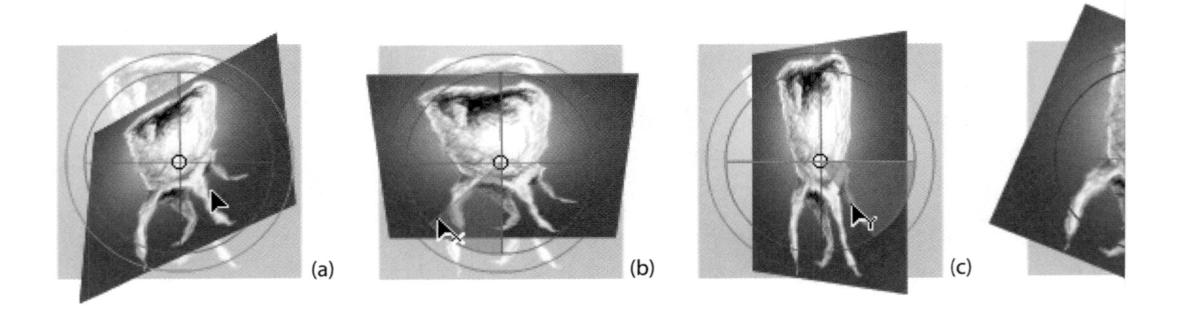

Figure 8-6. Rotating an image around each 3D axis, and freely

The labels in Figure 8-5(a) show that the red line controls ro x-axis, the green line controls rotation around the y-axis, aı circle controls rotation around the z-axis. A cursor is also pr orange circle in the lower-left corner of the illustration. T controls *free rotation*, or rotation that is not constrained alor Using this feature, you can drag your mouse around and ma *z* rotations simultaneously, as shown in Figure 8-6(a). Wh more immediate freedom of movement, it is more difficult ı your movie clip to a particular orientation.

For increased control, you can rotate around each axis individually. For example, Figure 8-5(b) shows rotation only around the x-axis. When you click the red line and drag it in a clockwise direction, the *x* rotation value increases and the image rotates toward you, shown in Figure 8-6(b). Figures 8-5(c) and 8-6(c) show positive rotation along the y-axis, and Figures 8-5(d) and 8-6(d) show positive rotation along the z-axis.

A gray wedge indicates the starting and stopping points of the rotation angle, allowing you to estimate the degree of rotation without consulting a panel. Because a circle consists of 360 degrees, a wedge that appears to occupy one quarter of the circle represents approximately 90 degrees of rotation along one axis (holding down the Shift key while using the 3D Rotation tool to snap rotation to 45-degree increments, making it easy to reach exactly 90 degrees, if desired).

The wedge display assists with *x*, *y*, and *z* rotation. However, because free rotation can include changes along any axes, the wedge cue is not helpful and is omitted for that tool adjustment.

WARNING

One thing to bear in mind when using this visual feedback is the way Flash handles rotation angles. Flash improves performance by using positive and negative rotation angles. That is, 270 degrees is equivalent to –90 degrees, and it's more efficient for Flash to use –90. As such, Flash sees rotating from 0 to 360 as rotating from 0 to 180 and then from –180 to 0. The wedge feedback in the 3D Rotation tool behaves the same way. When it becomes a semicircle (having traveled 180 degrees), the wedge flips to the other side of the tool and begins to diminish as you approach 0.

Transforming Multiple Objects in 3D Space

Thus far, this chapter has focused on transforming single objects in 3D space. However, you can also manipulate multiple objects using the 3D Translation and 3D Rotation tools. For example, Figure 8-7 shows two movie clips rotating around the y-axis at their mutual center.

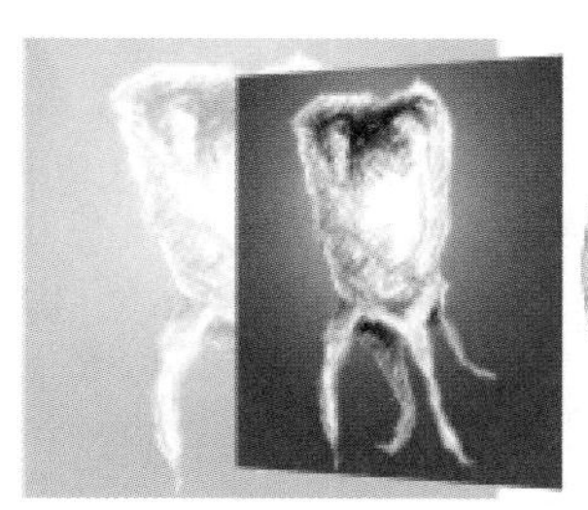

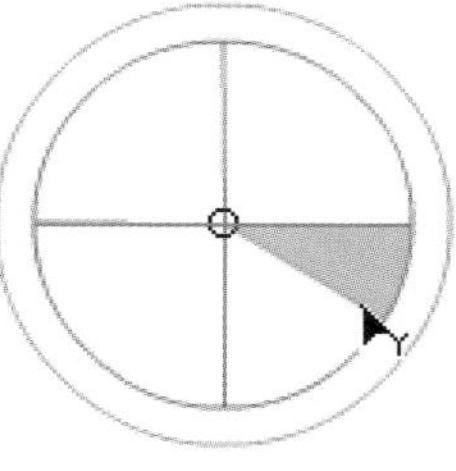

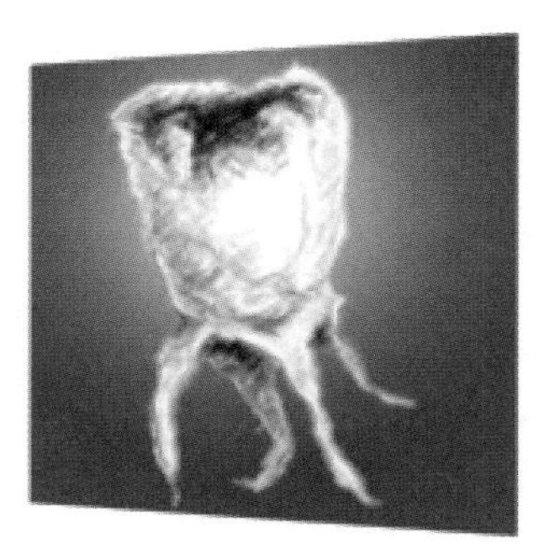

Figure 8-7. Rotating multiple objects in 3D space

NOTE

Once a movie clip has been moved into 3D space by a 3D transformation, some 2D properties no longer apply and can't be edited in the Properties panel.

To transform more than one movie clip in 3D space, you must first select all instances that you want to transform (by clicking with the Shift key pressed or by dragging over the objects) and then select the 3D Translation or 3D Rotation tool. Thereafter, all selected objects will be moved or rotated as a group. To rotate the objects around a common center point, double-click in the center of the chosen 3D tool, as described in the next section.

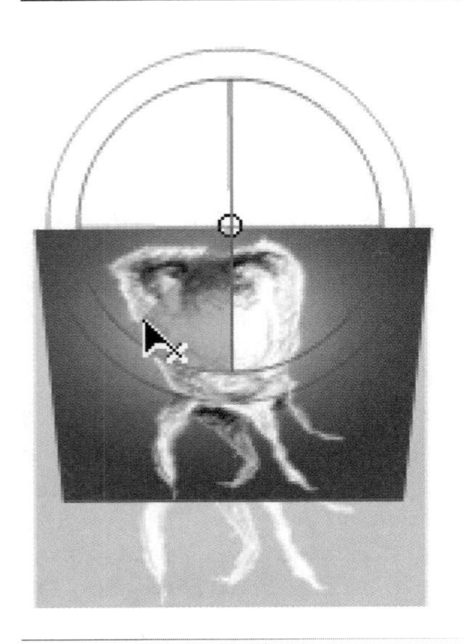

Figure 8-8. Rotating an object around the x-axis after changing its 3D center point

Changing the 3D Center Point

Just as you can change the 2D transform point using the Free T
you can also change the location of the 3D center point. Wh
movie clip with the 3D Rotation or 3D Translation, the whit
center of either tool's on-Stage interface is the 3D center point.
point, simply drag around the white circle.

Figure 8-8, for example, shows the 3D center point moved tc
movie clip. By rotating the movie clip around the x-axis at i
point, the movie clip appears to swing from its top—much lik
from a support outside a dentist's office.

To reset the 3D center point to the center of a selected movie
click on the white circle.

You can also change the 3D center point using the Transform pa
may be quicker and more intuitive to drag the center point aroun
y-axes on Stage using the 3D Rotation or 3D Translation tool, th
panel adds the ability to easily move the 3D center point along th

Global Versus Local 3D Transformations

By default, 3D rotations and translations are made in global spac
all movie clips with 3D transformations share the same global co
system. That is, when using global transformations, you are manip
a movie clip relative to the Stage. However, it's also possible to trans
movie clip in local 3D space using coordinates that are relative to its p
movie clip or, when no parent exists, the main timeline.

For example, imagine a movie clip on the Stage. Adjusting its global *z* position would move the movie clip perpendicular to the Stage, as described earlier in the "Translating 3D Objects" section. Now imagine rotating the movie clip 90 degrees on the y-axis. The global z-axis is still perpendicular to the Stage, but the movie clip's local z-axis, after the 90-degree rotation, is parallel to the Stage. Adjusting the movie clip's global *z* position still moves the movie clip in and out of the Stage. Adjusting its local *z* position, however, moves the movie clip left and right.

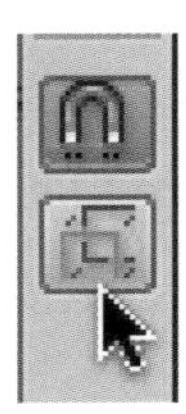

Figure 8-9. Rotating an object in local 3D space

To toggle between local and global 3D transformations, use the Global Transform button at the bottom of the Tools panel when a 3D tool is active (Figure 8-9). When you enable this option, movie clips will be transformed relative to the Stage. Otherwise, local coordinates will be used for transformations.

The visual feedback provided by the 3D tools updates to reflect this change. Previous figures, including Figures 8-3 and 8-5, show the global 3D tool on-Stage interfaces, while Figure 8-10 shows the local interfaces. The 3D Rotation tool (top) acquires a sphere-like quality, and the orientation of the 3D Translation tool axes (bottom) change with the orientation of the movie clip. Following the preceding z-axis discussion, note the blue z-axis in the bottom of Figure 8-10. Rather than a blue dot that points directly into the Stage, the local translation mode shows the z-axis clearly pointing somewhere behind you to the left.

Switching between global and local transformations when adjusting the location of a movie clip can make it much easier to set its position more accurately in 3D space.

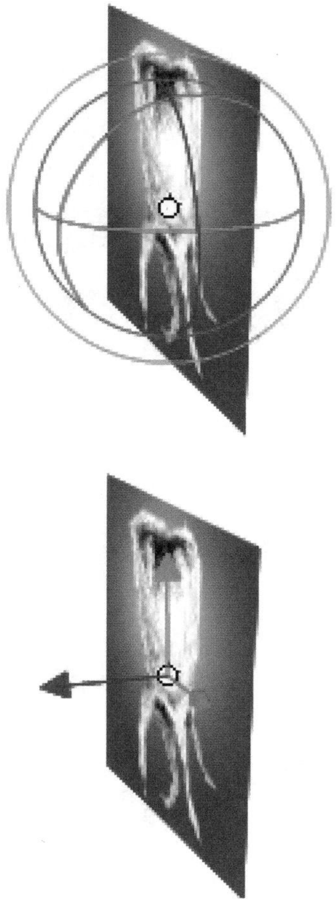

Figure 8-10. Rotating (top) and translating (bottom) a movie clip in local 3D space

Global 3D Environment Settings

Two important global 3D settings affect the way all 3D objects are rendered. The first is *vanishing point*, which determines the orientation of the z-axis. The second is *perspective angle*, which determines the viewer's field of view, much like a camera lens. You can adjust both settings in the 3D Position and View section of the Properties panel while a movie clip is selected. In Figure 8-11, a camera icon marks the perspective angle property and an icon of converging lines marks the vanishing point property.

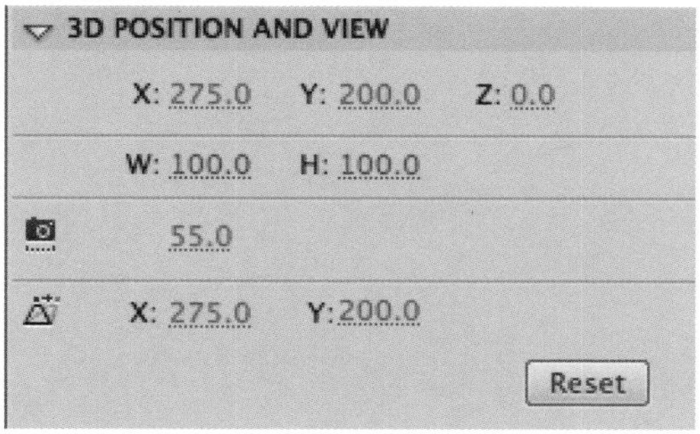

Figure 8-11. The default perspective angle (camera icon) and vanishing point (converging lines icon) settings in the Properties panel

Vanishing Point

The vanishing point is the point on the Stage to which the z-axis is aligned. That is, if you pushed a movie clip along the z-axis, it would approach the vanishing point. It is called the vanishing point because objects reduce in scale as they move farther away from the viewer, become smaller and smaller as they approach the distant point, and eventually vanish from view. In scenes with horizon lines, for example, such as looking down a highway running through the desert, the horizon usually goes through the vanishing point.

The default vanishing point is the center of the Stage. Figure 8-12 illustrates the point by showing lines emanating from all four corners of a movie clip, converging on the vanishing point.

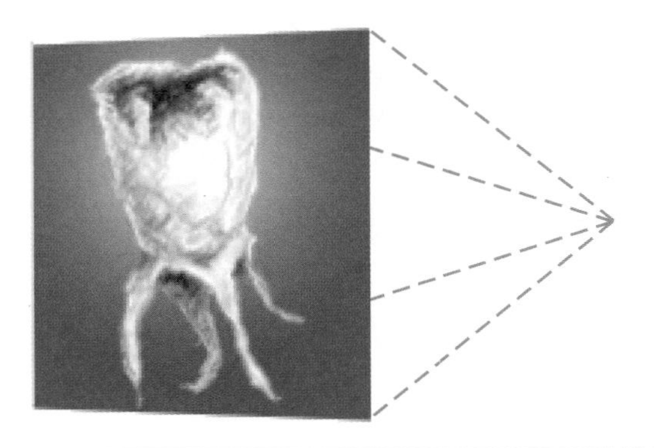

Figure 8-12. The default vanishing point, in the center of the Stage

You can change the global orientation of the z-axis in a 3D world, and thus the appearance of the scene, by changing the vanishing point. Figure 8-13 shows a before-and-after view of a vanishing point change. The original vanishing point is still depicted using purple dashed lines. Orange dashed lines indicate the new vanishing point.

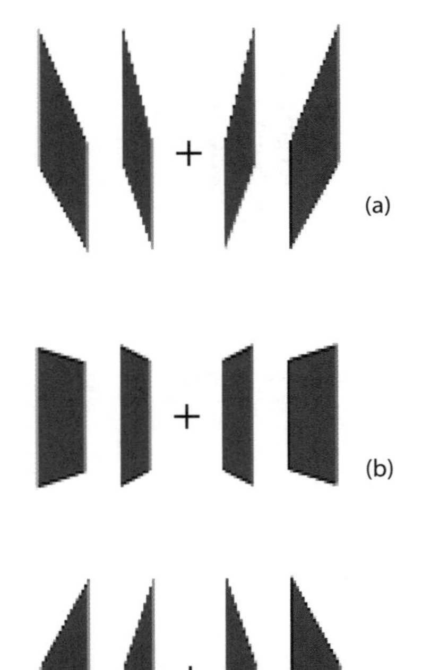

Figure 8-14. Three vanishing points applied to a 550×400 Stage: point (275, 350) (a), point (275, 200) (b), and point (275, 50) (c).

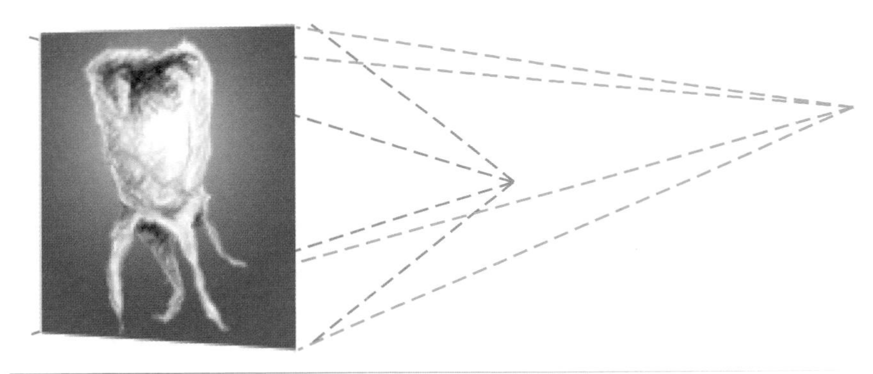

Figure 8-13. Changing the vanishing point to point (410, 170)

Note the change in the movie clip's appearance. It appears to have rotated a bit more on the y-axis. You can see where the previous corners of the movie clip were by looking at the ends of the purple dashed lines. In actuality, the movie clip's rotation has not changed. Instead, the visual update was caused by the shift in the vanishing point.

You can use the vanishing point to great effect to alter the viewpoint of a camera or viewer. For instance, Figure 8-14 shows three different examples of vanishing points applied to a 550×400 Stage. Figure 8-14(a) shows a vanishing point near the bottom of the Stage. This orients objects downward and gives the impression that the viewer is below ground, perhaps looking up at the ceiling. Figure 8-14(b) shows a vanishing point in the center of the Stage. This vantage point makes the viewer feel he or she is at ground level. Finally, a vanishing point at the top of the Stage, shown in Figure 8-14(c), implies that the user is aboveground, maybe looking down from a tall tower.

Perspective Angle

Choosing a perspective angle for a 3D scene is similar to choosing a lens for a camera. If you need to include a wide field of view in a picture, you will need a wide-angle lens. If you need a narrower field of view, to focus on something farther away, you will need a telephoto lens. A normal, or standard, lens falls somewhere in between.

Because different angles of view are optimized for different distances of viewing, changing the angle can often change the perspective of objects seen through a lens. For this reason, the field of view in a virtual 3D camera is often referred to as the *perspective angle*.

Figure 8-15 shows three examples of perspective angles using three rows of movie clips. In each example, four vertical rectangular movie clips are displayed with a 3D *y* rotation of 90 degrees. In the horizontal center of the screen, they would be invisible, as you would be looking at the very edge of the movie clip. Because the vanishing point is in the center of the four movie clips, you can see more of the sides of the rectangles that are farther to the left or right of center.

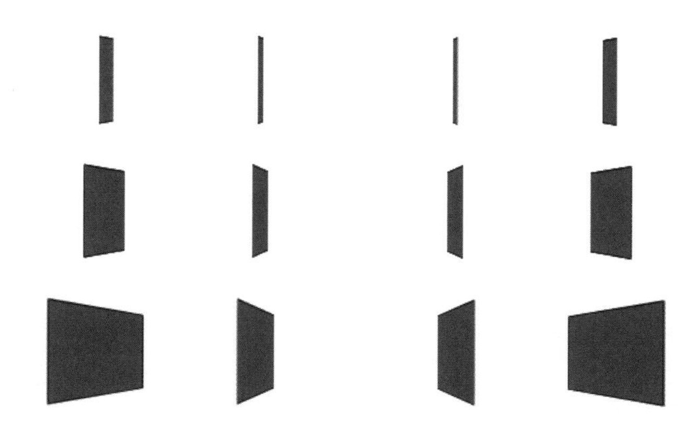

Figure 8-15. Four movie clips displayed using perspective angles of 15 (top), 55 (middle), and 105 (bottom) degrees

The top row of Figure 8-15 shows a perspective angle of 15 degrees. This is equivalent to a 100mm telephoto lens. The middle row shows a perspective angle of 55 degrees. This is Flash's default perspective angle and is equivalent to a standard 35mm lens. The bottom row shows a perspective angle of 105 degrees, which is equivalent to a 20mm wide-angle lens.

NOTE

Perspective angles in camera lenses can be measured horizontally, vertically, or diagonally, yielding different values. When making comparisons between Flash's perspective angle and camera lenses, it is helpful to consider the lenses' horizontal view angle values.

Changing the perspective angle when changing the stage size

It's important to remember that the perspective angle affects all 3D transformations in your file, and changing the Stage size can change the way objects are affected by the angle. Making the Stage much wider, for example, without changing the perspective angle, can make objects at the edges of the Stage seem shallower in the field of view.

Flash can automatically compensate for this effect if you change the Stage size. If your file includes 3D transformations and you need to change the Stage size (Modify→Document), enable the *Adjust 3D Perspective Angle to preserve current stage projection* option. Flash will automatically calculate a new perspective angle, based on the new stage size, which will preserve the appearance of your 3D objects.

Taking Advantage of Global 3D Settings with Movie Clip Containers

Because perspective angle and vanishing point settings affect all 3D objects, and because global 3D transformations are based on these values, Flash intelligently updates the appearance of every 3D object when any of these values change. You can take advantage of this by placing related 3D objects into container movie clips. By adjusting the location of the container movie clip, all its 3D children also update.

Figure 8-16 shows a movie clip with three 3D children being dragged up and away from the vanishing point. Changes caused by the leftward movement are evident in the horizontal skewing. In particular, the left square is becoming broader and the right square is becoming shallower. The upward movement is affecting the vertical skewing. The angles at all corners are more acute, and the center object appears flatter.

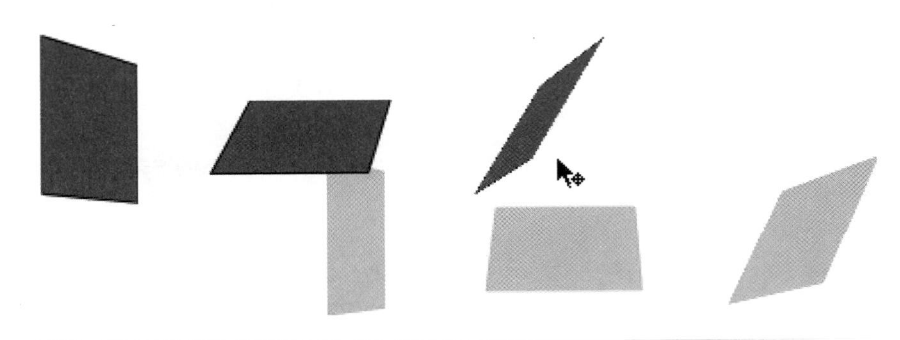

Figure 8-16. Moving a movie clip with children in 3D space

The basic example shown in Figure 8-16 can't convey the full impact or simplicity of manipulating 3D children just by positioning the parent container.

Any rotated or translated child will be affected by a change to its parent. If a child appears small and far away, for example, due to a positive *z* position, rotating the parent 180 degrees automatically brings that child far into the foreground. You'll see this effect of rotating children with multiple depths when you learn more about parallax scrolling later in the chapter.

The important thing to take away from this discussion is that you don't need to select objects as a group, reposition the 3D center point, or switch between global and local transformations using this technique. Once the children are positioned where you want them inside the parent container, all you need to do is transform the parent, and you're done.

Animating 3D Properties

One of the best parts of having 3D properties integrated into the Flash CS4 Professional interface is that you can animate them as easily as you can animate 2D properties—even in tweens. Figure 8-17 shows a movie clip rotating 180 degrees around the x- and y-axes (the faded frames have been inserted in the illustration to show you the progress of the tween).

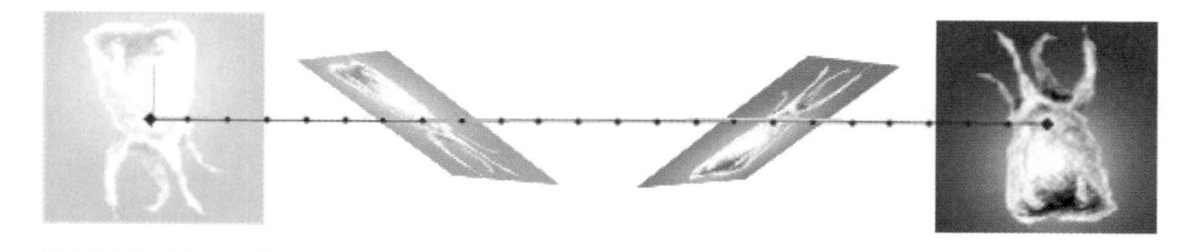

Figure 8-17. An animation rotating an image 180 degrees around the x- and y-axes

You can use the Motion Editor panel (discussed in Chapter 5) to achieve this animation effect. Before applying the 3D transformations, set up a simple motion tween that moves the movie clip from one side of the stage to another:

1. Open *tween3d_01.fla* from the companion source files. This file has one movie clip on the Stage at point (70, 200).
2. Right-click (Windows) or Control-click (Mac) the movie clip and select Create Motion Tween from the pop-up menu. 23 frames are added to the tween span, and the playhead should be in frame 24. Right-click or Control-click again and enable 3D Tween from the same menu.
3. Select the movie clip on the Stage with the Selection tool and, using the Properties panel, change its *x* position to `480`.

As you did in Chapter 5, you have completed a 2D motion tween, and you can see the path taken by the movie clip. You can even scrub through the Timeline to watch it move. If you like, compare your work to *tween3d_02.fla*, then move on to add the 3D transformations.

4. Open the Motion Editor and scroll down to the *Rotation X* and *Rotation Y* properties. As usual, the first keyframe in any tween is provided.

5. Scroll horizontally until you can see frame 24. Click the frame numbers in the Motion Editor panel to place the playhead in this frame.

6. In the Keyframe column, click the center diamond to add a keyframe in the *Rotation X* and *Rotation Y* rows of the Motion Editor.

7. In the same frame (frame 24), use the Value column to set both properties to **180** degrees. Your Motion Editor panel should look like Figure 8-18. The graph for both *Rotation X* and *Rotation Y* properties should slant up and to the right between frames 1 and 24. Figure 8-18 shows the playhead midway through the tween, when both rotation values have just passed 90 degrees.

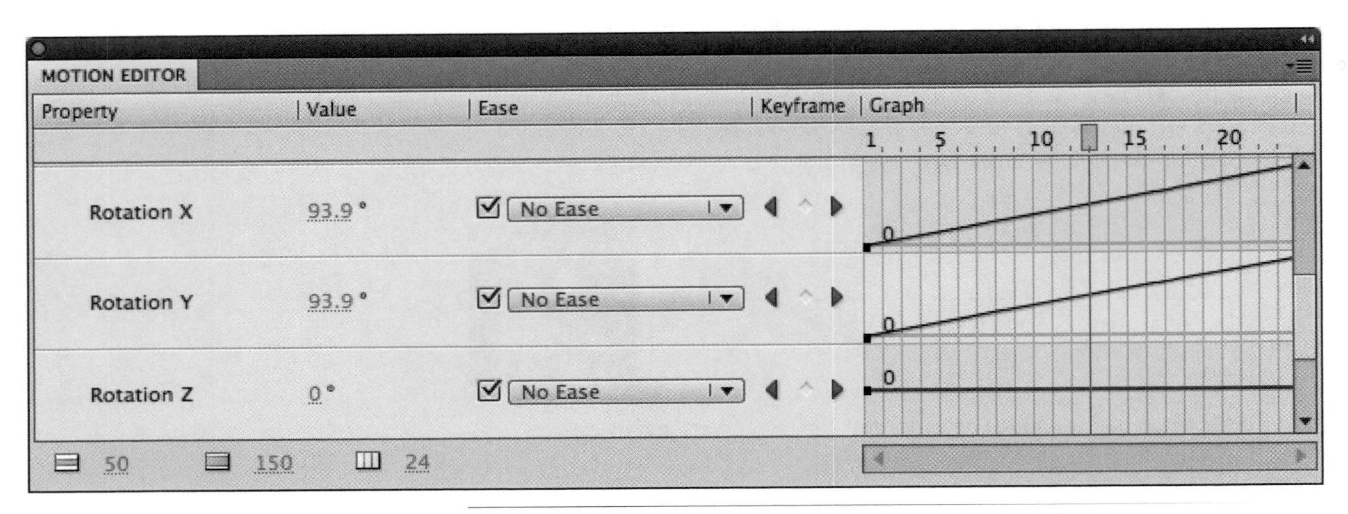

Figure 8-18. The Motion Editor showing x, y, and z rotation values

Save your work if desired (it will not be used in the portfolio project) and test your movie. Your movie clip should spin around both x-and y-axes as it moves across the Stage, as depicted in Figure 8-17 and demonstrated in *tween3d_03.fla*.

Based on this experience, you can probably see the benefit of having Flash's 3D features—basic though they may be—integrated directly into the authoring interface. While you certainly will have more control over 3D objects when using ActionScript, at least you can accomplish a few simple effects without knowing a lick of code.

Depth Management

There are drawbacks, however, with Flash's integrated 3D feature set. The biggest shortcoming is the lack of automatic *z-sorting*, or sorting movie clip stacking order based on the value of their *z* positions, called *z-depth* in most z-sorting discussions. Because the z-axis represents depth in the 3D world, z-sorting restacks elements when they overlap, based on their depth along the z-axis.

This is easily demonstrated by placing your hands in front of your face. Hold both hands up about a foot or so in front of your face, and about a foot apart from each other. This represents two movie clips with the same *z* position, or depth. Now move your right hand another foot away from your face. Your right hand is now farther away, or has a larger *z* value. Now move your hands horizontally so they overlap in your line of vision. Your left hand is in front of your right hand because its depth is closer to your point of view. It would obviously be wrong if your right hand were in front of your left, because your right hand is further away.

Unfortunately, Flash doesn't handle the z-sorting for you. If you place one movie clip each in layers 1, 2, and 3, no change in the z-depth of any of the movie clips will change the visual stacking order imposed by the layer in which each appears. This limitation doesn't apply only to layer order, either. If you set up the same three movie clips correctly in the same layer, so the stacking order matches your desired z-depths, everything will be fine. If you then select the frontmost movie clip and send it behind the others (Modify→Arrange→Send to Back), the z-depth will remain the same, but visually, the movie clips will overlap in the wrong order.

Figure 8-19 illustrates this point. Figure 8-19(a) shows three movie clips in the same layer, stacked in an arrangement that matches their z-depths. Figure 8-19(b) shows three similar movie clips, each in its own layer, but layered in a manner that does not reflect their z-depths. Even though the movie clip in layer 4 has the highest z-axis value, it does not appear behind the other movie clips, because it resides in a higher layer. Finally, Figure 8-19(c) shows a similar arrangement, but this time the movie clips are in the same layer and their stacking order doesn't match their z-depths.

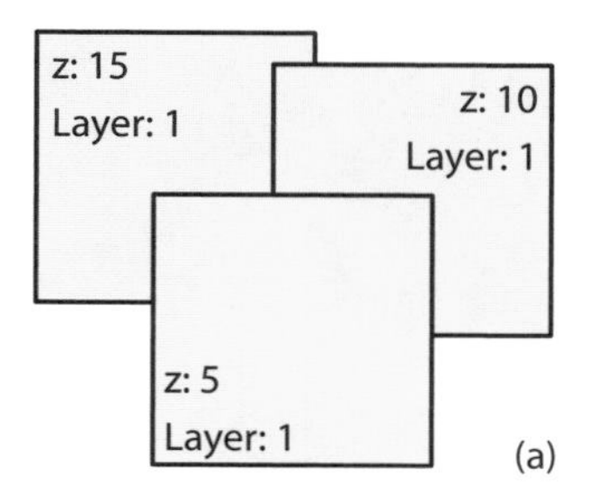

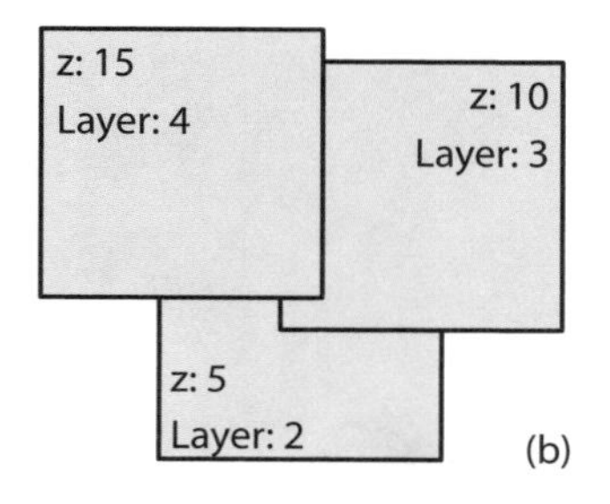

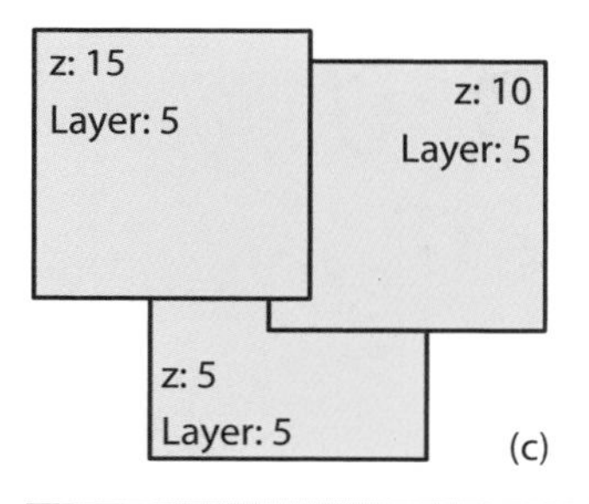

Figure 8-19. z-depth, correct in example (a), does not override layer order (b) or conflicting single-layer stacking order (c)

Unfortunately, there are only two ways to conquer this problem. The first way is to manually manage your stacking and layering orders so they don't conflict with the desired depths of your 3D objects. This is fine for work in the Timeline, but not helpful for dynamic or interactive scenarios. In these cases, you must use custom ActionScript to sort and reorder your display objects.

In the "Using ActionScript to Change 3D Properties" section, later in this chapter, you will learn how to explicitly set the *z* value of any movie clip. However, maintaining a running inventory of all movie clips, and properly sorting their depths based on when they overlap, requires quite a bit more effort and is beyond the scope of this book. The book's companion website, however, has more information on this topic, including links to solutions from top-notch coders. As your ActionScript skills improve, you may be able to use a script prepared for this purpose, and integrate it seamlessly into your project—or even write your own solution!

In the meantime, remain alert to awkward z-ordering and try to compensate by shuffling the layer or stacking order of your assets. When motion tweens require depth changes, impossible to achieve in a single tween, you can usually adjust your work by splitting your tweens into multiple segments, each in its own layer and using layer order to provide better depth management. This technique, too, is covered on this book's companion website.

Parallax Scrolling

Parallax scrolling is an effect that shows the appearance of depth during movement. The principle behind parallax scrolling is simple: as you move past objects at different depths, closer objects will seem to rush by faster.

You've probably experienced this effect yourself on a road trip or train ride. Imagine you are sitting in a window seat in a car or on a train as it embarks on a trip across the countryside. Outside the window you see rocks at the edge of the road or track. Some distance beyond the rocks is a row of bushes, and further beyond the bushes is a row of trees. Finally, far off at the horizon is a mountain range.

Figure 8-20. Four individual layers intended for a scrolling movie clip

As the car or train picks up speed, the rocks rush past your window. Although moving quickly by, the bushes don't seem to be traveling as fast as the rocks. The distant trees are moving slower still, and the mountains appear to be stationary.

This effect is also used in 3D to give the illusion of depth. Figure 8-20 shows the rocks, bushes, trees, and mountains previously described. If you placed these elements in their own layers in the Flash Timeline and animated them in a normal 2D manner, they would all scroll at the same time. However, if you push each element further back along the z-axis, all four elements will reside in different depths and scroll by at different speeds.

Figure 8-21. Scrolling assets, after z-translation is applied; this clip is rotated on the y-axis to illustrate parallax effect

Figure 8-21 shows such an arrangement, but rotated in 3D space so you can clearly see the depth between layers. Because of scaling along the z-axis, objects closer to the viewer are wider than the smaller objects in the distance. As such, when all layers scroll by the viewer's eye, the foreground layers move faster than the background layers.

You will use this technique for the Gallery screen of the portfolio project. The foreground elements will eventually be components (added in the next chapter) that will load in external assets. The graphics behind the component layer will scroll through the gallery at different speeds, giving the illusion of depth.

Using ActionScript to Change 3D Properties

Manipulating 3D objects with ActionScript is, in many ways, as simple as transforming display objects in two dimensions. Using a handful of properties, you can do just about everything the 3D Flash interfaces can do. For example, translating a movie clip in 3D space uses *x*, *y*, and *z* properties. Assuming you're transforming a movie clip with the instance name *mc*, here are examples of all three properties:

```
mc.x = 275;
mc.y = 200;
mc.z = 100;
```

Look familiar? It should. The *x* and *y* properties are the same properties you've been using for 2D manipulations, and the new *z* property is consistent in use.

Rotation is not much different. In 2D space, there is only one rotation property. In 3D space, the three axes around which a movie clip can rotate require three properties, but the syntax is still similar. Instead of the single *rotation* property for 2D space, the 3D properties are:

```
mc.rotationX = 10;
mc.rotationY = -20;
mc.rotationZ = 5;
```

Even the perspective angle and vanishing point properties are easy to use, albeit a bit verbose. The ActionScript properties themselves are straightforward enough—perspective angle is called *fieldOfView* and vanishing point is called *projectionCenter*, but they are used like this:

```
root.transform.perspectiveProjection.fieldOfView = 25;
root.transform.perspectiveProjection.projectionCenter = new Point(0,0);
```

Changing these values requires a bit more syntax because they are part of a larger transformation object. They're collected in an object called *perspectiveProjection* (which also contains other 3D transformation properties) that is part of the more encompassing *transform* instance of applicable display objects, like movie clips.

In the preceding example code, the transform object belongs to the *root* of the FLA's display list, the main timeline, which itself is a big ol' movie clip. This change affects all global transformations, just as you witnessed when making changes to the perspective angle and vanishing point controls in the Properties panel.

NOTE

As with the Properties panel, the perspective angle is measured in degrees, and the vanishing point unit is a point.

Practical Demonstrations

Here are a few examples of the ActionScript at work. A companion source file is provided to demonstrate each property. In all cases, a single frame event listener links mouse movement with a 3D property that manipulates a movie clip with an instance name of *mc*.

Rotation

The *as_3d_rotation.fla* example sets the *y* rotation of a movie clip to the horizontal mouse position. As you move your mouse left and right, it looks like you're spinning a placard on a nail.

```
1   this.addEventListener(Event.ENTER_FRAME, onLoop);
2   function onLoop(evt:Event):void {
3       mc.rotationY = mouseX;
4   }
```

Translation

The *as_3d_translation.fla* file will set a movie clip's z-depth to the mouse's vertical location (0 to 400)—moving the movie clip into and out of the Stage.

```
1   this.addEventListener(Event.ENTER_FRAME, onLoop);
2   function onLoop(evt:Event):void {
3       mc.z = mouseY;
4   }
```

Vanishing point

The *as_3d_vanishing_point.fla* file will move the vanishing point to wherever the mouse is, causing objects to reorient in 3D space. First, the location of the mouse is converted into a point (line 7), and the *projectionCenter* property is then set to that point (line 8). To see the effect work, 3D rotations have been applied to the movie clip in lines 1 through 3.

```
1   mc.rotationX = 10;
2   mc.rotationY = -20;
3   mc.rotationZ = 5;
4
5   this.addEventListener(Event.ENTER_FRAME, onLoop);
6   function onLoop(evt:Event):void {
7       var mouseLoc:Point = new Point(mouseX, mouseY);
8       root.transform.perspectiveProjection.projectionCenter =
    mouseLoc;
9   }
```

Perspective angle

The *as_3d_perspective_angle.fla* source file will change the field of view using the mouse's vertical location. Because the *mouseY* value is divided by 4, and the height of the stage is 400, the initial value is between 0 and 100. However, the perspective angle cannot be 0, so 40 degrees is added to the initial value. The result is a demonstration of perspective angles between 40 degrees (a normal lens) and 140 degrees (an extra-wide-angle lens). The movie clip is also translated and rotated to enhance the effect.

```
1    mc.x = 140;
2    mc.rotationY = -45;
3
4    this.addEventListener(Event.ENTER_FRAME, onLoop);
5    function onLoop(evt:Event):void {
6        root.transform.perspectiveProjection.fieldOfView =
     mouseY / 4 + 40;
7    }
```

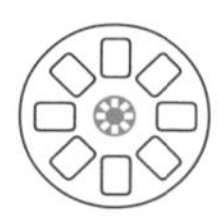

Project Progress

In this chapter, you will create a parallax scrolling effect that will serve as the interface for the gallery screen. The gallery interface will consist of three components added in the next chapter. Components are precreated combinations of code and graphics designed to complete tasks of varied complexity, but with little to no additional ActionScript provided by you. The gallery will ultimately present two external graphics and one video, and visitors will scroll between the three displays with the click of a button.

For this exercise, you will need a provided source file, *gallery_01.fla*. This file differs slightly from a new FLA derived from the book template you've used in prior chapters. In this case, the art required for the gallery interface is already in place. The illustrations, shown in Figure 8-22, are typical 2D movie clips, and it will be your job to manipulate these assets in 3D space.

Figure 8-22. Scrolling assets before z-depth is applied

Inside the content movie clip, renamed to *GalleryScreen* in the Library and given an instance name of *gallery*, are the parts required to create a parallax scrolling interface. The content of this movie clip has three layers. Two layers are static and contain an image of a mountain range with a grassy plain and a button that will advance through the gallery. The third layer is a movie clip called *parallax*, and this will be the focus of your attention.

In the *parallax* movie clip are four layers. The background is a movie clip of trees, substantially wider than the stage and thus suitable for horizontal scrolling. Above the trees are two movie clips containing bushes and rocks, respectively. Finally, the foreground movie clip contains artwork of three frames, each labeled with the name of a component that you will add in later chapters.

Adding Depth for Parallax Scrolling

To prepare your movie clip for parallax scrolling, you will adjust both the z-depths of layers in the movie clip and the position of the vanishing point:

1. Open *gallery_01.fla* from the companion source files.
2. Double-click the *gallery* movie clip to edit its contents.
3. Double-click the *parallax* movie clip to edit its contents.
4. Ignoring the foreground layer that contains the artwork of three empty frames (which will remain unchanged until next chapter), select the movie clip in the *trees* layer and give it a *z* value of `1000` in the Properties panel. Select the movie clip in the *bushes* layer and set its *z* value to `600`. Finally, select the movie clip in the *rocks* layer and change its *z* value to `300`. With all layers pushed away along the z-axis, the assets are resized and repositioned to achieve a very basic appearance of 3D depth.
5. Looking at the effect of the depth changes, the collection of movie clips is sitting a bit high on the stage, causing a few elements to appear as if they are floating in air. Changing the vanishing point will adjust the position of all movie clips that have *z* values. Select any movie clip on stage and, in the Properties panel, change the Vanishing Point *y* value to `320`. This will move the movie clips down a bit, all moving relative to one another.
6. Using the Edit Bar above the Stage, click on GalleryScreen to leave the *parallax* movie clip and go back to its parent clip, *gallery*.
7. Click once on the *parallax* movie clip to select it and, using the Properties panel, set its 3D *x* coordinate to `1235`. This is the starting point of the parallax scrolling interface and the first thing viewers will see when visiting the Gallery screen.
8. Save your work. If you want to check your progress, compare your file to *gallery_02.fla*.

Animating the Gallery

Now it's time to create a tween and animate the scroll. You will extend your timeline to 100 frames, change the location of the *parallax* movie clip over time, and add a few frames of ActionScript:

1. Continue with your file or, if you prefer, open *gallery_02.fla* from the companion source files and save that to your own file as your starting point.
2. Double-click the *gallery* movie clip to edit its contents. Don't worry if the guide layer is no longer visible. This is just a little Flash quirk. When double-clicking a 3D element, the symbol is not edited in place. This is equivalent to editing any symbol by double-clicking in the Library.
3. Right-click (Windows) or Control-click (Mac) on the *parallax* movie clip and select Create Motion Tween from the pop-up menu.
4. In the Timeline panel, drag the end of the *parallax* layer's tween span out to frame 100.
5. In the *actions*, *button*, and *mountains* layers, click in frame 100 and press F5 to add frames. When you are finished, all four layers should span to frame 100.
6. In the actions layer, create empty keyframes (F7) in frames 35 and 70. You will later add stop actions to these frames.
7. Move the playhead to frame 35, click on the *parallax* movie clip to be sure it's selected, and, using the Properties panel, set the *x* value under 3D Position and View to `365`. The ScrollPane frame artwork should be visible. This will automatically add a property keyframe to your tween, reflecting the movie clip's change in *x*.
8. Move the playhead to frame 70, make sure the *parallax* movie clip is still selected, and set the 3D *x* value to `-435`. The FLVPlayback frame artwork should be visible.
9. Move the playhead to frame 100, make sure the *parallax* movie clip is still selected, and set the 3D *x* value to `1235`. The UILoader frame artwork should be visible. This last frame provides an animated return to the first item in the gallery.
10. Save and test your movie.

The animation should scroll through the three gallery items, showing UILoader, ScrollPane, and FLVPlayback frames, and then returning to frame 1. If you look at the assets in the animation, you will see that the frames, rocks, bushes, and trees all move at a different rate, creating the parallax scrolling effect. This is achieved automatically because Flash's 3D depth management handles the scrolling of each layer independently.

Adding ActionScript Control

Now that you've seen the parallax scrolling effect, it's time to prepare your file for ActionScript control. You will first add stop actions to stop the animation when the three component frames come into view, and then you'll add a script to react to user button clicks.

1. In the *actions* layer, add a stop action (**stop();**) to frames 1, 35, and 70. Frame 100 does not receive a stop action because you want the movie clip to loop back to frame 1 on its own. Figure 8-23 shows a detail of the Timeline panel, depicting the layer structure, parallax tween, and added actions.

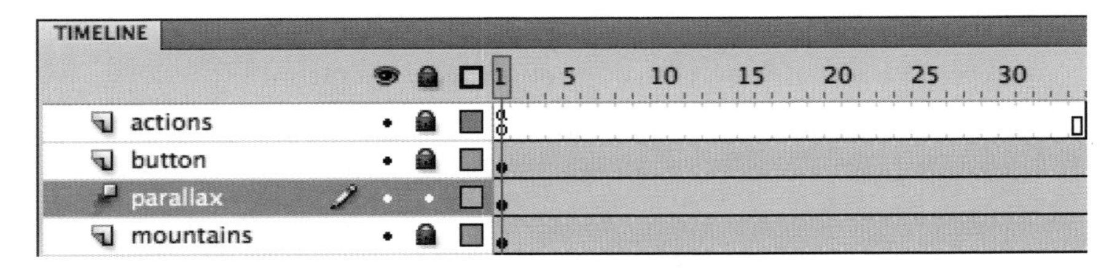

Figure 8-23. A detail of the gallery movie clip Timeline panel, showing the parallax tween

2. Test your movie once again, and you'll see that the movie clip doesn't scroll. This is correct, and is due to the stop action in frame 1.
3. Save your work and compare your file to *gallery_03.fla*.

With the stop actions in place, you will need to add a small script to continue to preview your file. You'll add this script in the main timeline, rather than inside the gallery movie clip, to make it easier to add it later to the main project file.

4. Using the Edit Bar above the Stage, click on Scene 1 to return to the main timeline. In the actions layer, add the following script, save your work, and test your file:

```
1   gallery.more.addEventListener(MouseEvent.CLICK, onGalleryClick);
2   function onGalleryClick(evt:MouseEvent):void {
3         gallery.play();
4   }
```

This script adds a mouse-click event listener to the *more* button inside the *gallery* movie clip. The main thrust of the script is line 3, which plays the movie clip. This animates the gallery interface after it's been stopped by the stop actions in frames 1, 35, and 70. When testing your file, you should be able to click the *more* button and advance to the next gallery item. When the final item (the FLVPlayback component frame) is in view, clicking the *more* button will cause the movie to return to the first item (the UILoader component frame).

After you have finished testing, compare your file with *gallery_04.fla* from the companion source files.

Adding a Layer Mask

When testing the interface, you may notice that the *parallax* movie clip is visible to the left of the viewing wheel depicted in the guide layer. When you integrate your gallery assets into the main project file, this will be a problem, as the gallery interface will appear from under the wheel as you scroll. To prevent this, you'll add a layer mask to show the *parallax* movie clip only within the desired frame.

1. Continue with your gallery file or use *gallery_04.fla*.
2. From the main timeline, double-click the *gallery* movie clip to edit its contents.
3. Right-click (Windows) or Control-click (Mac) on the *parallax* layer name or icon and select Insert Layer from the pop-up menu. Name the new layer `mask`.
4. Select the *mask* layer and, using the Rectangle tool, draw a rectangle anywhere on the stage. Using the Properties panel, set its *x* location to `80`, its *y* location to `20`, its width to `520`, and its height to `400`.
5. Right-click (Windows) or Control-click (Mac) on the *mask* layer and select Mask from the pop-up menu. This will turn the *mask* layer into a layer mask and automatically mask the underlying layer, *parallax*. It will also lock both layers so you can see the mask in effect in authoring mode. Figure 8-24 shows a Timeline panel detail of the *gallery* movie clip after adding the mask.

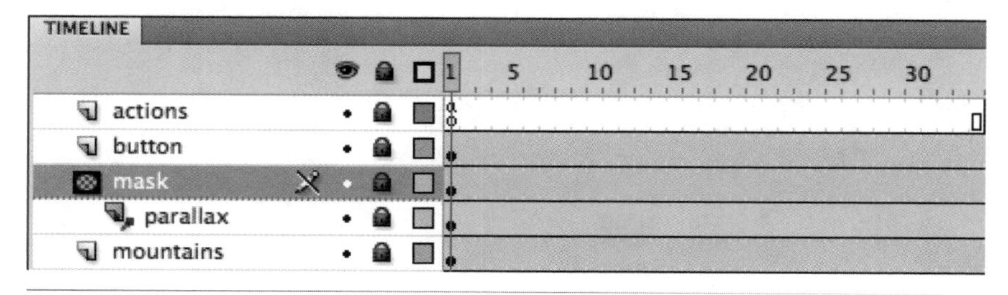

Figure 8-24. A detail of the gallery movie clip Timeline panel, showing the masked parallax tween

6. Test your movie and scroll the gallery. The art should now be visible only within the masked area.
7. Save your work and compare your file to *gallery_05.fla*.

Adding the Gallery to the Main Project File

Once your parallax scrolling effect is working, you'll need to add the necessary content to the main project file:

1. Continue with your gallery file or open *gallery_05.fla*.
2. From the main timeline, select and copy the *gallery* movie clip.
3. Open the main portfolio project file that you've been building over the past several chapters.
4. Scroll to the gallery section of the portfolio project file's main timeline. Select frame 144 in the content layer and, using Paste in Place (Edit→Paste in Place), paste the gallery movie clip that you copied from your gallery file.
5. Switch back to the gallery file and copy the script from the actions layer.
6. Switch once again to the main portfolio file, select frame 147 in the actions layer, and paste the script after the existing stop action.
7. Save your work and test your file. The gallery should now work correctly in your main portfolio file (Figure 8-25).

Figure 8-25. The Gallery screen, after integrating the assets into the main project file

The Project Continues...

In the next chapter, you'll add two of the three components destined for the Gallery. You will add the UILoader component, which loads external images and SWFs, and the ScrollPane component, which displays details of larger content within a scrolling pane.

CHAPTER 9

COMPONENTS

IN THIS CHAPTER

Introduction

Components are wondrous little widgets that add functionality to your projects with little to no programming required. Used by ActionScript neophytes and veterans alike, their feature sets range from simple to complex. Usually, they contain a graphical front end and integrated code to assist in or complete their purpose. However, code-only components are also sometimes used as ActionScript libraries to expand the capabilities of the language.

Components make it possible to accomplish goals without having to reinvent the wheel. That is, when you need a video player, a component can save you the need to create one on your own. This is especially true of oft-used items such as buttons, scroll bars, menus, and other user interface controls. In this chapter, you will look at all of these kinds of components and more, and you will add some to your portfolio project. Components will enhance the portfolio by giving users the ability to load external content and display details of images too large to fit on the screen. In Chapter 13, you will complete your portfolio project by adding video to the Gallery screen using a video component.

There are, of course, limitations that go along with having these features provided to you in a precreated package. For one thing, because you didn't create the component yourself, it may not match the look of your project. This is usually not a big problem for applications, but the more unique your project looks, the more components can stick out. To get around this design issue, you will learn later in this chapter how to *skin*, or change the outward appearance of the components that are shipped with Flash CS4 Professional.

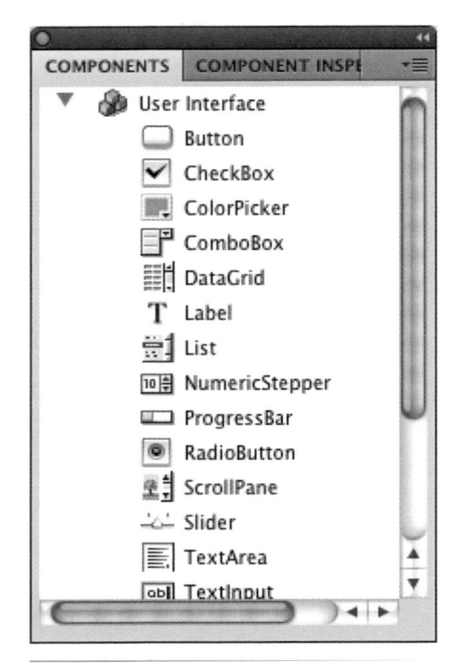

Figure 9-1. The Components panel

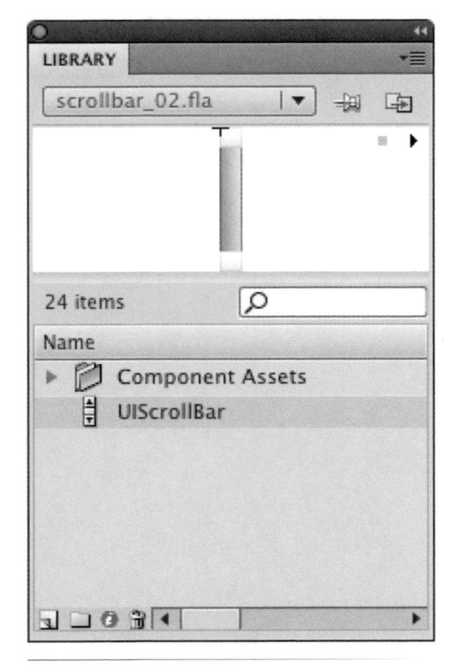

Figure 9-2. The Library after adding a UIScrollBar component

Another drawback is that components contribute to file size. This is not unexpected because they add both code and design elements to your files. However, if you are really concerned about file size—such as when developing small advertising banners—this can be an issue. In that context, the weight a component adds to a FLA can be significant if the design or ActionScript it uses is complex.

In this chapter, you'll focus primarily on the user interface components that ship with Flash CS4 Professional, each of which adds approximately 15 to 30 KB to your file. As another example, the video component that ships with Flash CS4 adds 50 to 65 KB to your file. In most cases, this is a drop in the bucket when it comes to the overall size of the average SWF. Just keep this in mind if your files are on a diet.

Fortunately, Flash CS4's user interface components have been optimized to be as slim as possible. Not only are most of them smaller than their counterparts from prior Flash versions, they are often designed so that graphics are external to the component and shared by several components. In fact, in some cases, components use other components for specific functionality. A menu component might use a text label component, for example, to display the text in each menu item. Shared assets that are used by more than one component are not added to your file again. So, with each added component, the contribution to file size diminishes.

NOTE

There are hundreds of third-party components available. Three good places to start looking are the Adobe Exchange (http://www.adobe.com/exchange), Jumpeye Components (http://www.jumpeyecomponents.com), and Flashloaded (http://www.flashloaded.com).

Adding and Configuring Components

The first step in using a component is to add it to your file. The easiest way to do this is by dragging a component from the Components panel (Figure 9-1) to the Stage. The component then appears on the Stage and in your Library. Depending on how the component was built, an accompanying folder of assets may also appear in the Library (Figure 9-2). Typically, you won't have to work with this folder at all. As described earlier, it may contain assets that are shared with other components, or simply external graphics that you can customize through editing.

You can often use the Component Inspector panel (Figure 9-3) to configure a component, thereby controlling its functionality. This panel contains editable text fields, menus, and other user entry controls. You can select from available options, enter text strings (such as button labels or URLs), and even access a file browser to locate external files. Once configured, the component may exhibit an immediate visual change, while other times you will have to test your movie to see the asset in action.

Functionality Without Programming

Especially for ActionScript beginners, components are very attractive because some can enhance the functionality of your project without a single line of code. This section shows how easy it can be to produce a scrolling text field—no programming required.

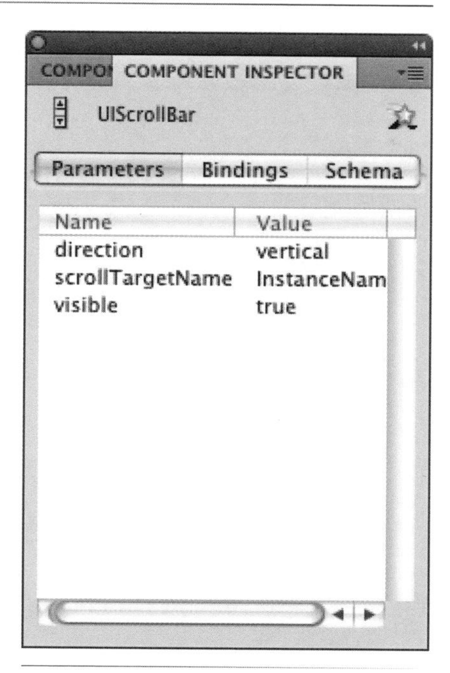

Figure 9-3. Configuring a UIScrollBar component using the Component Inspector panel

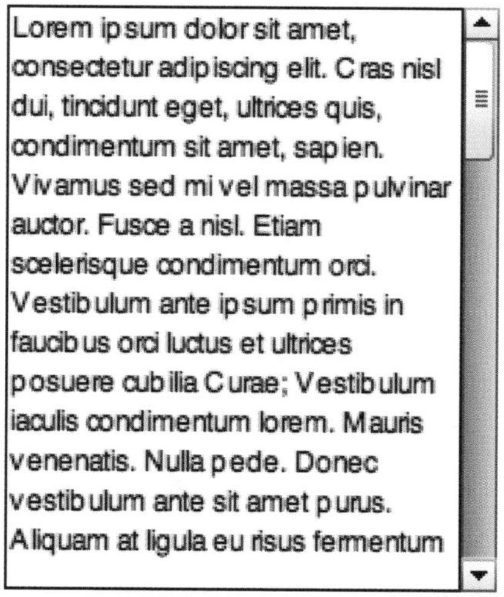

Scrolling Text

You can add the *UIScrollBar* component to a simple Dynamic text field to give the field scrolling capabilities. The outcome of this exercise is shown in Figure 9-4.

1. Open *scrollbar_01.fla* from the companion source files.
2. On the left side of the Stage are a few paragraphs of text. You'll copy and paste this text in a moment.
3. On the right side of the stage is a smaller Dynamic text field. You learned in Chapter 2 that Flash Player renders Static text fields as graphics. Dynamic text fields, on the other hand, are optimized for ActionScript control. Using ActionScript behind the scenes, the UIScrollBar component will control your text field for you, so the field type used here must be `Dynamic` (you'll learn about programming text in Chapter 11).
4. Select the field with the Selection tool and look in the Character section of the Properties panel. Choose a type family and size appropriate for scrolling, such as `_sans`, `Arial`, or `Times`, `12` point. The sample file uses `_sans`, which is the default sans serif typeface used by your computer's operating system.
5. Note that the Dynamic text field has a width and height of `140` and `200`, respectively. The size really isn't important, but the field needs to be big enough to accommodate a scroll bar and too small to contain the text you will eventually paste into it.
6. Copy the large block of text into the clipboard. Before pasting the text into the small field you created, select the field and use the Text→Scrollable menu command to make the text scrollable. This will fix the field size at its original dimensions and prevent overflowing text from resizing the field.
7. Paste the text into this field. If the field changed height to accommodate the pasted text, the field is either not dynamic or is not set to scrolling. Revisit steps 2 and 3 to check your work.

Lorem ipsum dolor sit amet, consectetur adipiscing elit. Cras nisl dui, tincidunt eget, ultrices quis, condimentum sit amet, sapien. Vivamus sed mi vel massa pulvinar auctor. Fusce a nisl. Etiam scelerisque condimentum orci. Vestibulum ante ipsum primis in faucibus orci luctus et ultrices posuere cubilia Curae; Vestibulum iaculis condimentum lorem. Mauris venenatis. Nulla pede. Donec vestibulum ante sit amet purus. Aliquam at ligula eu risus fermentum

Figure 9-4. A UIScrollBar component in use

8. Open the Components panel and locate the UIScrollBar in the User Interface category of components. Drag it to the Stage and place it directly on top of the text field. This may seem odd, but this step automatically associates the scroll bar with the text field so the code inside the component knows which field to control without any additional intervention from you.

9. Move the scroll bar to the right side of the text field and set its height to **200** using the Properties panel.

10. Test your movie and scroll the text. The area used when dragging the scroll bar automatically resizes itself to indicate how much text is scrollable. A very small drag interface element (often called a *thumb*) means a lot of text is scrollable. Conversely, a very long scroll thumb means there may only be a few additional lines of text eclipsed from view.

Look at Figure 9-3 again, and you will see the UIScrollBar options that you can configure in the Component Inspector panel. When you dropped the scroll bar onto the text field in step 5 of this exercise, the component automatically assigned the field a generic instance name (*InstanceName_1*, for example). Instead of using this approach, you could manually assign the field an instance name and type that name into the Component Inspector's *scrollTargetName* property. You could also make the scroll bar invisible and even choose to orient the scroll bar horizontally instead of vertically.

Adding a Pinch of ActionScript

Some components can behave self-sufficiently to a point, but don't have enough functionality to be useful without a nudge from ActionScript. The *Button* component is just such a component. The button will display multiple states upon mouse interaction and allows you to customize its appearance, but it won't actually trigger any response from your application unless you assign an event listener in ActionScript.

Triggering Actions with Buttons

Figure 9-5. A Button component in use, showing the button Down state

In this exercise, you'll add a *Button* component to the Stage, adjust its text label, and add a few simple lines of ActionScript to make it work. When clicked during testing, the button will place text into the Output panel. Figure 9-5 shows the button with its new label in action.

1. Create a new file using File→New. This file will not be used in the final project, so you don't need to use the book template and you don't need to save your work.

2. Drag the Button component from the Components panel to the stage.

3. Select the component and, in the Component Inspector, click the label value field and replace the default label with `Trace it!`, as shown in Figure 9-6.

4. Test your movie and check to be sure the button displays the correct label and multiple button states based on mouse interaction. Review the previous steps, if needed.

You probably noticed that the button behaved as it should, but didn't do anything when clicked. For additional functionality you need ActionScript and, as the button label implies, you will write a short script that traces a message to the Output panel.

5. Select the button and, using the Properties panel, give the button an instance name of `myButton`.

6. Open the Actions panel, enter the following script, and test your movie. "Button clicked" should appear in the Output panel.

```
1   myButton.addEventListener(MouseEvent.CLICK, onClick);
2   function onClick(evt:MouseEvent):void {
3       trace("Button clicked");
4   }
```

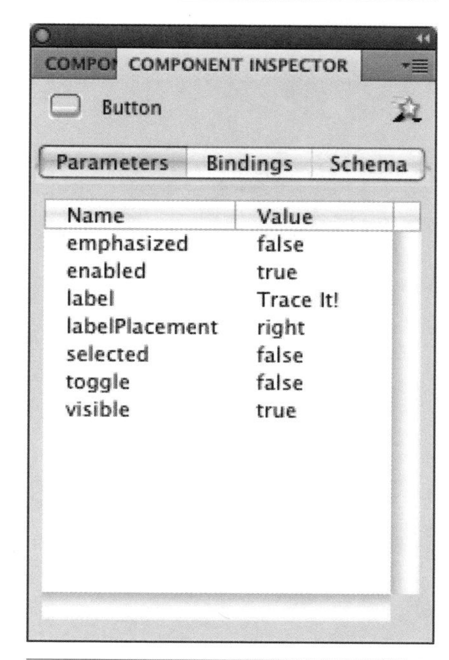

Figure 9-6. Configuring a Button component using the Component Inspector panel

This script adds an event listener to the button and listens for the mouse click event. When the event is received, the listener calls the **`onClick()`** listener function and traces the message to the output panel. If desired, you can compare your work to *button_01.fla* in the companion source files.

Other features that you can configure in the Component Inspector include whether or not the button is *visible*, *enabled* (disabled buttons are not interactive and have a muted appearance), *emphasized* (visually), and initially *selected* (showing a visual change but not automatically triggering any event listener function). You can also align the label you edited using *labelPlacement*, and turn the button into a *toggle*—a button that stays pressed until it is clicked again. This is in contrast to a standard push button, which releases as soon as you let go of the mouse.

Working Hand-in-Hand with Code

So far, the main point of this chapter has been to stress that components can be strong allies to nonprogrammers. That doesn't mean, however, that their usefulness is limited to this area. The following section will walk you through the creation of a pop-up menu, or *ComboBox*. You will use increasing quantities of ActionScript until the steps culminate in a code-only solution that adds the menu to an empty Stage.

Navigating with Menus

This exercise will power two menus with which the user can navigate through the timeline shown in Figure 9-7. Briefly, the four frames at 1, 10, 20, and 30 each have a *stop()* action, content, and corresponding label. The *home* frame (1) is empty, and the *one* (10), *two* (20), and *three* (30) frames have illustrations of the numbers 1, 2, and 3, respectively.

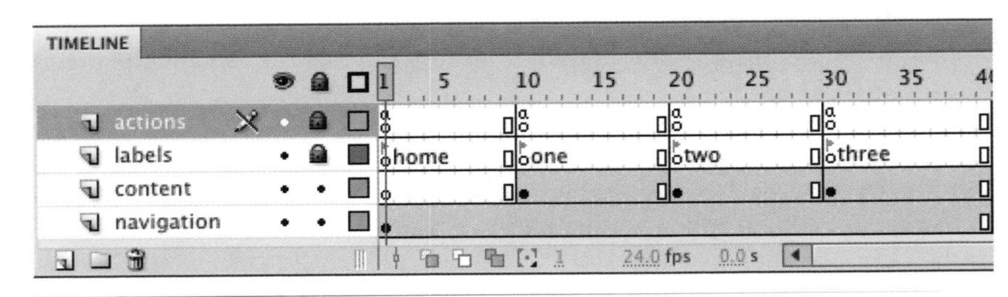

Figure 9-7. A simple timeline through which a ComboBox will navigate

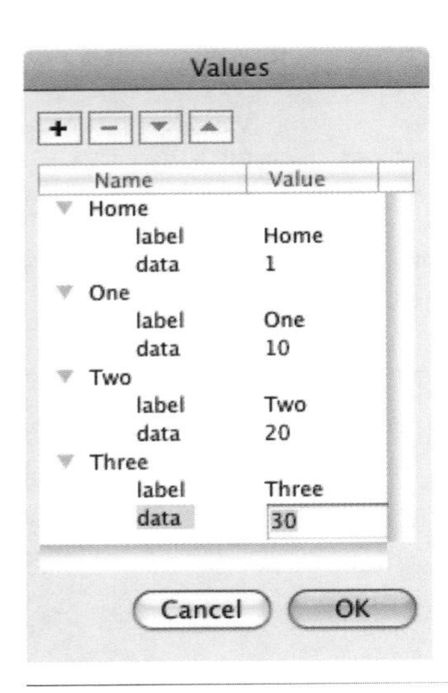

Figure 9-8. Configuring a ComboBox component using the Component Inspector panel

On the Stage through all the frames (in the *navigation* layer) are two empty menus. You will populate these menus—using the Component Inspector for one and ActionScript for the other—and then write a script that reacts to a user choice from either menu and sends the playhead to the desired frame.

1. Open *comboBox_01.fla* from the companion source files.
2. Select the top *ComboBox* menu and, using the Properties panel, give it an instance name of `inspectorMenu`.
3. Open the Component Inspector panel and double-click the *dataProvider* value. A dialog called Values will open. If the dialog doesn't open, click the small magnifying glass icon to open the dialog.
4. In the upper-left corner of the dialog, click the plus sign. This will add a menu item to the ComboBox that consists of two entries: a text label and a corresponding data value. In the *label* value, type `Home`. This is the text that will appear in the menu for this item. The name of the item will update to reflect this change. In the *data* value, type `1`. This corresponds to the frame you want to visit in the Timeline when selecting this menu item. This is a simple example use of the data property. You learned in Chapter 5 that it's usually better to use frame labels rather than frame numbers for navigation, and you'll do that, too.
5. Repeat step 4 three more times to create the values shown in Figure 9-8. When you're finished, click OK and note that the *dataProvider* value has been populated with the information you just entered (Figure 9-9).
6. Next, click the *prompt* value and type `Navigate`. This is the word displayed at the top of the menu prior to its first use. Thereafter, the current menu selection will be displayed in this location.
7. Click the *rowCount* value and type `4`. This setting limits the number of rows shown below the menu when the menu is clicked.

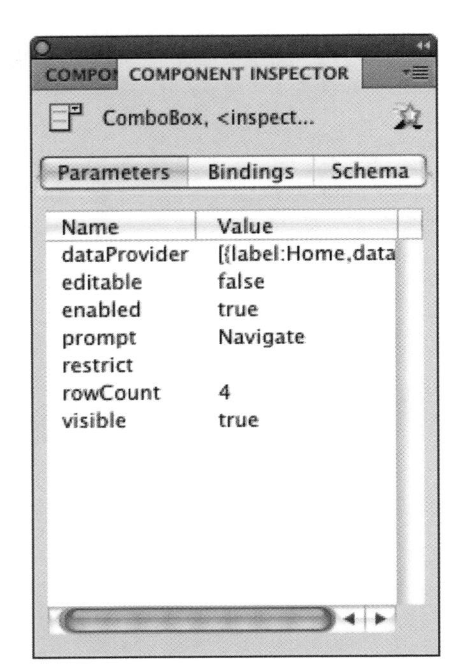

Figure 9-9. Configuring a ComboBox component using the Component Inspector panel

8. Test your movie, and you will see that the ComboBox displays menu items as you would expect it to when interacting with the mouse (Figure 9-10), but it requires ActionScript for any navigation to occur.

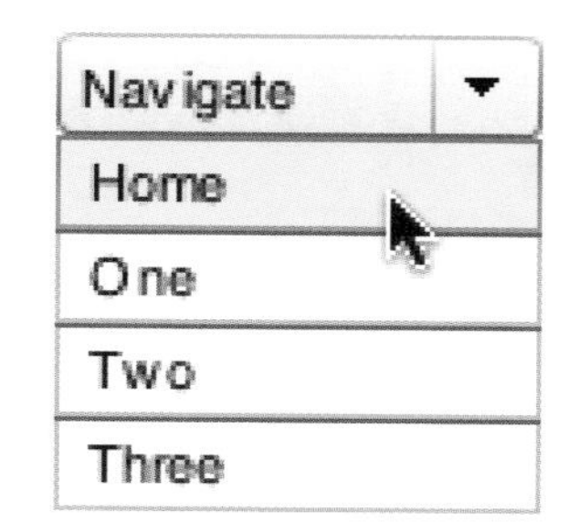

Figure 9-10. A ComboBox in use

Other options that you can configure through the Component Inspector panel include whether or not the menu is *visible* and *enabled*, whether it is *editable* (allowing the user to change the text that appears in the menu) and, if so, which characters the *restrict* property limits the user to when typing a new menu item.

Now it's time to add ActionScript to the first menu. When you're finished, test your movie. When picking any frame but *Home*, you should see an illustration of the corresponding number. When selecting *Home* from the menu, the playhead will return to the empty first frame:

9. Select the first frame in the actions layer and open the Actions panel. A **stop()** method should already be in the frame script, so add lines 3 through 6 to the script:

```
1   stop();
2
3   inspectorMenu.addEventListener(Event.CHANGE, onChange);
4   function onChange(evt:Event):void {
5       this.gotoAndStop(evt.target.selectedItem.data);
6   }
```

This script creates an event listener that will listen for the event fired when the menu is changed (line 3). When this event is received, the main timeline will go to a specified frame and stop (line 5).

Take a close look at how this line works. The navigation code is straightforward. The **this** identifier references the main timeline because that's the scope of the script. The **gotoAndStop()** method moves the playhead. The *data* property contains the frame number used by the method, but where does this information come from? The *data* property is pulled from the menu's *selectedItem*—the menu item the user selected when last using the menu. Note, however, that the menu is not specified by instance name. Instead, the *target* of the event is queried. This means the function can work with more than one menu.

NOTE

If necessary, review Chapter 6 for more information on event listeners.

Now it's time to build the second menu, this time with ActionScript.

10. Select the second menu and give it an instance name of **asMenu**.

11. Add the following to your script in frame 1:

```
7   asMenu.prompt = "Navigate";
8   asMenu.addItem({label:"Home", data:1});
9   asMenu.addItem({label:"One", data:10});
10  asMenu.addItem({label:"Two", data:20});
11  asMenu.addItem({label:"Three", data:30});
12  asMenu.addEventListener(Event.CHANGE, onChange);
```

Notice that you don't need to use the Component Inspector at all in this stage; the ActionScript replaces that labor. Line 7 of the script creates the menu text visible prior to the menu's first use, and line 12 adds an event listener, listening for the menu's change event. Because your prior listener function doesn't restrict its use to a specific menu, both menus can use the same function. The **evt.target** property will automatically determine which menu was used.

Lines 8 through 11 replace the work you did in the Values dialog by adding menu items and populating their **label** and **data** properties. The syntax used for each menu item is an associative array. Look back over Chapter 6 for a refresher of arrays if this form is not clear.

The final step in this exercise is to rely solely on ActionScript to not only populate the menu, but create it as well. As long as the component is in your library, you can create the menu dynamically the same way you can create a movie clip dynamically.

12. Without manually adding another menu to the stage, add the following to your script in frame 1.

```
13  import fl.controls.ComboBox;
14  var dynamicMenu:ComboBox = new ComboBox();
15  this.addChild(dynamicMenu);
16  dynamicMenu.x = 40;
17  dynamicMenu.y = 270;
18  dynamicMenu.prompt = "Navigate";
19  dynamicMenu.addItem({label:"Home", data:1});
20  dynamicMenu.addItem({label:"One", data:10});
21  dynamicMenu.addItem({label:"Two", data:20});
22  dynamicMenu.addItem({label:"Three", data:30});
23  dynamicMenu.addEventListener(Event.CHANGE, onChange);
```

Focus on the first five lines of this script. Lines 18 through 23 are essentially the same as the last script you added, but they have been included here for clarity because the reference to the menu has changed.

Line 13 includes an **import** statement, which is an important structure new to your work in ActionScript thus far. Up to this point, you've restricted your exercise work to using only ActionScript that the Flash compiler already knows about. When using components, however, you must tell the compiler where to look to find their classes so the compiler can validate and compile your code. Without this **import** statement pointing the way to the **ComboBox** classes, your script would generate errors telling you that the compiler doesn't understand the code.

Line 14 creates a new **ComboBox** and stores a reference to the menu in the **dynamicMenu** variable. This reference is akin to a manually created instance name. Line 15 adds the menu to the display list as a child of the main timeline, and lines 16 and 17 set the components position.

Test your work and compare it to *comboBox_02.fla*.

Skinning UI Components

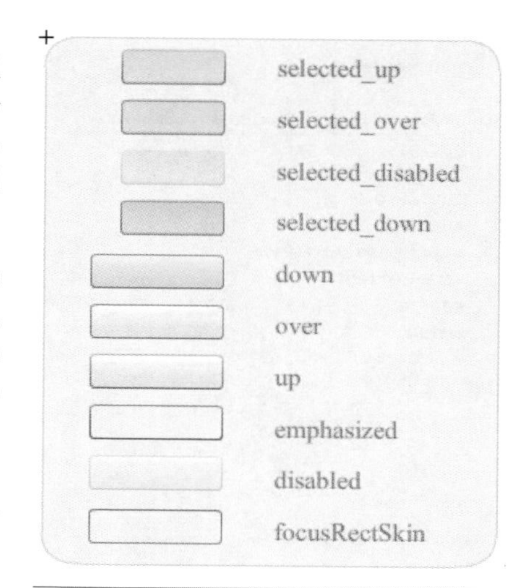

Figure 9-11. Editing the Button component skin

Depending on the design of a given project, you may not always be satisfied with the appearance of a component. Some components allow you to modify their appearance through ActionScript (a process typically called *styling*, similar to the way you style text with ActionScript) or by modifying the visual elements used by the component. The latter process is called *skinning*.

The user interface components you've been concentrating on thus far are really easy to skin. All you have to do is double-click a component to edit it, just like you were editing a movie clip that you created yourself. Inside the component, you'll find a guide layer that contains all the separate movie clip parts the component uses for its various display forms.

Figure 9-11 shows the *Button* component skin. In this figure you can see the various button states for both normal use and when the button's *selected* property is true. You can also edit the art used when the button is emphasized, disabled, or focused (meaning the user is interacting with the button either with the mouse or keyboard).

Each skin element is a movie clip and can be double-clicked for editing. When doing so, remember that you're editing the symbol. Every instance of the component will be updated and, because some assets are shared between components, more than one component may be updated.

Project Progress

In this chapter, you will add two components to the Gallery page. The first will load a JPG at runtime and the second will display a detail of a large image (allowing the user to scroll the image around to see more). You will add the third component in Chapter 13 when you add video to the project. These components are all very easy to use and require no ActionScript.

Loading an External Graphic at Runtime

Figure 9-12. The project UILoader component in use

The *UILoader* component is optimized to load external display objects such as bitmaps (JPG, PNG, or GIF) and other SWFs. It will load ActionScript behind the scenes and includes a few nice optional features, such as loading the asset automatically, scaling the content to fit the size of the component, and maintaining aspect ratio during scaling. The component also contains no graphical elements to get in the way of the appearance of the loaded asset. Figure 9-12 shows the component in use.

1. This is the first time you're loading external assets in the portfolio project, so it's a good idea to make sure your relative file locations are consistent. This will eliminate the need to change scripts or component configurations later on. In the same folder as your main portfolio FLA, create a

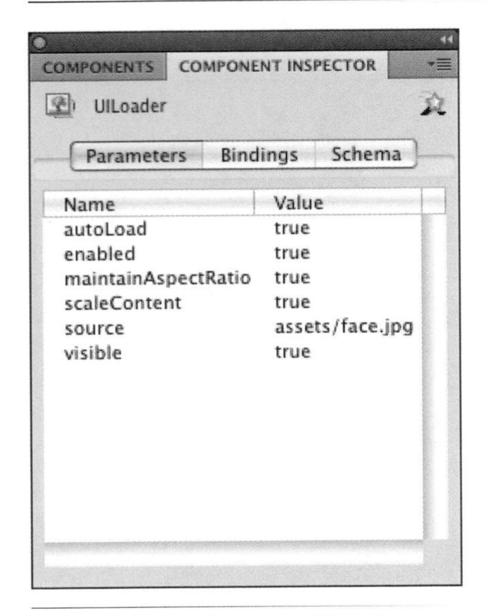

Figure 9-13. Configuring a UILoader component using the Component Inspector panel

folder called *assets*. In that folder, place the *face.jpg* image from the furnished source files.

2. Open the main portfolio FLA and open the Library panel. Look inside the *Gallery Assets* folder and double-click the *foreground* movie clip.
3. Scroll to the far left of the movie clip so you can see the frame in which the *UILoader* component will reside. Select the *components* layer and drag the *UILoader* component from the Components panel to the frame on the Stage.
4. Using the Properties panel, set the size of the component to a width and height of `200` and `200`, respectively, and position the component neatly within the frame.
5. With the component selected, open the Component Inspector (Figure 9-13). Click the *source* value field and enter `assets/face.jpg`.
6. Test your movie, go to the Gallery screen, and see if the image loads. If not, review your steps. If you can't resolve the issue, you can compare your file to the chapter final portfolio file, *portfolio_09_final.fla*.

Display Details of a Larger Image

On occasion, you may run into scenarios in which you must display an image that is too large to fit on the screen, thereby showing only a detail of the image. A good example of this is showing a large map and displaying only a portion of the map at any one time.

One easy way to handle this is to use the *ScrollPane* component (Figure 9-14). This component allows you to scroll the detail area vertically and/or horizontally so you can see more of an image when you need to.

Figure 9-14. The project ScrollPane component in use

1. Place the *words.jpg* image from the furnished source files into your *assets* folder.
2. If you're not already there, return to editing the *foreground* movie clip.
3. Scroll to the center of the movie clip so you can see the frame in which the *ScrollPane* component will reside. Select the *components* layer and drag the *ScrollPane* component from the Components panel to the frame on the Stage.
4. Using the Properties panel, set the size of the component to a width and height of `300` and `200`, respectively, and position the component neatly within the frame.
5. With the component selected, open the Component Inspector (Figure 9-15). Click the *source* value field and enter `assets/words.jpg`.
6. Test your movie, go to the Gallery screen, advance to the *ScrollPane* position using the Next button, and see that the image loads. Use the scroll bars to scroll the image both vertically and horizontally. If the image doesn't load, review your steps.
7. Close the SWF to try the optional feature of scrolling the image with your mouse. With the component selected, change the *scrollDrag* setting to `true`.
8. Test your movie again, go back to the Gallery screen, and see that you can now scroll the image with your mouse, as well as the scroll bars. If any problems remain, compare your file to the chapter final portfolio file, *portfolio_09_final.fla*.

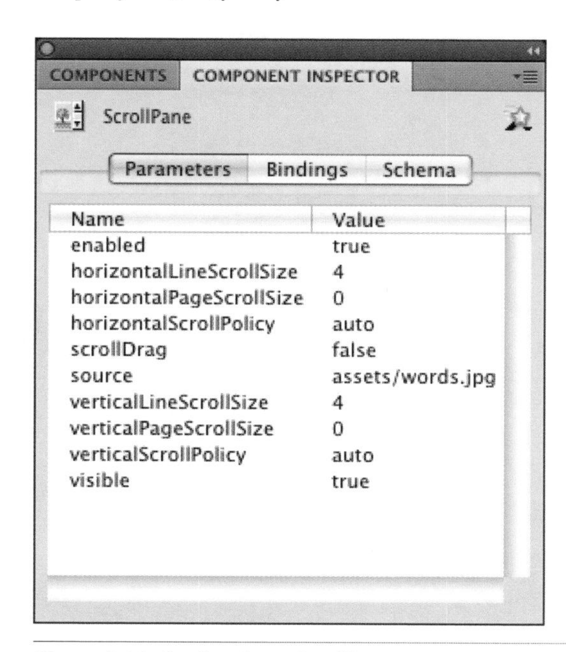

Figure 9-15. Configuring a ScrollPane component using the Component Inspector panel

Additional available options include the ability to choose in which directions the user can scroll and how much is scrolled at one time. The scrollable direction is controlled by the *horizontalScrollPolicy* and *verticalScrollPolicy* settings, which can each be *on*, *off*, or *auto* (the latter only scrolls when needed). The *horizontalLineScrollSize* and *horizontalPageScrollSize* properties, as well as their vertical counterparts, set the size of scrolling. The line values are the amount scrolled when clicking on the scroll bar arrow, and the page values are the amount scrolled when clicking on the scroll track.

The Project Continues...

In the next chapter, you'll learn how the project video featured in this chapter was created. You'll learn about *armatures* and *inverse kinematics*. In short, you'll be animating elements by tying them together like the bones of a skeleton, and moving the bones to create poses. Plus, Scaly returns!

CHAPTER 10

INVERSE KINEMATICS

Introduction

IN THIS CHAPTER

Programming the motion of a robotic arm is no small feat. It requires knowledge of trigonometry and *kinematics*, the study of motion. Moving a robotic arm is complex because you must calculate the angles of each joint required to place the mechanical hand of the robot at its destination. This is typically called *forward kinematics* because the motion goes forward through the structure. The angle of the shoulder positions the upper arm and elbow, the angle of the elbow positions the forearm and wrist, the angle of the wrist positions the hand, and so on.

Unless you're a math wiz, this won't help you much when animating jointed structures in Flash. Fortunately, Flash CS4 Professional introduces an easier approach to the problem; an animation technique called *inverse kinematics* (IK). Common in higher-end 3D and animation packages, IK is the opposite, or inverse, of forward kinematics. Using IK, you start from the end and work backward. You specify a pose of a jointed structure by positioning its end segment, and let the animation software determine all the joint positions and angles required to create that pose. In Flash, for example, you can drag the hand of an articulated character into position, and Flash then calculates the location and angles of the wrist, forearm, elbow, and shoulder joints.

In the art world, these jointed skeletal structures are called *armatures*. Armatures are very much like your own skeleton in that they provide the structural foundation for surrounding objects. In sculpture and stop-frame animation, for example, figures in the physical world are often built over armatures to give them support. Jointed armatures are especially important for posable figures, as you can adjust the armatures many, many times without breaking them. Computer animation tools also use armatures. Rather than containing the armatures inside physical models, they are typically superimposed on the figures they control.

In this chapter, you will learn how to create and animate armatures using Flash's Bone tool. Found in the Tools panel, the Bone tool lets you connect multiple movie clips to form a single posable framework. The tool is very easy to use, requiring little more than a process akin to connecting the dots. To create an armature, you click on one movie clip of the figure to be animated and drag to the next movie clip in the chain. As you'll see later, the Bone tool connects them together as part of a new skeletal structure.

You can also create armatures that serve as skeletons inside a single shape. For example, you can draw multiple bones with the Bone tool, connecting one point to another, inside a snake-like shape. The armature can then be animated and the shape will deform according to influence exerted by the bones. When working with shapes, you can even control to which bone each anchor point of the shape is connected. When a shape deformation is incorrect, you can use the Bone tool's close companion, the Bind tool, to change the bone to which a shape anchor point is bound.

Once you've completed your armature, you can animate it with the Timeline, control it with ActionScript, or even enable it for runtime manipulation by end users. First, however, it will help to understand how skeletons are made.

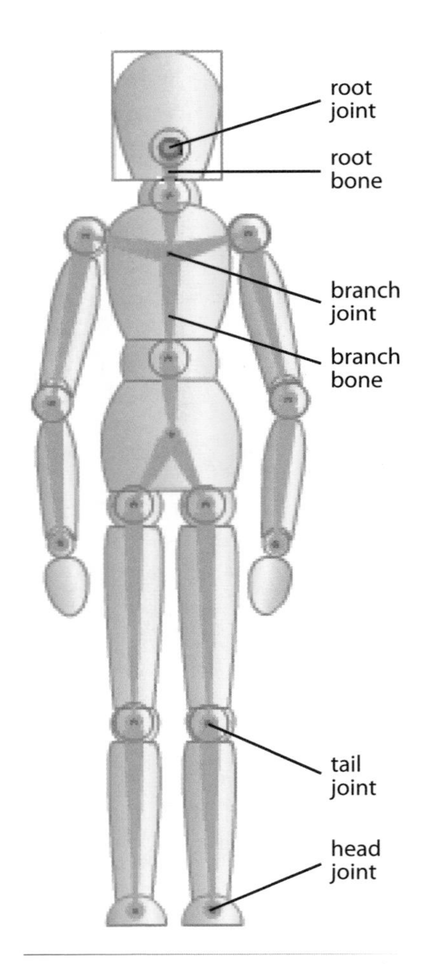

Figure 10-1. A sketch mannequin with visible armature

Anatomy of an Armature

Inverse kinematic structures begin with a *root joint*—the point around which the entire armature may rotate. For an arm, the shoulder is the logical root joint, however, any skeletal starting point can serve in this capacity. The head, for example, is the root joint of the basic full-body skeleton in Figure 10-1.

A *root bone* then spans from the root joint and terminates at the first *child joint*, such as an elbow. A *child bone* then spans from that joint to another, and so on, until the armature is complete. Each bone in the armature can have two joints. The *tail joint* is closest in line to the armature root and the *head joint* is closer to the end of the armature. Just like your skeleton, the last bone in an armature progression has no head joint. For example, following your arm from shoulder to fingertip, the knuckle of your last finger segment has a tail joint but no head joint.

When a skeletal structure requires more than one linear segment as part of the armature, you can introduce additional joints and bones, called *branches*. Although the arms and legs in Figure 10-1 contain linear progressions of bones, the skeleton is not complete without branch bones at the sternum and pelvis. These branches help unite the left and right sets of bones into one skeleton.

Because the purpose of inverse kinematics is to move the bones and joints of the armature together to create poses, you can't drag around pieces of the armature the way you can drag normal display objects. In simple terms, if bones link display objects, you can move them only in directions constrained by the armature. For example, by default, you can't pull off the head of the sketch mannequin depicted in Figure 10-1 simply by dragging its head away from its body. Similarly, a joint can rotate in 360 degrees unless you take steps to limit its movement to a more natural range. You'll learn later how to transform bones and constrain joint rotation, but the default limits imposed on display objects by an IK armature already make it easier to pose your objects.

Using the Bone Tool with Movie Clips

Now that you know a little about armatures, it's time to create one and apply some basic manipulations. Your first couple of armatures will be for practice, so there's no need for extensive detail, and the book template is not required.

1. Create a new FLA (File→New). If you prefer to start with existing assets, open *beads_01.fla* from the companion source files and skip to step 4.
2. Using the Oval tool, create a small circle on the Stage and then make it a movie clip. Do not give it an instance name.
3. Duplicate this movie clip by selecting it with the Selection Tool and holding down Alt (Windows) or Option (Mac) when dragging it on the Stage. Place the copies adjacent to one another in a horizontal line like a string of beads, shown in Figure 10-2.
4. Using the Bone tool from the Tools panel, click the center of the leftmost circle and drag it to the next in line, then let go to connect the two circles with a bone, shown in Figure 10-2.
5. Repeat this process, connecting the second circle to the third circle, then connecting the third circle to the fourth circle, and so on, until all the circles are connected with bones (Figure 10-3).
6. Save your work as *beads.fla*. If you wish, compare your file to *beads_02.fla* from the companion source files.

Figure 10-2. Connecting movie clips with the Bones tool

Figure 10-3. The completed armature

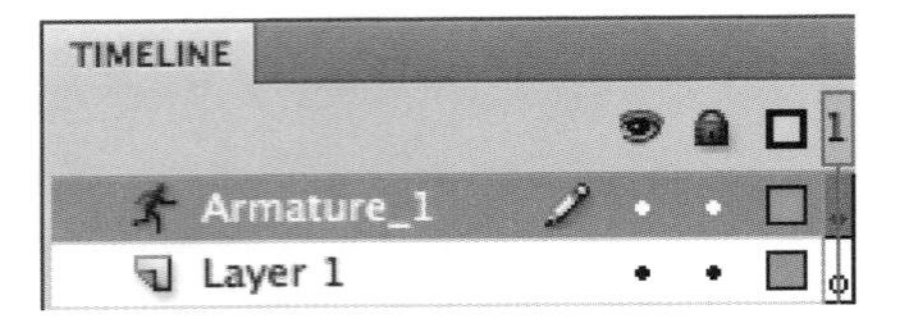

Figure 10-4. All pieces of the armature have been moved to a dedicated Armature layer

You've now completed an IK armature connecting all the circles. If you look at your Timeline (Figure 10-4), you will see that a new layer type, an armature layer, has been created and all the movie clips that became a part of the armature have been moved to this layer.

Although you needn't bother to try this now, you will eventually find that you can't add anything else to this layer unless it becomes a part of the layer's associated IK form. Just like a tween layer, the content of an armature layer is restricted to one armature and no unrelated assets.

Now that you have a working armature, it's time to put it to work. There are two basic modes for using inverse kinematics: *Authortime*, in which you can pose armatures during authoring mode for animation purposes, and *Runtime*, which allows the user to manipulation the armature freely while the SWF is running.

Authortime Mode

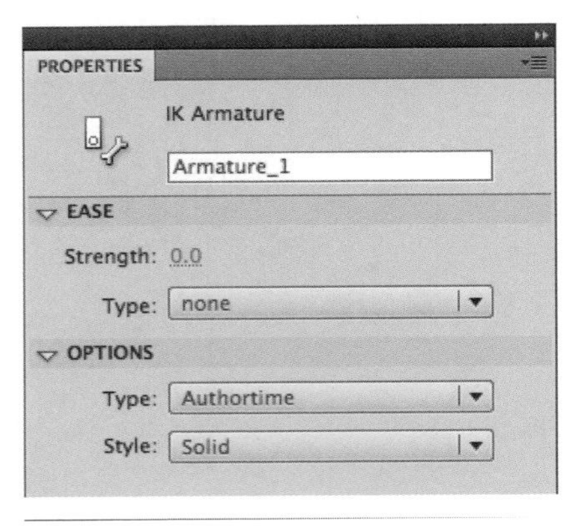

Figure 10-5. An armature set to Authortime mode in the Properties panel

To set an armature for authortime manipulation, first select the armature by clicking on its layer in the Timeline panel. Open the Properties panel (Figure 10-5) and set the *Type* property of the Options section to Authortime.

Using the Style setting in the same section, you can switch between solid, wire, and line. Solid displays normal solid bones, as shown in Figure 10-1. Wire displays the same bone shapes, but in outline form. Finally, line displays each bone as a single thin line. Switching between these settings is a matter of preference, but can be useful if the superimposed armature obscures small details in the object you're trying to animate.

Once you have set the armature Authortime mode, you can drag it around freely on the stage. Figure 10-6 shows the string of beads you created earlier being dragged upward from its original position. You can use Authortime mode to position armatures in static poses, but you can also animate armatures.

Figure 10-6. Dragging the armature around in Authortime mode

Tweening Armatures

To animate IK armatures, you just need to follow the same simple process described in Chapter 5 for tweening symbol instances.

1. Continue with the *beads.fla* file you created earlier.
2. Extend the frame span of the armature layer in the Timeline panel. To do this, grab the end of the armature layer (the end of frame one, in this case) with your mouse and drag it out to frame 100 (Figure 10-7).
3. Every 20 or 30 frames or so, move the armature on the Stage until you are satisfied with a pose. You will end up with four or five poses, approximately 20 to 30 frames apart.
4. Save your movie and test it (Control→Test Movie). You should see the beads come to life and move around a little bit like a worm. Compare your file with *beads_03.fla*.

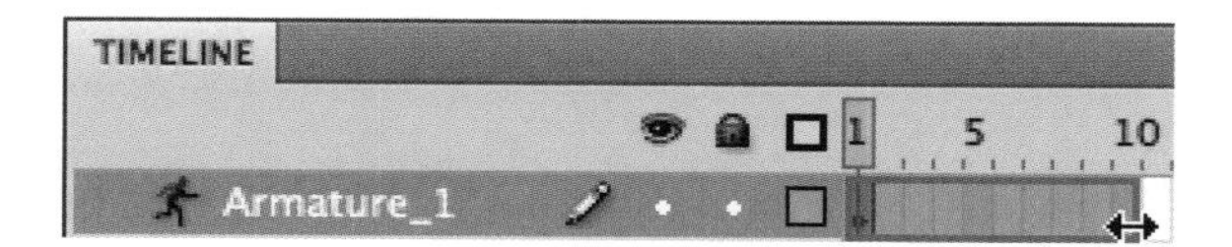

Figure 10-7. Extending an armature layer in the Timeline

Using Onion Skinning

To get help with poses, or to preview multiple frames of the animation as you scrub through the Timeline, enable *onion skinning* in the Timeline panel. This feature can display semitransparent images of the content in the frames before and after the current frame currently in view. The images will appear on the stage while the feature is enabled, making it appear as if you were looking at the content through layers of onion skin. The farther away a frame is from the current frame, the more layers of metaphorical onion skin you're looking through, and the fainter the art becomes.

To enable onion skinning, look for the Onion Skin button in the row of icons beneath the frames in the Timeline (Figure 10-8). The playhead will change to show a gray span of frames before and after the playhead. You can change these gray spans to reveal more or fewer frames in either direction. For example, by dragging the right gray span leftward until it stops at the playhead, and the left gray span leftward until five frames are covered, you will be seeing five frames before the current frame wherever you drag the playhead.

Figure 10-9 shows onion skinning applied for an example animation of the sketch mannequin waving. Two prior frames are visible as the current pose is being manipulated.

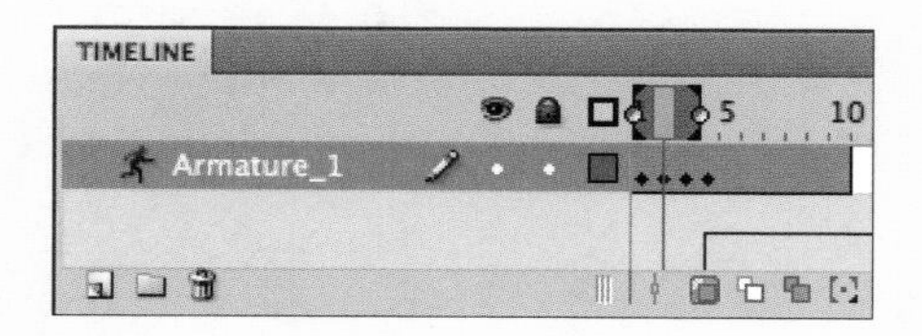

Figure 10-8. Onion skinning enabled in the Timeline

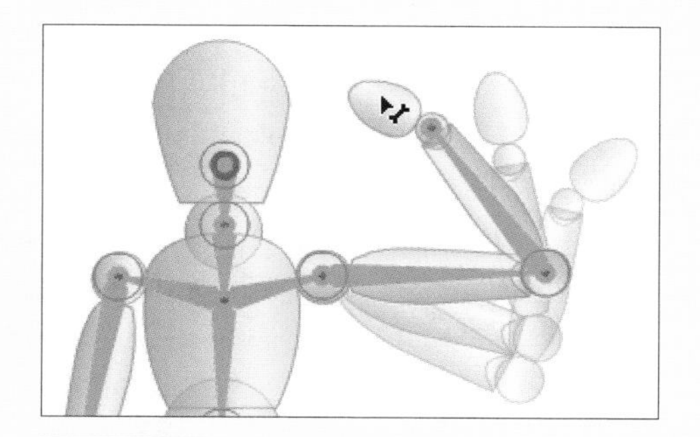

Figure 10-9. Using onion skinning to assist in posing an armature

Armature Easing

While previewing your animation, you might notice that it seems a bit expressionless, and perhaps even a bit rigid when the armature changes direction. You may recall from Chapter 5 that you can improve this by applying *easing* to the tween. Easing is so named because the motion eases into or away from keyframes.

IK animations do not use the Motion Editor, however, so you must apply easing using the Properties panel. The benefit of this approach is that you can apply unique easing between keyframes very easily. Adding easing to IK animations mirrors your experience in Chapter 5 when applying easing to the wheel rotation in the project interface. In that case, easing was also applied in the Timeline panel on a keyframe-by-keyframe basis because you used a classic tween.

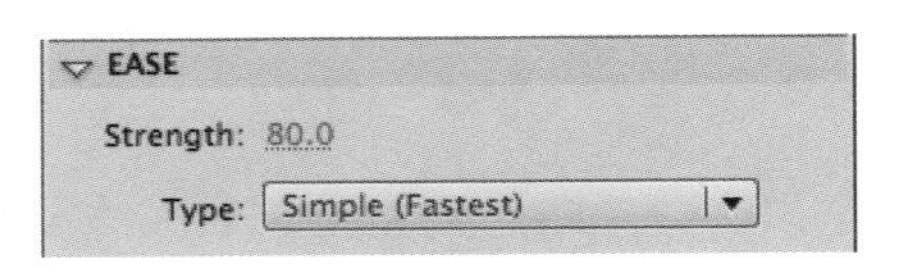

Figure 10-10. Setting an armature's easing values in the Properties panel

To add easing, select a pose keyframe in your armature tween. Any easing applied will affect the frames between the selected pose and the next pose. With the pose keyframe selected, open the Properties panel and look at the Ease section of the armature properties (Figure 10-10).

The *Strength* setting, which ranges from –100 to 100, determines if you will ease into or out of a pose keyframe. Negative values slow the acceleration of motion as you *leave* a keyframe. Positive values slow the acceleration as you *approach* a keyframe. That is, –100 slows the motion closest to the current keyframe, 0 causes no easing, and 100 slows the motion closest to the next keyframe.

The *Type* setting includes Simple, as well as Stop and Start. Simple slows the motion relative to *one* pose (either leaving the current pose keyframe or approaching the next pose keyframe, depending on the *Strength* setting).

The Stop and Start setting slows the motion relative to *two* poses—both current and next keyframes. When using the Stop and Start easing type, negative Strength values slow the motion at both keyframes, while positive values slow the motion between the two keyframes.

Both easing types have variations called Slow, Medium, Fast, and Fastest. Slow applies the subtlest effect, while Fastest applies a more pronounced affect.

Practice applying easing with your own files, as well as by replicating a sample source file. Using this precreated file, in addition to any of your own experiments, will allow you to compare your results to a matching final version. Try to imagine what will happen from the text descriptions and then consider how well the results matched your expectations.

1. Open the *beads_03.fla* file included in the companion source code. This file contains the beads animation you worked with earlier, to which you will now apply easing.
2. Select the pose keyframe in frame 1. Skew the easing 50% toward the starting keyframe and make the effect relatively pronounced by setting *Strength* to `-50` and set *Type* to `Simple (Fast)`.
3. Select the pose keyframe in frame 35 and apply the same settings (set *Strength* to `-50` and set *Type* to `Simple (Fast)`).
4. Select the pose keyframe in frame 50 and use the same type of tween as in the prior two frames, but this time affect all of the easing at the very start of the animation segment by setting *Strength* to `-100` and *Type* to `Simple (Fast)`.
5. Select the pose keyframe in frame 70 and apply an ease that slows motion to the greatest extent at the first keyframe, and uses a subtle overall effect. Set the *Strength* to `50` and set *Type* to `Stop and Start (Fast)`.
6. Test your file and compare it to *beads_04.fla*. The movement of the beads should seem more expressive, particularly as the movement changes direction.

Runtime Mode

While Authortime mode allows you to pose armatures for tweening, any resulting animation is permanent once you compile your file to SWF. Runtime mode, on the other hand, allows users to manipulate armatures at runtime, *without any ActionScript*. For example, an animation created in Authortime mode could show a marionette performing a short dance. However, enabling Runtime mode for the marionette could allow a user to control any of its movements.

To enable Runtime mode, simply select the armature layer and, in the Properties panel, choose Runtime from the Options→Type menu. You can try Runtime mode with your own files or use *beads_02.fla* to experiment.

There are some significant limitations to Runtime mode. First, you can only enable armatures with a single pose keyframe (in other words, no IK tweening animation) for Runtime mode, so preplanning is required. You must either animate your armature using tweening or enable the armature for user manipulation. Second, without ActionScript, only armatures in a FLA's first frame will work with Runtime mode enabled. Later in this chapter, you will learn the minimum ActionScript required to enable armatures in other frames to function.

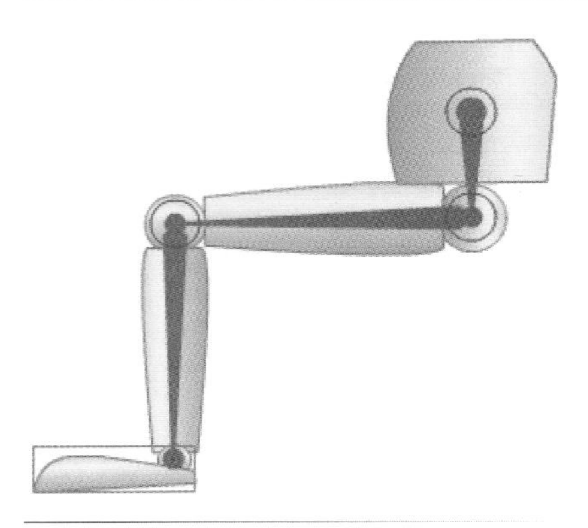

Figure 10-11. A posed leg armature

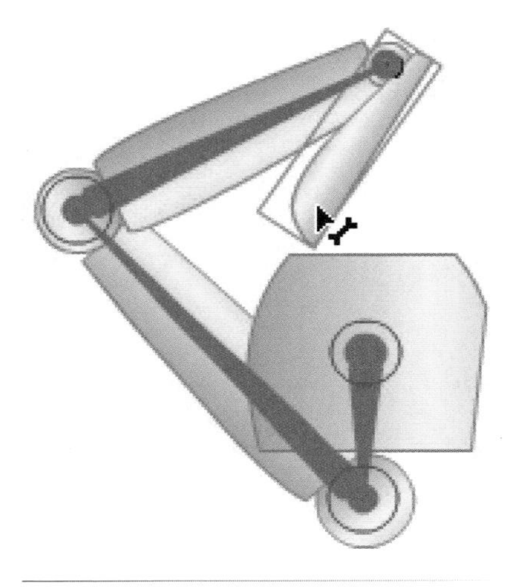

Figure 10-12. Creating an awkward pose using an unconstrained leg

Joint Rotation and Translation

By default, armature joints can rotate freely in 360 degrees. Figure 10-11 shows a sketch mannequin in a standard leg pose, as if sitting on a chair. Notice that a complete circle surrounds each joint: hip, knee, and ankle. This visual cue indicates at a glance that the joint can rotate a full 360 degrees.

360-degree freedom of rotation provides for nearly limitless posing, but can also be a little too much of a good thing at times. Think about using armatures for humanoid character animations, for example. Restricting joint movement to mimic human skeletal capabilities will make your animation more realistic and easier to create.

You'll also end up with fewer awkward poses like the one shown in Figure 10-12. Dragging the foot in a clockwise motion around the pelvis created the pose depicted in this figure. The leg has consequently angled up in front of the body, the knee has bent backward, and the foot is where the spine of a complete skeleton might be. Clearly, this pose is not possible for the average human.

Constraining Joint Rotation

To prevent the ambulatory disaster shown in Figure 10-12, you can constrain the minimum and maximum rotation of each joint in the armature to simulate the range of motion afforded by a human hip, knee, and ankle. To accomplish this, you'll first add another bone to the foot and then enable the rotation constraint feature for all the joints in the armature. You'll then set the degree of allowable rotation for each joint and see how these changes more closely mimic natural motion.

Adding a bone

The first step in being able to fully constrain the movement of the sketch mannequin's leg is to create a foot bone. This is a very important concept because it falls outside the default behavior of Flash IK armatures, and is not intuitive. Knowing how and when to take this step will help smooth your use of IK as an animation or interactivity tool.

As you may remember from the "Anatomy of an Armature" section earlier in this chapter, a bone only has a head joint if it is connected to another bone. In Figure 10-11, the armature ends at the ankle because the foot does not connect to an additional bone segment. Trying to draw a bone *within* the foot, from ankle to toe, for example, will fail because there is nothing to which the bone can connect.

The lack of a foot bone does not limit manual posing, because the point of rotation for the foot is the ankle. As Figure 10-12 demonstrates, you can manipulate the foot by hand just like any other bone. The trouble begins when you attempt to constrain the movement of the foot or, as you will learn later, try to pose the foot using ActionScript. Because there is no *tail joint* at the ankle, you cannot restrict the rotation at the ankle or set it with code.

The solution to this problem is to create a placeholder display object at the end of the final shape, to which a bone can connect. The top of Figure 10-13 shows a small movie clip placed at the end of the foot. The alpha value of the placeholder is set to 0 so it doesn't add a visual distraction to the animated object. The bottom of Figure 10-13 shows a bone connected from the ankle to the "toe placeholder," which then creates a joint at the ankle that you can constrain. The source file *leg_01.fla* shows a typical leg armature without a foot bone, while *leg_02.fla* makes use of this technique to add a foot bone ready for constraint.

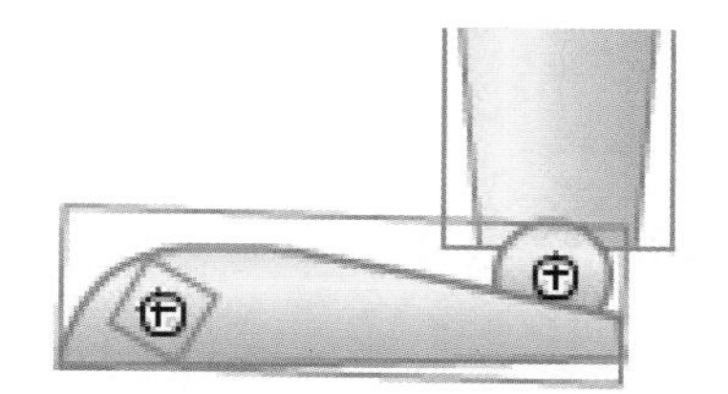

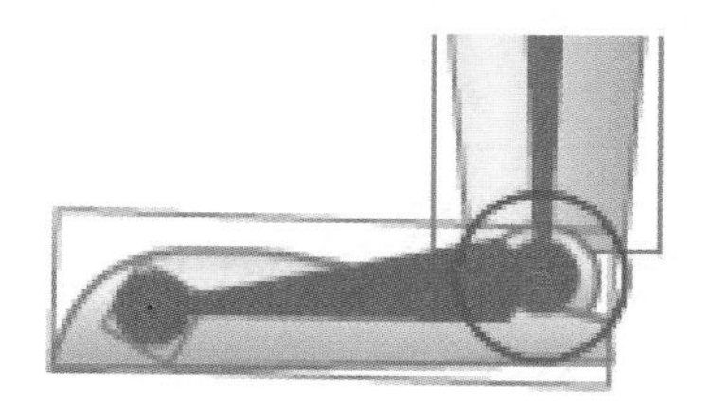

Figure 10-13. Adding a small movie clip to an armature (top) to create a bone without a visible head joint

Setting minimum and maximum rotation angles

To constrain the rotation of a tail joint, you must select a bone, expand the Joint:Rotation section of the Properties panel, and enable Constrain (Figure 10-14). By default, you will be presented with a Minimum rotation value of –45 degrees and a Maximum rotation value of 45 degrees. The desired movement of your bone and joint, however, determines the final values of these settings.

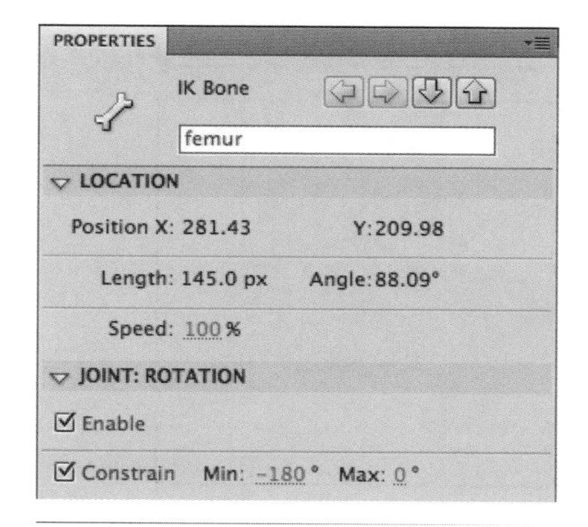

Figure 10-14. Constraining joint rotation to 180 degrees

Think of the constraint values of the sketch mannequin's thigh bone (femur), for example. To mimic the profile view of a human leg, the bone must be capable of rotating approximately 180 degrees at the hip—from parallel to the floor in front of the body to parallel to the floor behind the body.

NOTE

In addition to constraining rotation, you can alter the speed at which the bone can rotate to simulate weight. A Speed setting of 100% is the baseline speed of movement, so a speed setting of only 25% will make the bone seem much heavier.

Selecting the actual numeric values for the Minimum and Maximum constraint depends on the orientation of the bone. The best way to determine the angles is to look at a close-up of the joint when the Constrain setting is enabled and adjust the values based on visual feedback.

Figure 10-15 shows detailed views of each joint in the leg. Shown from top to bottom are hip, knee, and ankle. In each case, the circle that indicated unconstrained rotation has been replaced with wedges of different shapes, within which a protruding, handle-clad stem sits. The stem represents the current angle of the bone, and the wedge is the span of allowable rotation. In other words, the ends of the wedge stop the stem from moving any further, and the joint rotation is constrained.

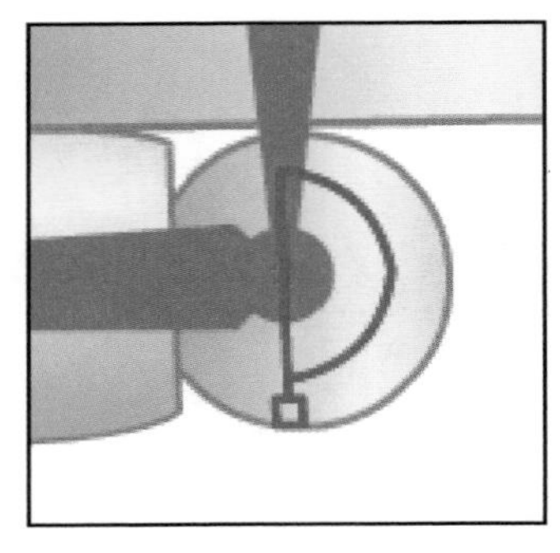

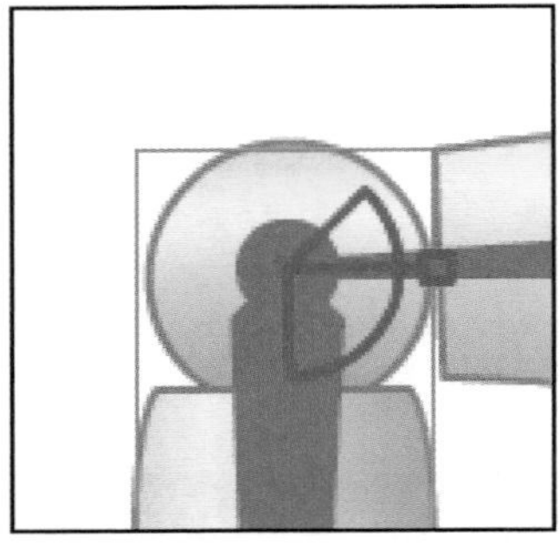

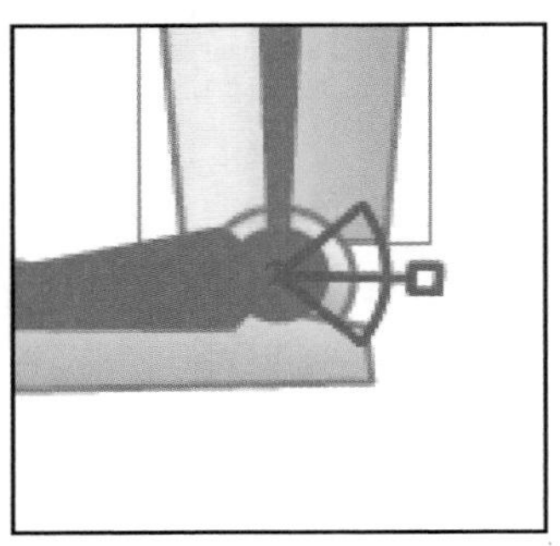

Figure 10-15. Joint rotation constraints at (from top to bottom) hip, knee, and ankle

The average human operates under similar constraints. The hip can rotate about 180 degrees, so its wedge is a semicircle. The knee can rotate a bit less than 180 degrees (between full leg extension and contraction when calf meets thigh). The ankle can move about 45 degrees forward (pointing up toward the knee) and 45 degrees backward (pointing away from the knee).

Determining numeric values for rotation constraint settings

As previously mentioned, the visual feedback at the armature joints is fairly clear, but the numeric values of the constraint angles depend on the orientation of the bone and can take some getting used to.

To begin with, Flash angles of rotation start at 0, which is due east, and move clockwise until reaching 180 degrees, or due west. At that point, the angles change to negative values. Continuing clockwise, the angles span from –180 degrees to 0 degrees. This approach was adopted because it is easier for the computer to rotate to –90 degrees than to rotate to 270 degrees—both of which are due north. You saw this in practice in Chapter 5 when rotating the viewing wheel of the portfolio project interface.

Furthermore, the original location of the stem shown in each joint in Figure 10-15 is determined by the original orientation of the bone. Based on the initial bone position, the stem is created at angle 0, or due east. If you repose the armature, this position changes. Looking at Figure 10-15, for example, the numeric values of the ankle's *Minimum* and *Maximum* constraints are –45 degrees and 45 degrees, respectively. Similarly, the knee constraints are –45 degrees and 90 degrees.

The hip constraints, however, are somewhat unexpected (–180 degrees and 0 degrees). Upon further consideration, this makes sense because the angle values reflected in the wedge overlay are relative to the stem protruding from the wedge. In Figure 10-15, the *Maximum* rotation has been reached, so the joint can rotate 0 degrees further. The leg can be folded under the hip, however, and it can rotate halfway around the circle to its minimum angle of –180 degrees.

The values of –180 degrees and 0 degrees appear in the Properties panel because the armature was originally created while the sketch mannequin was standing up. If you open the source file *leg_03.fla* (used for Figures 10-12 and 10-13) and rotate the leg joints so the mannequin is standing, you will see that the hip joint shows an angle of 0 and logical constraint values of –90 degrees and 90 degrees.

Seeing the constraints in action

Once you have applied the rotation constraints, no armature joint will rotate further than allowed, and the poses will be much more natural. Figure 10-16, for example, shows the result of rotating the leg in the same manner as for Figure 10-12: grabbing the foot and rotating clockwise around the hip. Notice that the result is very different from when the joints were unconstrained; now the constraint angles limit the leg to a natural stopping point at full extension.

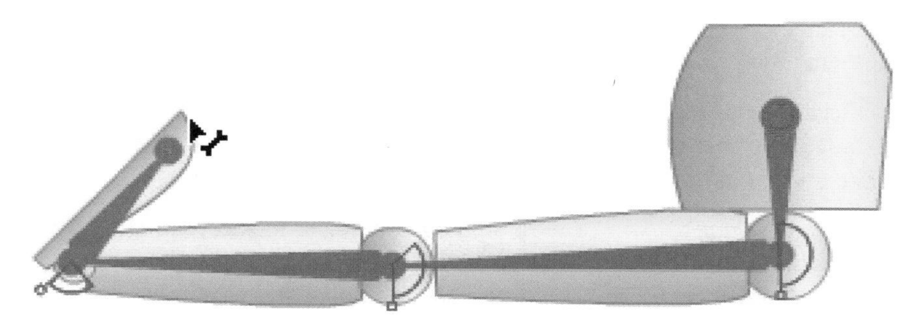

Figure 10-16. Testing posing limits of a constrained leg

Enabling Joint Translation

By default, every bone in an armature has a fixed length at the time of creation. Although bones can be rotated around joints, changes to their locations cannot alter the length of a parent bone. However, by enabling *joint translation*, you can move a bone along the x- or y-axis, changing the length of a parent bone in the process.

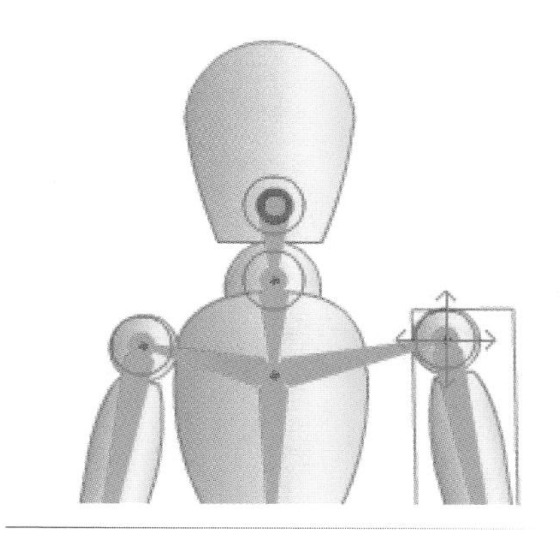

Figure 10-17. x and y joint translation enabled for one shoulder

Figure 10-17 shows both x and y translation enabled for the right shoulder of the sketch mannequin. The arrows that run parallel to the bone (currently facing up and down due to the orientation of the bone) indicate that y translation is enabled. The arrows perpendicular to the bone (currently facing left and right) indicate that x translation is enabled.

Enabling joint translation is a simple matter of turning on the Enable feature in the Properties panel (Figure 10-18), just as you did for joint rotation. You can also constrain the distance of translation allowed in either direction. If constrained, the arrows shown in Figure 10-17 will change to lines indicating the distance of allowable movement in either each direction.

▽ JOINT: X TRANSLATION

☑ Enable

☐ Constrain Min: 0.0 Max: 0.0

▽ JOINT: Y TRANSLATION

☑ Enable

☐ Constrain Min: 0.0 Max: 0.0

Figure 10-18. x and y joint translation enabled in the Properties panel

NOTE

If you enable both the x and y translation properties of a bone, it is easier to position the bone by temporarily disabling its rotation.

Transforming Movie Clips and Armatures

You can transform movie clips, even when they are part of an armature. For example, you can use the Free Transform tool to rotate a movie clip used in an armature without altering the properties of the armature itself. Transforming movie clips associated with an armature is no different than transforming any other display object. The only caveat is that you need to be sure you're selecting one or more movie clips rather than bones. This may affect the armature, too. For example, if you enlarge a movie clip, the attached bone will lengthen.

NOTE

If you transform movie clips that are part of an armature animation, these changes will be affected by keyframes like any other tweened property. So, be sure to apply such transformations to any and all relevant keyframes.

You can use this technique to add expression or humor to an animation. For instance, you might subtly rotate and scale arm and leg parts of a robot to make it seem in ill repair as it moves.

If you only want to *move* a bone or movie clip, while still maintaining its relationship to an armature, you can hold down the Alt (Windows) or Option (Mac) key when dragging the object. This is a quick and easy way to change the length of a bone. It's also a convenient way to move an entire armature. To do so, just select all pieces of the armature before dragging.

Finally, you can change the visual stacking order of movie clips after they have become a part of the armature. By default, each movie clip that is added to an armature will move to the top of the stack. If you want to change this stacking order, just use Modify→Arrange to reorder like you would with any other display objects overlapping in the same layer.

Using the Bone Tool with Shapes

You can also use the Bone tool to add armatures to shapes. Rather than altering the location or rotation of separate objects joined by bones, as was the case when working with movie clips, manipulating a shape armature will deform the shape itself.

To demonstrate adding an armature to a shape, this discussion focuses on a single simplified bat wing. Figure 10-19 depicts the wing in dark gray, atop an outline of a bat to provide visual context. A three-bone armature will originate in the lower-right corner of the wing, where it joins the body of the bat. To follow along with the discussion, open *bat_wing_01.fla*.

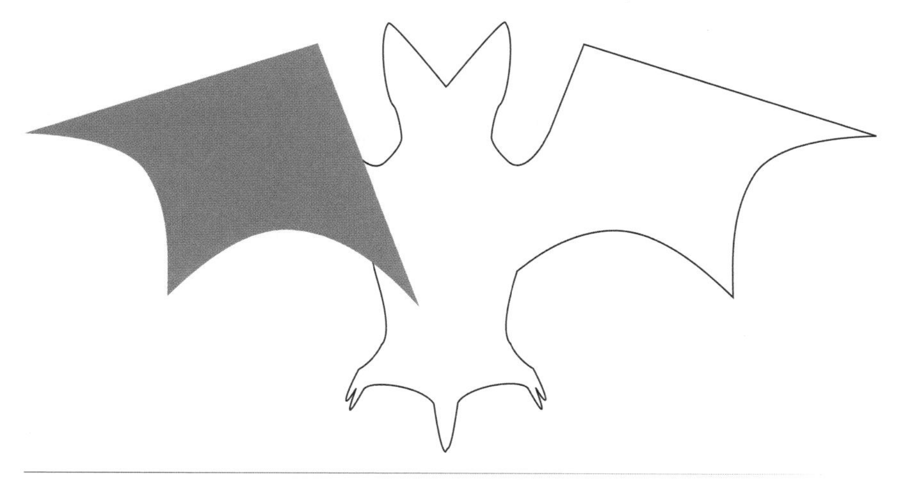

Figure 10-19. Original wing shape prior to creation of armature (outline of bat for visual context)

Creating an armature in a shape is no different than linking movie clips with an armature. Using the Bone tool, start where you want your root joint to be and draw bones end to end. Figure 10-20 shows the root joint in yellow, in the lower-right corner of the wing, and the first bone spans up to the top corner of the wing. From that top corner, two additional bones are drawn down to left and mid corners of the web-like shape.

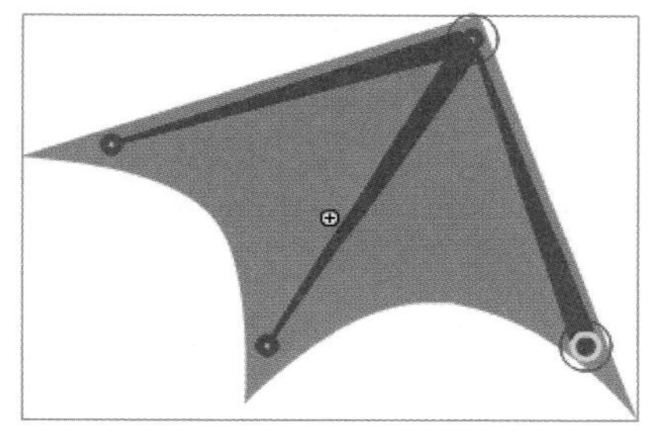

Figure 10-20. Three bones added to wing shape; root joint shown in yellow

If you are following along, your file should now resemble *bat_wing_02.fla*.

Dragging the bones outward will deform the shape to an extent that will make it appear like the wing span is elongating and its webbing is becoming taught. Dragging the bones inward will fold up the wing the way a real bat wing might fold in to its body.

However, Figure 10-21 shows that the default deformation of the wing begins to appear uncharacteristic as the left and center bones are moved inward to fold up the wing. This is because anchor points in the wing shape are associated with specific bones by default and their movement of the points doesn't reflect all of the adjustments made to the armature.

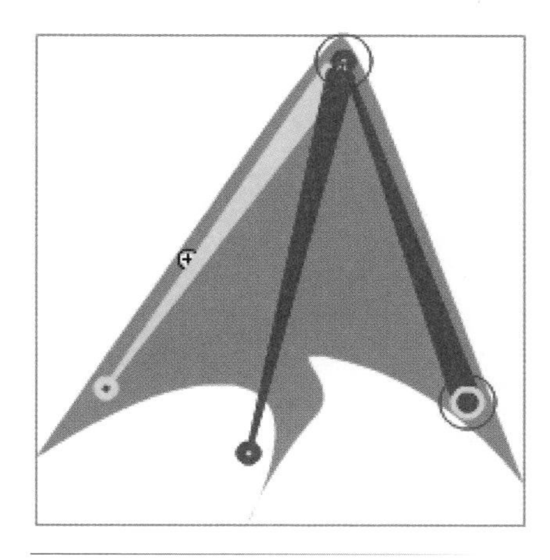

Figure 10-21. Moving the outer bone (at left) shows unpleasant deformation of default wing

For example, the webbing in the left half of the wing looks OK, but it is being pushed too far to the right because it is not considering the position of the center bone. The webbing in the right half of the wing is deforming in an unnatural-looking way because there are too many points in the shape being affected by the bones.

To correct these problems, you will use another IK tool, called the *Bind* tool, to associate points in the shape with relevant bones. You'll also use the Pen tool to remove extraneous points to simplify the morph as a whole.

Using the Bind Tool

Focusing on the left half of the wing to start, notice that a single point sits at the top of the webbing's arc. Clicking on this point with the Bind tool changes the point to a larger red rectangle and highlights the left bone with a yellow line (Figure 10-22). These two visual cues indicate that the point is bound to the bone highlighted in yellow. Because the point is only bound to one bone, it is being pushed beyond the center bone during the shape's deformation. To fix this, you must bind the point to both bones so it can update its position intelligently.

Figure 10-22. Anchor point in wing's left web is originally bound only to the leftmost bone, indicated by square marker

If you click on the point with the Bind tool and drag it to a bone, it will unbind the point from its previous bone and bind the point to the new bone. If you hold down the Shift key when dragging, it will bind the point to the new bone in addition to any existing bindings. Figure 10-23 shows this process in action, binding the point to the center bone as well as to the left bone. After a point is bound to more than one bone, it will display a red triangle when you select it with the Bind tool (Figure 10-24).

Figure 10-23. Using the Bind tool to bind the anchor point to the center bone in addition to the left bone

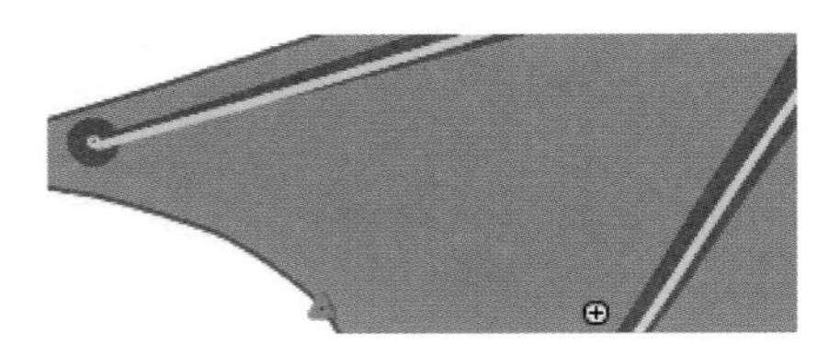

Figure 10-24. A triangular marker indicates that a point is bound to more than one bone

If you test the file again and move both left and center bones inward, the deformation of the left half of the wing will be more realistic. You can compare your work to *bat_wing_03.fla* to check your progress.

Adjusting Points

The right half of the wing must be bound in this way, too, but there is a more pressing problem. The arc in the right webbing has three points instead of one, and they are moving independently. To remove the extraneous points, switch to a Pen tool variant, the Delete Anchor Point pen tool, and click to remove the left and right points (Figure 10-25). Only the center point, at the top of the arc, should remain.

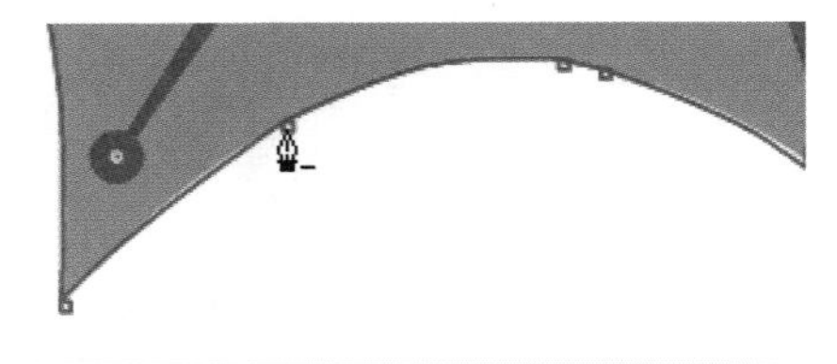

Figure 10-25. Deleting a shape anchor point with the Delete Anchor Point tool

After removing the extraneous points, switch to the Subselection tool, click on the remaining point, and adjust its position and control handles to reshape the bottom of the wing into a nice arc (Figure 10-26).

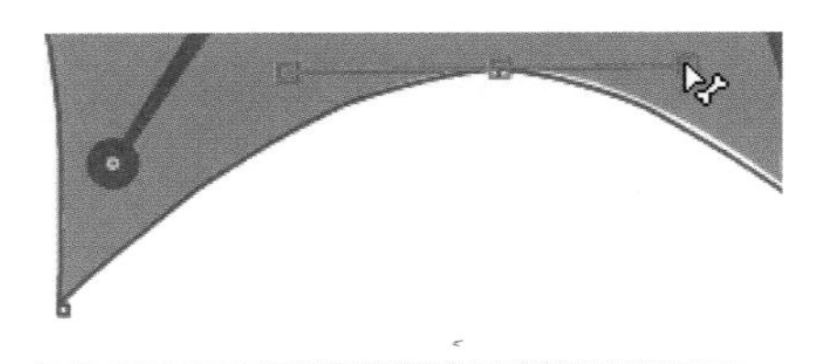

Figure 10-26. Adjusting a shape's anchor points and control handles

Once you are satisfied with the wing's shape, you can bind the center point to the center and right bones so it will take the position of both bones into consideration when morphing the shape. Figure 10-27 shows the folded wing after removing the extraneous points and binding the remaining points to the appropriate bones.

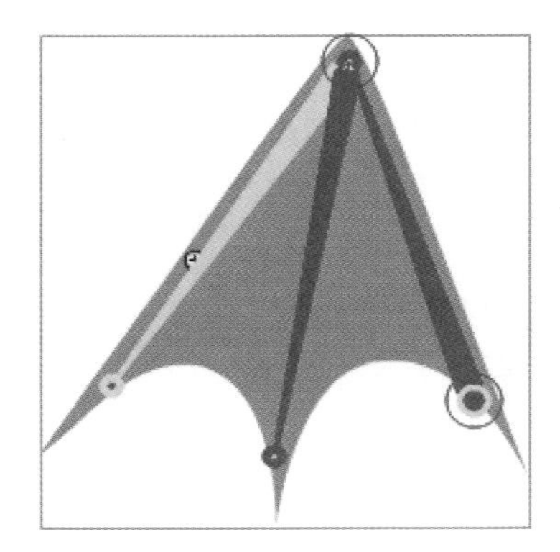

Figure 10-27. The wing folds correctly after cleaning up and binding anchor points

Basic ActionScript

The ActionScript required to exert any substantial control over IK armatures is more suited to experienced ActionScript coders. However, you can accomplish two important tasks with minimal code: supporting an armature's Runtime mode in frames other than frame 1, and posing an armature.

Supporting Runtime Armature Use Beyond Frame 1

One of the biggest weaknesses of codeless Runtime manipulation of IK armatures is that, by default, they only work in frame 1 of a FLA. Any armatures that you place in other frames will work in Authortime mode, and thus can be used to pose for IK tweens, but they will not work with Runtime mode enabled.

Although ActionScript is still not required to control these armatures at runtime, a small script is needed to register them with the player so it knows the armatures are available for user tinkering.

1. To try this script, open *beads_second_frame_01.fla*. This file has been set up with an armature and **`stop()`** action in frame 2, and the armature has already been set to Runtime mode.
2. Test your movie and see that you can't manipulate the armature at runtime.
3. Add the following script to the actions layer of frame 2, below the existing **`stop()`** method.

```
1   import fl.ik.*;
2
3   var armtr:IKArmature = IKManager.getArmatureByName("Armature_1");
4   armtr.registerElements(this.stage);
5   IKManager.trackIKArmature(armtr, true);
```

4. Save your file as *beads_second_frame.fla* and test again. The armature will now work in frame 2.

Line 1 of this script imports all the IK classes using a wildcard (*) so the Flash compiler knows where they are when compiling your SWF.

Line 3 creates a reference to an armature by name. Armature_1 is the default name for the first armature created. If your armature is not named the same way, you can either change the script to match the name you are using or rename the armature. You can rename the armature by changing the armature layer name or by clicking on the armature layer in the Properties panel.

Line 4 registers the armature within the scope of the Stage of your FLA so ActionScript can control it, and Line 5 tracks the armature so you can update it as needed.

Flash will automatically register the armature for you when it is in frame 1, but anytime you need to place an armature in another frame, the preceding script is required. You will need to place this code into your portfolio project, too, as you will have an armature on the Help screen.

Posing an Armature with Code

It's also possible to control armatures exclusively with code. This is preferable when you want to display animations that must vary from the permanence of the Timeline, without relying on user intervention.

NOTE

For an example of using ActionScript to move an armature over time, visit the book's companion website.

To display motion over time requires ActionScript that is a bit beyond the scope of this text. However, striking individual poses is a relatively straightforward process. To learn the process initially, you'll control the movement of a simple single-segment armature. Later, in the "Project Progress" section, you will apply what you've learned to pose a three-segment armature.

1. To try this script, open *single_bone_armature_01.fla*. For simplicity, an armature with one bone has been created in this file, and the armature has already been set to Runtime mode.

2. Test your movie. Notice that the bone's initial position (pointing north in the original source file) reflects the last pose you created in the authoring environment. Also notice that you can manipulate the armature at runtime.

3. Add the following script to the actions layer:

```
1   import fl.ik.*;
2
3   var armtr:IKArmature = IKManager.getArmatureByName("Armature_1");
4
5   var bone:IKBone = armtr.getBoneByName("ikNode_0");
6   var boneTJ:IKJoint = bone.tailJoint;
7
8   var ikMvr:IKMover = new IKMover(boneTJ, boneTJ.position);
9   ikMvr.moveTo(new Point(0,0));
```

4. Save your file as *single_bone_armature.fla* and test again. The armature will now initially point to the upper-left corner of the Stage. It also remains available to runtime control.

Lines 1 and 3 are the same as in the previous script. Line 1 imports all the required IK classes for the compiler, and Line 3 stores a reference to the armature with an instance name of *Armature_1*.

Lines 5 and 6 store a reference to the bone and joint you want to manipulate. Line 5 identifies a bone named *ikNode_0* in the previously referenced armature. You can change the names of a bone by selecting it and editing its instance name at the top of the Properties panel. Line 6 stores a reference to the tail joint of that bone.

Lines 8 and 9 create an instance of the **IKMover** class, designed to move bones, and attempt to move the bone to point (0, 0), or the upper-left corner of the Stage. Because the armature is anchored, it can't be moved to point (0, 0), and strains to get there. The result is that it ends up pointing in that direction.

You will initialize an armature in your portfolio project this way, but the armature will be in a movie clip. As such, the coordinates used will be easier to understand if they are translated from values relative to the movie clip to values relative to the Stage. To prepare for this adjustment, see the "Local and Global Coordinates" sidebar, next.

WARNING

The point used in the bone-moving script must be an instance of the Point class. You can't pass in separate x and y values.

Local and Global Coordinates

When working with display objects, many properties and methods are dependant upon the scope, or relative location, to which the code applies. For example, consider a movie clip with an instance name of **mc**, an upper-left-corner registration point placed on the stage at an x coordinate of 20 and a y coordinate of 60 (Figure 10-28).

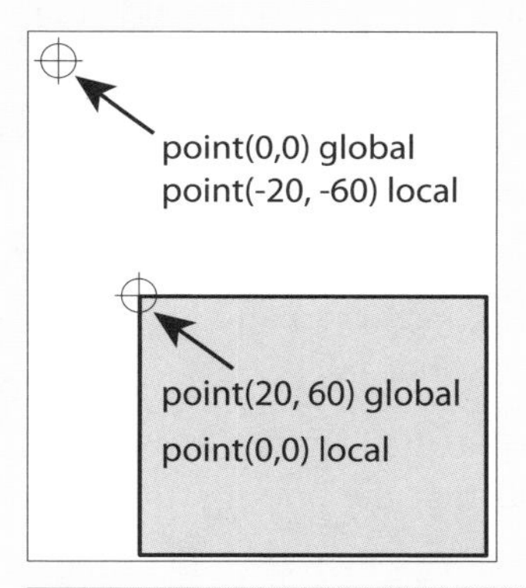

***Figure 10-28.** x and y coordinates relative to the stage (global) and a display object (local)*

From the Stage's perspective, that movie clip is at point (20, 60) and the upper-left corner of the stage is point (0, 0)—exactly as you'd expect.

However, from the movie clip's perspective, those points have different locations. The upper-left corner of the movie clip is no longer point (20, 60), but rather point (0, 0), because that position is now relative to the movie clip.

Similarly, the upper-left corner of the Stage isn't point (0, 0), but is point (–20, –60) instead. That is, the movie clip remains at the origin of its perspective and the upper-left corner of the stage is 20 pixels to the left and 60 pixels up.

On occasion, you may need to refer to x and y coordinates in another scope. To do so, you can use the *globalToLocal()* and *localToGlobal()* methods. These methods translate *Points* from one scope to another so you don't have to.

For example, if you wanted to reference the upper-left corner of the Stage from within the movie clip described previously, you would have to use point (–20, –60). However, that location isn't very intuitive. So, you can translate that information on the fly. You can use point (0, 0), which makes the most sense as the upper-left corner of the Stage, but the location will still be correct if you translate it first.

Using **globalToLocal()**, you are asking for a global coordinate (relative to the Stage) to be converted to a coordinate at the same location but relative to the movie clip:

```
trace(mc.globalToLocal(new Point(0, 0)));
//(x=-20, y=-60)
```

Using **localToGlobal()**, you are asking for a local coordinate (relative to the movie clip) to be converted to a coordinate at the same location but relative to the Stage:

```
trace(mc.localToGlobal(new Point(0, 0)));
//(x=20, y=60)
```

The companion source file, ***local_global_coordinates.fla***, shows a simple example of how this works, and you will also use the technique in this chapter's "Project Progress" section, next.

Project Progress

In this chapter, you'll add an armature to the Help screen so users can point to interface controls with Scaly's arm. In the next chapter, you'll populate a dynamic text element with help text based on which on-screen control Scaly's arm touches.

For this exercise, you'll need to open the accompanying source file *help_01.fla*. This file is very much like the default template you usually start with, but with two small exceptions. In the content movie clip, which has been renamed *HelpPage* for consistency with prior chapters, Scaly's torso has already been imported from Illustrator. More importantly, a new layer has been added in which Scaly's arm resides inside a movie clip.

Scaly's arm has been moved to a higher layer so that it can appear above the portfolio's viewing wheel when the Help screen loads. You want Scaly's torso to be cropped by the viewing wheel frame, just as the content from other frames has been cropped. However, his arm must be free to move outside the frame so it can interact with the navigation buttons and sound control at runtime.

You will modify the *HelpPage* movie clip in the next chapter when you add text. For now, you will focus on turning the arm movie clip into an IK armature.

1. Double-click the *arm* movie clip to edit its contents. You will see that the arm is composed of four pieces: upper arm, forearm, hand, and a small green dot at the end of the finger (Figure 10-29).
2. Select the Bone tool in the Tools panel. Enable the *Snap to Objects* feature by clicking the context-sensitive magnet icon at the bottom of the Tools panel.
3. Connect your first bone from the upper arm to the forearm. Click in the *upperarm* movie clip, at the shoulder, and drag to the *forearm* movie clip, at the elbow. Because you enabled *Snap to Objects*, the Bone tool will snap to the registration points of the movie clips if you are close enough.
4. Connect your second bone from the forearm to the hand. Click on the head joint of the previous bone, at the elbow, and drag to the *hand* movie clip, at the wrist. Don't worry if the *hand* movie clip is partially obscured; it will appear at the top of the armature when it is connected to the bone.
5. Connect your third bone from the wrist to the placeholder graphic. Click on the head joint of the previous bone, at the wrist, and drag to the *armature_placeholder* movie clip, at the finger. Don't worry if the *armature_placeholder* movie clip is partially obscured. It will appear at the top of the armature when it is connected to the bone.

6. Your armature is now complete. Click on the armature layer so the armature's properties are visible in the Properties panel. Change the Options Type to Runtime and test your movie. You should be able to drag the arm around the shoulder, unconstrained.

7. Close your SWF and return to the *arm* movie clip in the FLA. Now that you have moved the four movie clips that make up the armature to a dedicated armature layer, you can delete the remaining empty layer if desired.

8. Save your work as *help.fla*, and compare your file to *help_02.fla*. Continue with the exercise or, if you've had any problems thus far, continue with *help_02.fla*.

Figure 10-29. Four arm pieces prior to adding armature

Now that your armature is working, it's time to constrain its joints to move more like a human arm:

9. Double-click the *arm* movie clip to edit its contents.

10. Using the Selection tool, select the upper arm bone. At the top of the Properties panel, name the bone `humerus`. Under Joint Rotation, enable *Constrain*. Enter values of `-130` for *Minimum* and `110` for *Maximum*.

11. Repeat the process outlined in step 10 for the forearm and hand bones, using the following values:

    a. Name: `radiusUlna`, Minimum: `0`, Maximum: `140`

    b. Name: `metacarpal`, Minimum: `-60`, Maximum: `90`

12. The joints of your armature are now constrained. The final step in preparing your armature for inclusion in the project file is to hide the placeholder

at the end of the finger. Open the Library panel and double-click the *armature_placeholder* movie clip. Select the movie clip inside and set its *alpha* value to 0.

13. Test your movie again. When you drag the arm around, it should now be constrained at all three joints to more closely mimic human joint movement, and the placeholder at the end of the finger should now be invisible.

14. Close your SWF and return to the *arm* movie clip. If your arm doesn't behave as expected, look at the visible constraint wedges in Figure 10-30 and compare them with your own. If necessary, adjust the constraint values until the constraint wedge overlays look more like the ones in Figure 10-30.

15. Save your work and compare your file to *help_03.fla*. Continue with the exercise or, if you've had any problems thus far, continue with *help_03.fla*.

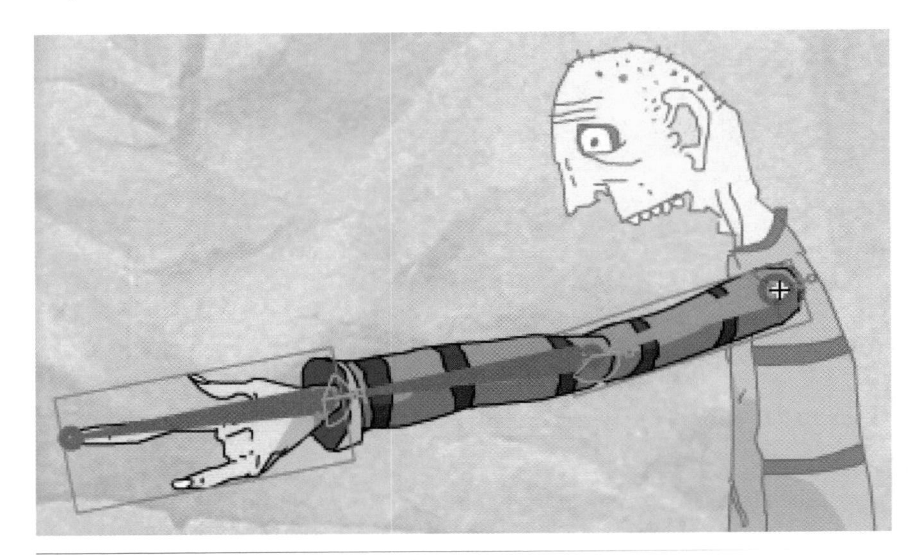

Figure 10-30. *A Scaly x-ray, showing joint constraints*

Your armature is now complete, and it's time to add it to the project file. You will copy the main content movie clip, *HelpPage*, from your *help.fla* file and paste it into the content layer of your main project file, as usual. However, this time you will also create a new layer in the project file, above all other interface layers, for the arm. This way, the arm can be dragged above the navigation buttons and sound control. When the layer drags over these elements, text feedback will display, based on a script that you'll add in the next chapter.

You'll end this chapter by adding ActionScript to the main project file that registers the movie clip for runtime operation and presets the arm's bone positions upon entering the frame:

16. In your help FLA, unlock the content layer, if necessary, and select the HelpPage movie clip by clicking on Scaly's torso, then copy it to memory.
17. While your *help.fla* file remains open, open your main project file. Scroll to the *help* section at the end of the Timeline.
18. In the *content* layer, at frame 224, add the *HelpPage* movie clip by using Edit→Paste in Place.
19. Select the *sound* layer and add another layer to the Timeline using the New Layer button at the bottom of the Timeline, or by using the Insert→Timeline→Layer menu command. Name the layer *helpArmature*.
20. Add empty keyframes (F7) to the *helpArmature* layer at frames 224 and 231 to match the keyframe locations in the *content* layer.
21. Switch back to the *help.fla* file, select the *arm* movie clip, and copy it to memory.
22. Switch to the main project file again and paste the *arm* movie clip into the *helpArmature* layer using Edit→Paste in Place.
23. Double-click the *arm* movie clip to edit its contents.
24. Click the armature layer in the Timeline and, in the Properties panel, be sure Runtime is selected in the Options Type setting.
25. Save and test your movie. After the opening animation is complete, click the Help button and test your armature.

Theoretically, you should be able to use your armature, but it won't work. This is because runtime armatures will only work in frame 1, by default. You will need to add ActionScript to register the armature. You will also add ActionScript to pose the armature when entering the screen.

26. In frame 227, add lines 2 through 7 of the following script after the **stop()** action already in place in the *actions* layer. This script will register the armature for runtime operation. If explanation is required, reread the "Supporting Runtime Armature Use Beyond Frame 1" segment of the "Basic ActionScript" section, earlier in this chapter:

```
1   stop();
2
3   import fl.ik.*;
4
5   var armtr:IKArmature = IKManager.getArmatureByName("Armature_1");
6   armtr.registerElements(this.stage);
7   IKManager.trackIKArmature(armtr, true);
```

27. Test your movie again, and navigate to the Help screen. Test the armature by dragging it around. It should rotate in a constrained manner and be capable of dragging over the navigation buttons, sound control, and logo.

28. All that remains now is to initialize the armature every time you visit the *Help* screen. Add the following code to frame 227 at the end of the same script you edited in step 26:

```
8    presetBone("humerus", arm.globalToLocal(new Point(380, 350)));
9    presetBone("radiusUlna", arm.globalToLocal(new Point(270,
     300)));
10   presetBone("metacarpal", arm.globalToLocal(new Point(270,
     200)));
11
12   function presetBone(whichBone:String, pt:Point):void {
13       var bone:IKBone = armtr.getBoneByName(whichBone);
14       var boneTJ:IKJoint = bone.tailJoint;
15       var ikMvr:IKMover = new IKMover(boneTJ, boneTJ.position);
16       ikMvr.moveTo(pt);
17   }
```

29. To better learn about the script from the previous step, test your file to see its results. Everything will be as it was before, including the ability to drag Scaly's arm around the screen. However, each time you enter the Help screen anew, Scaly will be pointing up (Figure 10-31).

Figure 10-31. Scaly's default pose each time you visit the Help screen

As for the script described in step 28, it's very much like the script discussed in the "Posing an Armature with Code" segment of the "Basic ActionScript" section, earlier in this chapter. However, there are a few important differences.

First, the code is wrapped in a function in lines 12 through 17, and called in lines 8 through 10. This allows you to pose three bones efficiently, rather than repeating code three times. You're passing information into the function, including the name of the bone, and the location to which the bone should be set. The bones are given custom names (roughly corresponding to the names of human bones) for clarity only. In your own projects, you can stick with the default assigned names if you prefer.

Second, the armature is already stored in the `armtr` reference variable determined in the prior script.

Most importantly, however, the points passed into the function are translated from values relative to the *arm* movie clip to values relative to the Stage. During development, you can look at the Stage and estimate the location of the bones you want to use. However, these coordinates are relative to the upper-left corner of the stage, or point (0, 0). If you do not translate these points into values relative to the movie clip, they will all be offset from the movie clip's location and the armature will point southeast.

Conversely, if you have to use points relative to the movie clip, they will be difficult to determine when writing your script, and probably confusing later on. To avoid this issue, you can use points that are familiar and easy to determine based on the upper-left corner of the Stage, and then let Flash translate them on the fly to values relative to the movie clip. For more information about translating point values among scopes, see the "Local and Global Coordinates" sidebar, earlier in this chapter.

The Project Continues...

In the next chapter, you'll work with dynamic and input text elements on the Help and Lab screens.

TEXT

Introduction

The lion's share of the attention Flash receives is usually due to the eye candy that so many talented artists and programmers create every day. Over the last several years, however, Flash has become recognized more and more as a tool for creating applications in which text plays a major part.

Increasingly, Flash is used for everything from mortgage calculators to newsreaders to rich Internet applications of many kinds. Flash is even sometimes used as a front end for very text-intensive products, such as blogs, bulletin boards, and other dedicated content management systems.

As Flash matures, this trend increases. New versions of Flash and ActionScript continue to add text-related features, including improved components, application interface support for font families and styles, and even in-menu rendering that displays sample text in each font for an on-the-fly preview.

NOTE

At the time of this writing, Adobe was working on a new text architecture for the Flash Platform called the Text Layout Framework. Although still in beta, the framework boasts many dramatic text improvements, including reading and writing in vertical, right-to-left, and left-to-right languages; support for double-byte characters (used for Chinese, Japanese, and Korean, among other languages); flowing text across multiple columns and linked containers, as well as improved flow around inline images; improved typographical control, such as kerning, ligatures, and hyphenation; enhanced runtime cut, copy, paste, and undo; and much more. Check http://labs.adobe.com/technologies/textlayout/ and http://www.adobe.com/ for more information.

In this chapter, you'll learn how to create and populate text elements using components, the Flash interface, and ActionScript. You'll also learn how to format text using the interface and code, including native ActionScript objects and HTML and cascading style sheets (CSS). Finally, you'll learn about a great new workflow that lets you export layouts from Adobe InDesign and open them in Flash.

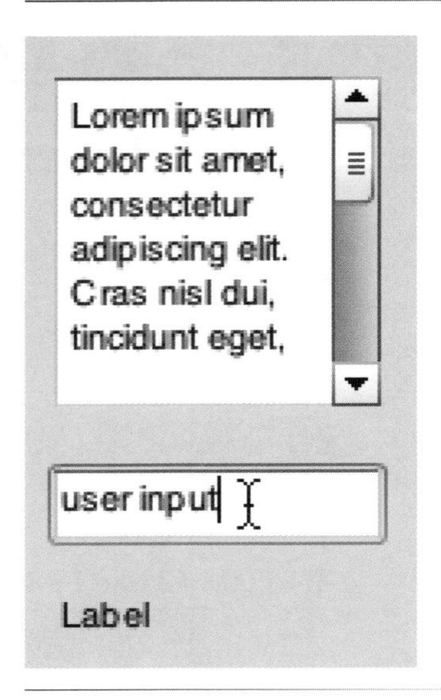

Figure 11-1. From top to bottom, TextArea, TextInput, and Label text components in use

Components

When you need to build user interfaces that require text, components are a quick and easy way to get started without using any ActionScript. In addition to the *UIScrollBar* component for easily customizable scrolling text (demonstrated in Chapter 9), three other text components ship with Flash CS4 Professional. Shown in Figure 11-1, from top to bottom, they are *TextArea* (used for scrolling text passages), *TextInput* (for capturing and processing user input), and *Label* (for basic text display).

Using the Component Inspector Panel to Configure Text Components

You can use the *TextArea* and *Label* components without ActionScript by configuring the components with the Component Inspector panel. You can also configure the *TextInput* component this way, but it typically requires additional ActionScript to validate the user's input. This section of the chapter will discuss use of the Component Inspector panel, and the next section will demonstrate simple ActionScript tasks for each component.

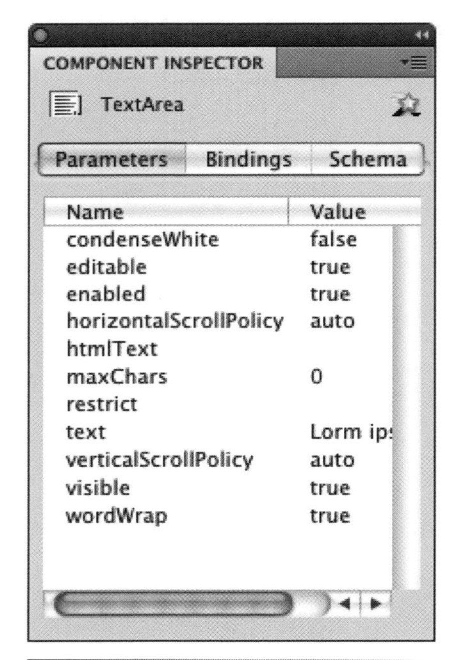

Figure 11-2. Configuring the TextArea component

Shown in Figure 11-2, the Component Inspector panel indicates that the *TextArea* component can be *enabled*, *visible*, and *editable*. The *wordWrap* property can wrap words onto multiple lines, and the *horizontalScrollPolicy* and *verticalScrollPolicy* properties can set horizontal and vertical scrolling, respectively, to on, off, or auto (based on length of text). The component can be populated using the *text* or *htmlText* properties and, if the latter is used, *condenseWhite* can remove extraneous spaces, tabs, and return characters from the HTML text. If editable, user input can be restricted to specific characters (using the *restrict* property) and/or a limited number of characters (using the *maxChars* property).

NOTE

The TextArea component makes use of the UIScrollBar component discussed in Chapter 9. This nicely demonstrates the efficient reuse of components that stems the increase in file size as additional related components are used. If you use the TextArea component, for example, adding the UIScrollBar component will not increase file size further.

Figure 11-3 shows that the *TextInput* component shares many of its properties with *TextArea*, including *editable*, *enabled*, *maxChars*, *restrict*, *text*, and *visible*. The lone unique property, *displayAsPassword*, automatically replaces user input with asterisks to obscure the entered text from prying eyes.

Like the *UIScrollBar* component, the *Label* component displays text inside other components. This is apparent in Figure 11-4, where you may notice that *Label* shares most of its properties with the other text components. Its only unique property is *autoSize*, which you can set to left, center, right, or none to automatically resize the label based on the quantity of text applied. The *Label* component is a viable alternative to creating, styling, and populating a text field using ActionScript. This is a handy option for users without programming experience, but it can't be changed at runtime. If you require a more dynamic solution, ActionScript is your answer.

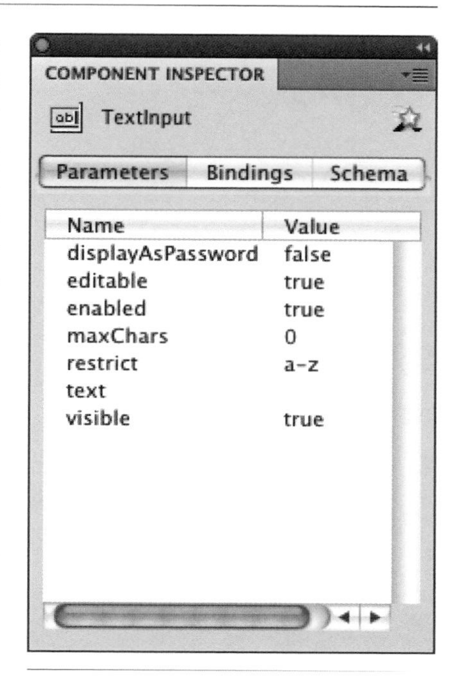

Figure 11-3. Configuring the TextInput component

Using ActionScript to Configure Text Components

As with most components that ship with Flash CS4 Professional, you can use the text components not only by dropping them onto the Stage and configuring them with the Component Inspector panel, but also by applying ActionScript. In most cases, the component's parameter names correspond with like-named ActionScript properties.

This section provides one simple ActionScript example for each component. Feel free to create a file (or files) or use the *text_components_01.fla* companion source file. For the following code to work, each component used must already be in your file's Library. As described in Chapter 9, dragging a component from the Components panel to the Stage also places that component in the file's Library. Thereafter, it is available for use as many times as you need it, just by dropping it from your Library onto the Stage.

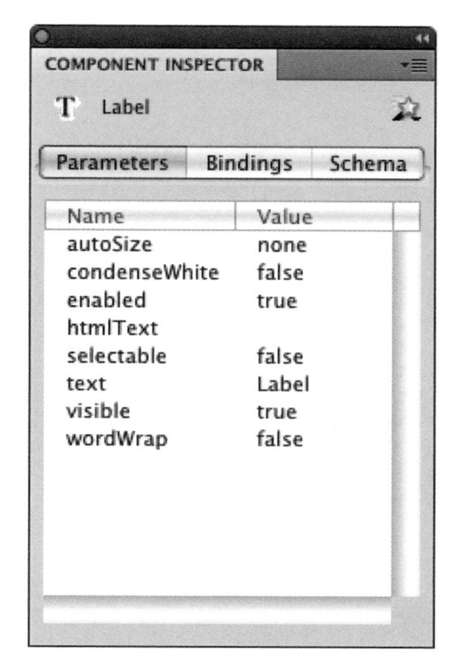

Figure 11-4. Configuring the Label component

Common coding

Before discussing the code specific to each component, let's look at the concepts that are common to all three. The first is the need for information that lets the Flash compiler know how to validate and check your code when creating a SWF. Not every Flash designer or developer will use components. As such, the compiler is not automatically aware of their classes the way it knows about other native classes (such as *MovieClip*). So, unlike for most ActionScript written in a frame of the file's Timeline, you must use an `import` statement to tell the compiler where to find the necessary classes. This first line in the following three examples doesn't actually import anything into your file. Instead, it provides the compiler with a path, or pointer, to the location of each component's class.

The next thing common to all three examples is the standard procedure for creating a display object, adding it to the display list, and positioning and sizing it on the stage. As you've seen in prior chapters, the `new` keyword creates an instance of the component, the `addChild()` method adds it to the display list, and the `x`, `y`, `width`, and `height` properties position and size the component.

The component-specific sections that follow omit these common elements, focusing instead on the properties specific to each component.

TextArea

This example varies the default behavior of the *TextArea* component by setting its **`editable`** property to **`false`**. It then populates the display-only component with text using the **`text`** property:

```
1   import fl.controls.TextArea;
2   var txtArea:TextArea = new TextArea();
3   addChild(txtArea);
4   txtArea.x = 200;
5   txtArea.y = 10;
6   txtArea.editable = false;
7   txtArea.width = 100;
8   txtArea.height = 100;
9   txtArea.text = "Lorem ipsum dolor sit amet, consectetur adipiscing
    elit. Cras nisl dui, tincidunt eget, ultrices quis, condimentum sit
    amet, sapien.";
```

TextInput

The *TextInput* component can restrict user input to specific characters and obscure visible text to hide password entry:

```
1   import fl.controls.TextInput;
2   var txtInput:TextInput = new TextInput();
3   addChild(txtInput);
4   txtInput.x = 200;
5   txtInput.y = 130;
6   txtInput.restrict = "a-z";
7   txtInput.maxChars = 8;
8   //txtInput.displayAsPassword = true;
9   txtInput.addEventListener(Event.CHANGE, onChange);
10  function onChange(evt:Event):void {
11      if (evt.target.text == "actionscript") {
12          trace("Password correct");
13      }
14  }
```

This example uses the **`restrict`** property of the *TextInput* component to limit user input to a *space* and letters *a* through *z*. It also limits the user to entering eight characters using the **`maxChars`** property. Optionally, you can set the **`displayAsPassword`** property to **`true`** to display asterisks instead of entered characters. This instruction is commented out in line 8 to let you see what you type. If you try this script, comment and uncomment this line a couple of times to see its effect on the component's behavior.

Finally, the text property is used again in line 11, as the event listener in lines 9 through 14 monitors the user's input. Every time the contents of the component change, the text is compared with the **`"actionscript"`** string. If the two compared strings match, an affirmative message is traced to the Output panel.

Label

This simple component adds a text label that you can easily set with code:

```
1   import fl.controls.Label;
2   var txtLabel:Label = new Label();
3   addChild(txtLabel);
4   txtLabel.x = 200;
5   txtLabel.y = 170;
6   txtLabel.width = 100;
7   txtLabel.autoSize = TextFieldAutoSize.RIGHT;
8   txtLabel.text = "Lorem Ipsum";
```

This example populates the **`text`** property of the *Label* component, and sets its **`autoSize`** property to the constant **`TextFieldAutoSize.RIGHT`**. This value aligns the text in the label to the right and automatically resizes the component down and to the left as needed. A string value of **`"right"`** would also work for *autoSize*, but, as you read in previous chapters (including Chapter 6), ActionScript 3.0 makes heavy use of public static constants to make sure these values are consistent and easy to maintain.

Configuring Text Fields with the Flash Interface

In Chapter 2 you learned how to create *Static* text fields and configure basic character and paragraph properties. You may recall that Static text fields are rendered as graphics at runtime and, therefore, can't be edited on the fly. If you need more flexibility, however, you can turn to Flash's Dynamic and Input text fields. *Dynamic* text fields are ActionScript programmable, and *Input* text fields are Dynamic fields with the added feature of accepting user input.

Creating each type of field follows the same process. First use the Text tool to draw a text field on the stage and then, with the field still selected, use the top section of the Properties panel to change its type. Position and size settings are shared among all text types, but what makes each type different are the values of the properties in the lower sections of the panel.

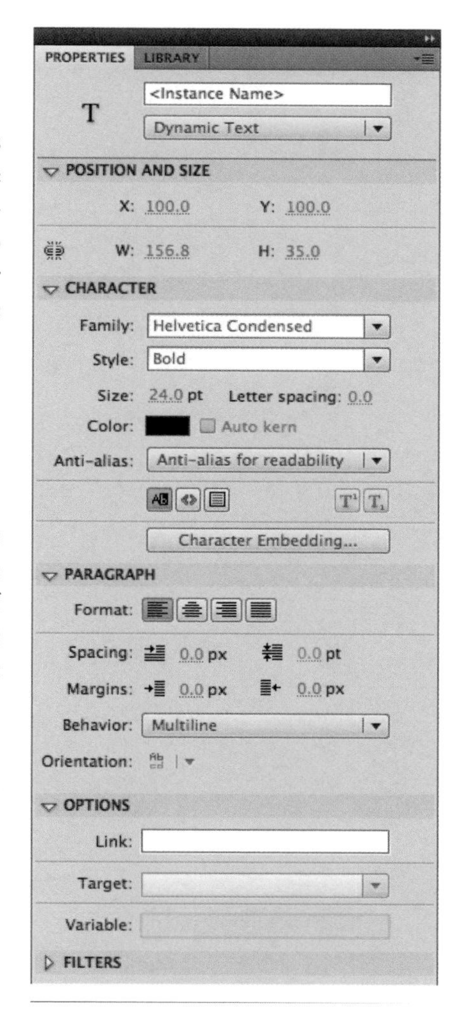

Figure 11-5. Properties of a Dynamic text field

Dynamic Text

Of the three text types, Dynamic fields include the largest number of features in the Properties panel, shown in Figure 11-5 and listed by section.

Character

The following text field properties control character-level manipulation but also affect attributes such as anti-aliasing, font embedding, and field appearance, among others:

Family

The *Family* property specifies the font family to use. To support ActionScript programming, Dynamic text fields are not converted to graphics when the SWF is compiled. As such, the font you select must either be commonplace among your audience or embedded to ensure proper display on any computer. Embedding fonts will be discussed later in the "Formatting Text" section of this chapter.

Style

The *Style* menu includes all font styles available for the chosen font. Styles are no longer limited to plain and faux font styles (skewing fonts for italics, and thickening fonts for bold). When found, faux styles are still included, but the menu also includes all family-specific styles, such as light, regular, book, medium, bold, heavy, and so on, where applicable.

Size

Size is the type size, in points.

Letter spacing

The *Letter spacing* setting controls the space between *all* the letters of the text field. This feature is sometimes called *tracking* in other applications. This is in contrast to *kerning*, which is the space between *two* letters.

Color

Color is the color of the text, and provides access to the pop-up color palette.

Auto kern

Disabled. *Auto kern* will use the kerning information built into the font to control the spaces between letter pairs. This option is only available to Static text fields.

Anti-alias

Anti-alias controls how the text is blended into its surroundings. This option contains four presets: *Use device fonts* (for operating-system-level anti-aliasing), *Bitmap text* (for no anti-aliasing), *Anti-aliasing for animation* (standard grayscale anti-aliasing composited by Flash Player), and *Anti-aliasing for readability*. This last setting is a unique anti-aliasing option that uses hints of color to blend better into any color background. This can usually reduce or eliminate the faint white halo that sometimes surrounds text when you use grayscale anti-aliasing over dark or color backgrounds. This menu also contains a custom anti-aliasing option through which you can control the thickness and sharpness of embedded fonts only.

Selectable

The final row of settings in the Character section of the Properties panel is a series of small buttons. The leftmost button controls whether the text is selectable at runtime. This is handy in some cases, such as for copy and paste support. However, if this is not a feature you want to include, it can be annoying because text-selection highlighting and/or the standard I-beam text cursor may appear and distract your users from design or animation. Typically, you should disable this setting.

Render text as HTML

The *Render text as HTML* option causes the field to parse text with limited HTML support when added to the field via ActionScript. The HTML features supported will be discussed later in the "Using HTML" section of this chapter.

Show border around text

The *Show border around text* option places a white background behind the text and surrounds the field with a thick black stroke. Without ActionScript, these colors are not configurable.

Subscript and Superscript

The last two buttons in the row below the *Anti-alias* option are the *Toggle the superscript* and *Toggle the subscript* buttons. These buttons convert selected characters to superscript or subscript but are active in Static text fields only.

Character Embedding

Character Embedding allows you to use custom fonts by embedding character outlines into the SWF during the compiling process. This means that custom fonts will appear consistent on all computers. Embedding fonts will be discussed later in the "Formatting Text" section of this chapter.

Paragraph

These text field properties control paragraph-level manipulation, including line wrapping and orientation, among other attributes:

Format

The *Format* property controls the alignment of the text within the field, and can be left, center, right, or full justified.

Spacing

The *Spacing* options contain paragraph-specific indent and leading (line spacing) values.

Margins

The *Margins* feature specifies the text indent within the field on the left and right sides.

Behavior

For Dynamic text fields, the *Behavior* menu controls the line-wrapping behavior. Options include single line, multiline, and multiline no wrap (line returns are processed but not line wrapping). Behavior options available to other text field types are explained in their respective sections.

Orientation

This feature is disabled for dynamic text fields. The *Orientation* property controls the direction of the text flow. Options include horizontal, vertical left-to-right, and vertical right-to-left. This feature is only available to Static text fields.

Options

These properties control how links applied to the text field behave.

NOTE

You can use additional ActionScript to trigger ActionScript functions from text links. This book's companion volume, Learning ActionScript 3.0: A Beginner's Guide *(O'Reilly), contains examples of this technique.*

Link

The *Link* property allows you to apply URLs to text without the need for HTML. For Dynamic text fields manipulated with the Properties panel, this property applies to the entire field. Static text fields allow the application of links on a per-character basis.

Target

The *Target* property, like its HTML counterpart, allows you to specify in which browser window the link will open.

Variable

Disabled in all text field types in ActionScript 3.0. In ActionScript 1.0 and ActionScript 2.0 only, the *Variable* option provides a direct link between a variable and a text field. You can therefore use it to easily display the contents of a variable on stage.

Filters

Filters were introduced in Chapter 7, and there is nothing new here that warrants a new discussion. It is worthy to note, however, that you can apply filters directly to text fields without first enclosing them in a movie clip.

Input Text

Input text fields are nearly identical to Dynamic text fields, with the important difference that they accommodate user input. Other subtle variances in behavior are cited in the following subsections.

Character

The *Selectable* feature is disabled, naturally, as the entire purpose of input text is for the fields to be editable by the user and, therefore, selectable.

Paragraph

A Password option has been added to the *Behavior* menu. When enabled, it substitutes asterisks for typed characters, obscuring the input from view.

The *Orientation* feature has been disabled. Vertical input is not supported in the shipping version of Flash CS4 Professional.

Options

The *Link* and *Target* features have been removed, as links cannot exist where user input has not yet been entered. Instead, a *Max chars* option has been substituted (Figure 11-6). This feature limits the maximum number of characters that can be entered into an Input field.

OPTIONS
Max chars: 0
Variable:

Figure 11-6. The Options section of the Properties panel with an Input text field selected

Static Text

As described earlier, Static text elements remain editable during authoring, but are rendered as graphics when compiled into your SWF. This means that they will render consistently on all computers, but it also means that you cannot control them with ActionScript. Accordingly, the first thing to notice is that no *Instance name* field exists at the top of the Properties panel while a Static text field is selected.

Position and Size

The *Height* option is disabled. Although you can set the width of a Static text field, its height is automatically set to the minimum size necessary to display the entire text.

Character

Auto kern is activated, allowing you to enable or disable the kerning information built in to each font. *Toggle the superscript* and *Toggle the subscript* are enabled, allowing users to apply superscript or subscript formatting to any character.

Render text as HTML and *Show border around text* are disabled, as neither ActionScript nor users can enter text or HTML into Static text fields.

Paragraph

The *Multiline* feature is disabled. Line returns and line wrapping are automatically supported by manual entry, but ActionScript control does not apply.

Orientation is enabled, allowing both horizontal and vertical directional alignment.

Options

The *Variable* option is removed because Static fields cannot display ActionScript variable values.

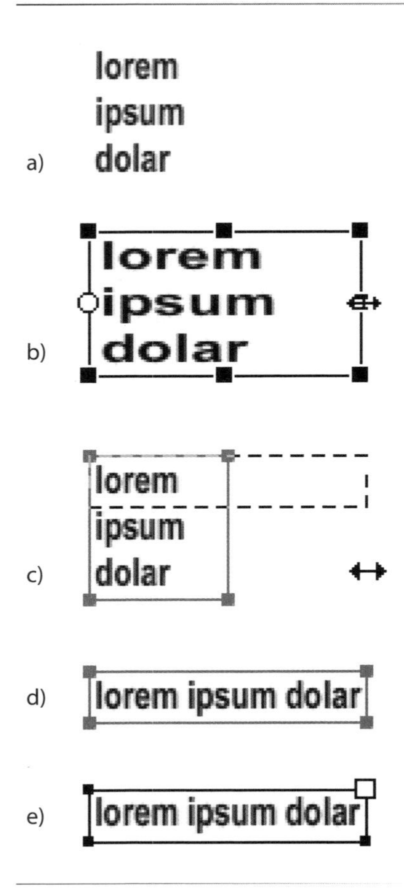

Figure 11-7. Resizing text: a) original text; b) scaling with the Free Transform tool; c) resizing using the text field handles; d) after resizing c), e) different handles displayed while editing the text at authoring time

Resizing Text Fields

Resizing text fields is a bit different than resizing other display objects. You can scale a text field using the Free Transform tool, just like you can scale a movie clip and the text characters will scale accordingly. Shown in Figure 11-7 b), this can result in distorted text if you are trying to resize the text field only, rather than scaling the characters within.

You can resize the text field without scaling the text in one of two ways. If you click on a text field with the Selection tool, four blue handles appear. These are the text resize handles, which you can drag to change the shape of the text field, as shown in Figure 11-7 c) and d). You can also double-click to edit the text field and, while editing, use the large solid-white handle in the upper-right corner of the field to resize without scaling.

Configuring Text Fields with ActionScript

Using ActionScript to create, configure, and populate text fields brings a heightened degree of control and freedom of runtime manipulation that is unavailable to text fields created solely with interface tools. This section contains one simple ActionScript example each for Dynamic and Input text fields. They will work together in the same sample file, and will not be used in the final project. If you prefer, you can test *as_text_fields_01.fla* to see how the script operates.

Dynamic Text

The following script not only configures the dynamic text field, but also uses code to populate the field with text:

```
1   var txtFld:TextField = new TextField();
2   addChild(txtFld);
3   txtFld.text = "Hello Scaly";
4   txtFld.x = 20;
5   txtFld.y = 20;
6   txtFld.width = 100;
7   txtFld.autoSize = TextFieldAutoSize.LEFT;
8   txtFld.selectable = false;
9   txtFld.multiline = true;
10  txtFld.wordWrap = true;
11
12  for (var i:int = 0; i < 25; i++) {
13      txtFld.appendText(" " + "word" + i);
14  }
```

Lines 1 and 2 create a text field, which is Dynamic by default, and add it to the display list. Line 3 adds a text string to an otherwise empty text field. Lines 4 and 5 position the field, and lines 6 and 7 size the field. The width is set explicitly in line 6, but line 7 anchors the height at the upper-left corner and automatically expands down as needed. Lines 8 through 10 disable the *selectable* option and enable the *multiline* and *wordWrap* features. Finally, lines 12 through 14 use the **appendText()** method to add text to an already populated text field. The **for** loop executes 25 times, adding a string containing three elements to the field. The string starts with a space and is followed by "word" and the number of the current iteration through the loop. For example, the first two additions are " word0", and this pattern continues on through " word24," where the process ends.

Input Text

Input text fields add control over user input, as well as styling options for border and background colors. This example script uses a simple conditional statement to check the accuracy of a user-typed password:

```
15  var txtFld2:TextField = new TextField();
16  txtFld2.type = TextFieldType.INPUT;
17  addChild(txtFld2);
18  txtFld2.x = 200
19  txtFld2.y = 20;
20  txtFld2.width = 100;
21  txtFld2.height = 20;
22  txtFld2.border = true;
23  txtFld2.borderColor = 0x990000;
24  txtFld2.background = true;
25  txtFld2.backgroundColor = 0xFFEFEF;
26  txtFld2.textColor = 0x990000;
27  txtFld2.maxChars = 10;
28  txtFld2.restrict = "0-9";
29  txtFld2.displayAsPassword = true;
30  stage.focus = txtFld2;
31  txtFld2.addEventListener(Event.CHANGE, onChange);
32  function onChange(evt:Event):void {
33      if (evt.target.text == "0123456789") {
34          trace("Password correct");
35      }
36  }
```

In this case, the first important change occurs in line 16 when the field is changed to an Input type. Lines 22 through 26 display a maroon border and text, with a light red background. Lines 27 and 28 restrict the user input to numbers only and limit it to 10 characters. Line 29 substitutes asterisks for characters typed by the user, and line 30 automatically places a text cursor in the input field awaiting user interaction. Finally, lines 31 through 36 duplicate the functionality of the *TextInput* component example from this chapter's earlier section,"Using ActionScript to Configure Text Components." If the user types the specified string, an affirmation is traced to the Output panel.

Formatting Text

So far, you've worked with very limited text formatting options that have focused on rudimentary field styling. In fact, the only text styling you've experimented with thus far is text color applied to all the text in a field. It is possible to use both the Flash interface and ActionScript to exert greater control over text formatting. The first thing to understand in this regard is how to specify a font, and how that impacts your work.

Fonts

In very broad terms, there are three basic categories of fonts in Flash. The first is a *device font*, which is an operating system-specific font that is sure to be available, even if it can differ slightly from computer to computer. The second category is a *custom font*, which may not display correctly if it is not available on a user's computer. Finally, the third category is a custom font that has been embedded into the SWF to ensure consistent display everywhere.

Device fonts

Flash targets three device fonts. The first is a font with *serifs*, or the small accents, tails, and flourishes added to each letter. Common examples of fonts with serifs include Times and Times New Roman. Next is a font without serifs, also called *sans serif* (sans means "without" in French). Examples of sans serif fonts include Helvetica and Arial. Finally, there is a typewriter font, which is a fixed-width font such as Courier and Courier New.

Device fonts save on file size because they are supplied by the operating system of each user and are relatively certain to display (barring a user's radical reconfiguration of his or her computer). However, the font chosen by the operating system may differ slightly, so there's no guarantee that everything will display the way you want it to.

All that is required to use a device font is to specify the names *_sans*, *_serif*, or *_typewriter* (note the preceding underscore characters) either in the Properties panel (Figure 11-8) or though ActionScript.

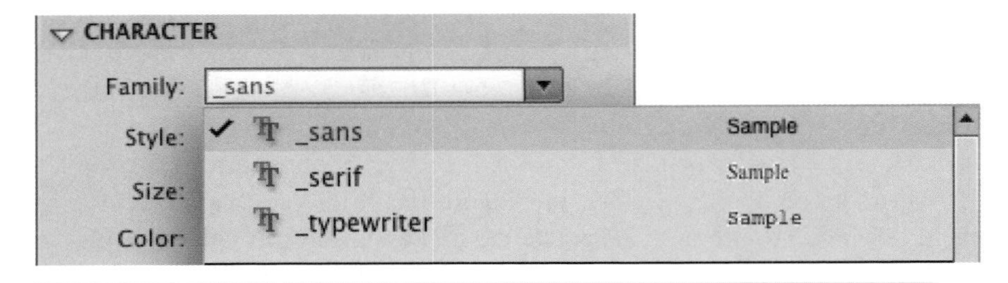

Figure 11-8. Choosing a device font from the font Family menu in the Properties panel

Custom fonts

As with device fonts, you specify custom fonts using the Properties panel or ActionScript. OpenType PostScript and TrueType font formats are supported. Instead of using a generic font reference (such as *_sans*, for example), you must specify the actual font name (Figure 11-9) and style (Figure 11-10) because there is no substitution by the user's operating system. As a result, there is no assurance that the end user will have the same font on his computer. For this reason, custom fonts are best used for Static text fields.

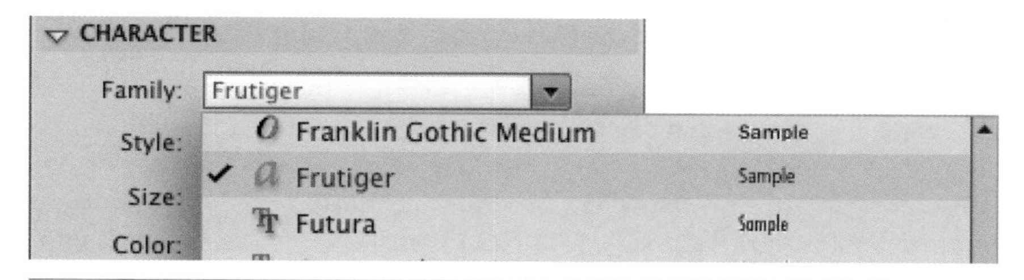

Figure 11-9. Choosing a custom font from the font Family menu in the Properties panel

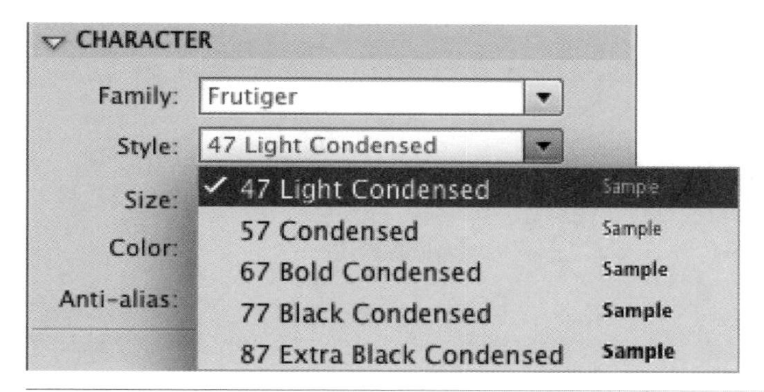

Figure 11-10. Choosing a style from the selected font family in the Properties panel

Embedding fonts

If you want to use a custom font for Dynamic or Input text fields, and expect it to be visible on all machines, you must embed the font. Flash Player also requires font embedding to properly render fonts when a field is transformed or composited in certain ways.

Up to and including Flash Player 9, for example, text created from nonembedded fonts would disappear in masked layers. These player versions also failed to apply transparency (*alpha*) settings when fields used fonts that were not embedded.

These issues have been corrected as of Flash Player 10, the player primarily associated with Flash CS4 Professional, but some circumstances still require embedded fonts. For example, even in Flash Player 10, if you apply a standard 2D rotation to a text field, you must use an embedded font to prevent the text from disappearing.

NOTE

If you want to rotate a text field without embedding a font, you can use a little-known trick. Place the field inside a movie clip and then apply 3D rotation along the z-axis. This works using the 3D Rotation tool, Transform panel, and ActionScript.

You can embed fonts using the Properties panel or by creating a font symbol and using ActionScript. When using the Properties panel, first select the field

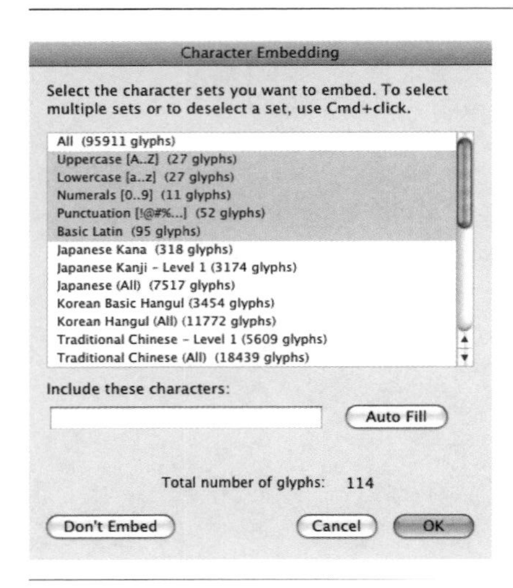

Figure 11-11. Embedding a subset of character glyphs using the Character Embedding option in the Properties panel

for which you want to embed fonts, specify the font of your choice, and then click the Character Embedding button in the Properties panel's Character section. The Character Embedding dialog will appear (Figure 11-11), and you can specify exactly which *glyphs* (font characters) or groups of glyphs to include. For example, Figure 11-11 shows the inclusion of uppercase and lowercase letters, numbers and punctuation, and basic Latin (special characters and symbols).

Every glyph you embed increases file size. If you are embedding fonts for a very isolated need, such as a single headline, you can even limit the glyphs you embed to specific characters by filling in the *Include these characters* input field shown in Figure 11-11. You can even have these characters automatically added to the Character Embedding dialog by clicking the Auto Fill button.

For ActionScript use, you must create a *font symbol*. Like other symbol types, you can use a font symbol throughout the entire file. To create a font symbol, choose New Font from the menu in the upper-right corner of the Library (Figure 11-12).

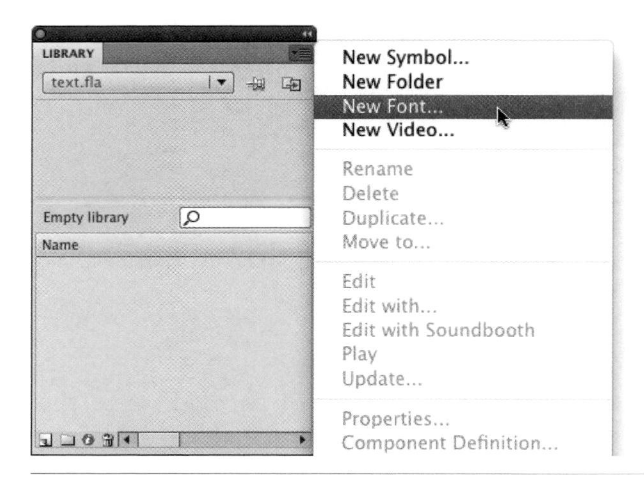

Figure 11-12. Adding a new font to the Library for use with ActionScript

This creates a font symbol and opens the Font Symbol Properties dialog (Figure 11-13). The creation of the font symbol from here on is similar in many ways to creating a movie clip symbol, for example. First, you should choose a meaningful name for your font symbol. This can be purpose-related, such as "Headline" or "BodyCopy," or you can label it as an embedded font by preceding the name with something like "emb" or by combining a font name and style. Figure 11-13 shows the embedding of Arial Regular; the chosen font symbol name is *ArialReg*.

Next, you should specify the *Font* and *Style* of your choice. If you want to use a bitmap font, rather than embedding font outlines that can be scaled to many resolutions, enable the *Bitmap text option* and indicate the size of the bitmap font you want to embed. This is fairly rare and is usually only used for bitmap micro fonts heavily optimized to look good at font sizes smaller than 9 points.

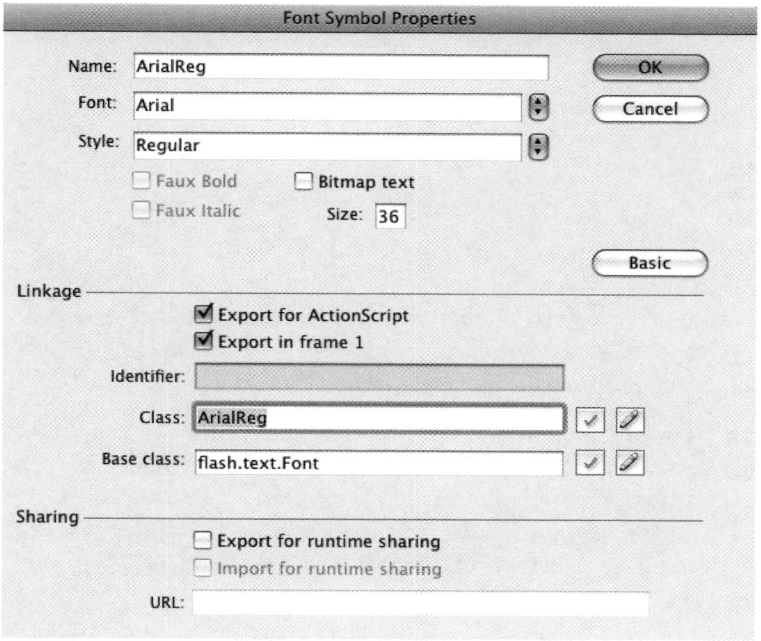

Figure 11-13. Assigning a linkage class to a new font symbol

Finally, just like with a display object, you need to export the font for use with ActionScript and give it a linkage class name. Flash automatically makes the base class `flash.text.Font` and, unless you're an experienced ActionScript developer writing your own classes for this purpose, you shouldn't change it.

NOTE

Remember from Chapter 6 that it is considered a best practice to start your class names with an uppercase letter.

Once you've filled out this dialog, click OK to create your font symbol. Flash embeds the font's entire set of glyphs, and you can now make use of the embedded font with ActionScript. The next section will discuss not only how to use embedded fonts with ActionScript, but also how to create native ActionScript objects to style the font without having to use the Properties panel.

ActionScript

To use ActionScript instead of the Properties panel for text formatting, you need to create and configure an instance of the *TextFormat* class. The following code example demonstrates how to format text this way, and also uses the embedded font discussed in the preceding section. This gives you the freedom to use ActionScript for runtime control of text formatting and also for use of custom fonts without fear of substitution on the end user's system.

This example can be found in the *text_formatting_01.fla* source file. The first step in this process is to create a custom font instance using your font symbol. Thanks to the consistency of ActionScript, you can use the same method that you would for creating any other new instance; this method is shown in line 1 of the script:

```
1   var customFont:Font = new ArialReg();
2   var txtFmt:TextFormat = new TextFormat();
3   txtFmt.font = customFont.fontName;
4   txtFmt.color = 0x000099;
5   txtFmt.size = 12;
6   txtFmt.leading = 4;
7   txtFmt.leftMargin = 6;
8   txtFmt.rightMargin = 6;
9   txtFmt.indent = 20;
```

Next you need to create and configure the `TextFormat` instance, in lines 2 through 9. The code is very straightforward and, for easy explanation, very closely mimics the property names used in the Properties panel.

The significant departure in this code is the specification of the font. Rather than citing the name of the font you used to create your symbol, you need to use the `fontName` property of the new `Font` instance you created. In this example, the linkage class given to the font symbol is `ArialReg`, and the instance is stored in the `customFont` variable. The font name is set in line 3 using `customFont.fontName`.

Next is some familiar code for creating and configuring a text field:

```
10  var txtFld:TextField = new TextField();
11  addChild(txtFld);
12  txtFld.x = 20;
13  txtFld.y = 20;
14  txtFld.width = 200;
15  txtFld.autoSize = TextFieldAutoSize.LEFT;
16  txtFld.textColor = 0x000099;
17  txtFld.border = true;
18  txtFld.selectable = false;
19  txtFld.multiline = true;
20  txtFld.wordWrap = true;
21  txtFld.embedFonts = true;
22  txtFld.defaultTextFormat = txtFmt;
23  txtFld.text = "Scaly was born in a bathtub, and grew so incredibly
    thin...";
```

The new information in this script segment is found in lines 21 and 22. Line 21 enables the use of embedded fonts, and line 22 sets the *default* text format for the field to the `TextFormat` instance you created earlier. You must apply a default text format *before* adding any text to the field. Note that line 22 sets the format and the text is added to the field in the next line, line 23. Note also that this approach uses the `defaultTextFormat` property and applies it to the entire field.

This is in contrast to the other way to apply a `TextFormat` instance, which uses the `setTextFormat()` method and must be used *after* the field already contains text. This approach is required when you need to apply a format only to a segment of the field. The final segment of this script creates another **TextFormat** instance and uses the **setTextFormat()** method to apply the format only to characters 0 through 5 of the field:

```
24  var txtFmt2:TextFormat = new TextFormat();
25  txtFmt2.color = 0xFF0000;
26  txtFmt2.underline = true;
27  txtFld.setTextFormat(txtFmt2, 0, 5);
```

It's also helpful to understand that **TextFormat** instances can work together. The **txtFmt2** instance specifies an underline style, as well as a color, but doesn't indicate a new font, for example. When applied along with an existing format, the text will retain any unchanged properties from the first format and take on new properties from the second format.

WARNING

It's important to know that when using embedded fonts, you must embed a separate version of each font style you want to use. For example, if you want to work with plain and bold versions of a font, you will need to create two font symbols. Similarly, if you want to embed the fonts using the Properties panel, you need to put placeholder text in the field and use plain, bold, and italic words to make sure they're all embedded. For more information, see the companion website.

Using HTML

Dynamic text fields, as well as *TextArea* and *Label* components, can display very simple HTML by using ActionScript. For basic applications, all you need to do is use the **htmlText** property instead of the **text** property when populating the object with text. However, Flash's HTML support is very limited. Table 11-1 shows the HTML tags supported, as well as qualifying notes.

NOTE

For more information about HTML, start with the World Wide Web Consortium tutorials at http://www.w3.org/MarkUp/Guide/.

Table 11-1. HTML tags supported by Flash Player

HTML tag	Notes
<font>	Supported attributes include **color**, **face**, and **size**
<b>	Bold version of font must exist to work
<i>	Italic version of font must exist to work
<u>	
<span>	Supported attributes include **class**
<p>	**multiline** must be enabled to work; supported attributes include **align** and **class**
 	multiline must be enabled to work
<li>	All lists are bulleted; ordered and unordered qualifiers are ignored
<img>	Supported attributes include **src**, **width**, **height**, **align**, **hspace**, **vspace**, and **id**; can embed external images (JPG, GIF, PNG) and SWF files with automatic text flow around source
<a>	Supported attributes include **href**, **event**, and **target**
<textformat>	Used to apply limited subset of **TextFormat** properties to enclosed text; supported attributes include **blockindent**, **indent**, **leading**, **leftmargin**, **rightmargin**, and **tabstops**

NOTE

Text elements with HTML support can also format text with cascading style sheets (CSS). Here again, Flash's support for CSS is very limited. For more information about CSS, start with the World Wide Web Consortium tutorials at http://www.w3.org/TR/CSS/.

Using CSS

Table 11-2 shows the cascading style sheet properties supported, their ActionScript equivalent, and any qualifying notes.

Table 11-2. CSS tags supported by Flash Player

CSS property	ActionScript property	Notes
`color`	`color`	Font color in 0xRRGGBB format
`display`	`display`	Controls display of item; values include **`none`**, **`block`**, and **`inline`**
`font-family`	`fontFamily`	Font name
`font-size`	`fontSize`	Font size in pixels
`font-style`	`fontStyle`	Font style; values include **`italic`** and **`normal`**
`font-weight`	`fontWeight`	Font weight; values include **`bold`** and **`normal`**
`kerning`	`kerning`	Turns kerning on or off; values include **`true`** and **`false`**
`leading`	`leading`	Font leading in pixels; not officially supported; similar to **`text-height`**; works well in internal style object, but may not be reliable in loaded CSS
`letter-spacing`	`letterSpacing`	Tracking in pixels
`margin-left`	`marginLeft`	In pixels
`margin-right`	`marginRight`	In pixels
`text-align`	`textAlign`	Aligns text; values include **`left`**, **`right`**, **`center`**, and **`justify`**
`text-decoration`	`textDecoration`	Underlines text; values include **`underline`** and **`none`**
`text-indent`	`textIndent`	First-line paragraph indent in pixels

There are two ways to apply CSS in ActionScript. The first is by creating your CSS data internally, as shown in the following example, which comes from the *internal_html_css.fla* source file. The second is by loading an external CSS document, which will be demonstrated in the upcoming section, "Loading Text."

This script will be discussed in two segments, beginning with the creation of the style sheet. The first step in creating the style sheet is to create an object for each style. In ActionScript, this is just like creating an instance of any other class and configuring its properties. If you have experience using CSS, you will find that this is similar to creating a CSS object, but the syntax is, necessarily, in ActionScript. When loading external CSS files, you can use traditional CSS syntax to the extent that it is supported within Flash.

There are a few things worthy of mentioning in this first script segment. The first style (lines 1 through 3) will be applied to the HTML **body** tag so that all text will use the specified font and paragraph indent. The second style (lines 5 through 11) is for a headline only, which will be applied using a CSS *class* in an HTML **span** tag. Because the headline should not be indented, the **textIndent** property is given a negative value (line 7) to compensate for the indent applied to all text.

Finally, the style sheet is created in lines 13 through 15 and each style is added to the style sheet using a process similar to standard CSS authoring. HTML tag styles are attributed to the tag name (line 14) and class style names are preceded with a dot (.), shown in line 15.

```
1   var body:Object = new Object();
2   body.fontFamily = "Verdana";
3   body.textIndent = 20;
4
5   var headline:Object = new Object();
6   headline.fontSize = 18;
7   headline.textIndent = -20;
8   headline.leading = 10;
9   headline.letterSpacing = 1;
10  headline.fontWeight = "bold";
11  headline.color = "#990000";
12
13  var css:StyleSheet = new StyleSheet();
14  css.setStyle(".headline", headline);
15  css.setStyle("body", body);
```

The next segment creates and configures a text field (lines 16 through 23), applies the style sheet (line 25), and then populates the field with HTML text (lines 25 through 27). It's very important to note that you must apply the style sheet *before* adding the HTML to the field. If this order is reversed, the text will not be styled.

```
16  var txtFld:TextField = new TextField();
17  addChild(txtFld);
18  txtFld.x = 20;
19  txtFld.y = 20;
20  txtFld.width = 500;
21  txtFld.multiline = true;
22  txtFld.wordWrap = true;
23  txtFld.autoSize = TextFieldAutoSize.LEFT;
24
25  txtFld.styleSheet = css;
26  txtFld.htmlText += "<body><span class='headline'>CSS Text Formatting
    Possible in Flash</span><br>";
```

```
27  txtFld.htmlText += "<p>Lorem ipsum dolor sit amet, consectetur
    adipiscing elit. Aenean facilisis luctus lacus. Mauris porta
    felis quis nulla. Vivamus sollicitudin magna vitae mauris. Nulla
    ultricies accumsan mi.</p></body>";
```

Loading Text

It is also possible to load text from an external file using ActionScript. The following simple script loads text from an external file and places it into a default text field. For this to work, create and save a plain-text file called *demo.txt* that contains a few words. For simplicity, the default text field size used here is only 100×100 pixels. The next example will create a larger field to demonstrate the process of loading HTML and CSS files.

Next, create a new FLA, save it into the same directory as your text file, and add the following script to frame 1. When you test your movie, your external text should appear in the small field in the upper-left corner of the Stage. If you prefer, you can test the *simple_text_loading.fla* source file.

```
1   var txtFile:URLLoader = new URLLoader();
2   txtFile.addEventListener(Event.COMPLETE, onLoadTXT);
3   txtFile.load(new URLRequest("demo.txt"));
4   function onLoadTXT(evt:Event):void {
5       var txtFld:TextField = new TextField();
6       addChild(txtFld);
7       txtFld.text = evt.target.data;
8   }
```

Line 1 of this script creates an instance of the **URLLoader** class, which is responsible for loading text from external sources. It can load text, server variables, and even binary data. In all cases in this chapter, you'll use it to load text, including simple text, HTML, and CSS data.

NOTE

A URL is not just used for web addresses. URL stands for universal resource locator and is, essentially, a pathname for any file.

Lines 2 through 8 are an event listener that will respond once the text has loaded. Line 3 loads the text and uses the **URLRequest** class to process the URL. This required class standardizes all external file references.

Finally, lines 5 through 7 create a default text field and populate it with the data coming in from the target of the listener. In this case, the target is the **URLLoader** instance, so the data coming in is the text that this instance loads.

You can combine this process with some of the CSS and HTML code used previously in the "Using HTML" and "Using CSS" sections of this chapter to load external HTML and CSS files in their native syntax. The following example requires that you save the HTML file named *demo.html* and a CSS file named *demo.css* (the text of which follows) into the same directory as your text FLA. These files are provided in the *external_html_css* directory of the companion source files.

HTML:

```
<body>
<span class='heading'>CSS Text Formatting Possible in Flash</span><br>
```

```
<p>Lorem ipsum dolor sit amet, consectetur adipiscing elit. Aenean
    facilisis luctus lacus. Mauris porta felis quis nulla. Vivamus
    sollicitudin magna vitae mauris. Nulla ultricies accumsan mi.</p>
</body>
```

CSS:

```
body {
    font-family: Verdana;
    text-indent: 20px;
}

.heading {
    font-size: 18px;
    font-weight: bold;
    text-indent: -20px;
    letter-spacing: 1px;
    color: #FF6633;
}

a:link {
    color: #990099;
    text-decoration: underline;
}

a:hover {
    color: #FF00FF;
}
```

FLA:

The only thing different about this example is that it uses two event listeners in sequence. The CSS is loaded first and, when the loading is complete, the HTML is loaded. When the HTML is loaded, the text field is then created and populated.

```
1   var css:StyleSheet;
2   var cssFile:URLLoader = new URLLoader();
3   cssFile.addEventListener(Event.COMPLETE, onLoadCSS);
4   cssFile.load(new URLRequest("demo.css"));
5   function onLoadCSS(evt:Event):void {
6       css = new StyleSheet();
7       css.parseCSS(evt.target.data);
8       var htmlFile:URLLoader = new URLLoader();
9       htmlFile.addEventListener(Event.COMPLETE, onLoadHTML);
10      htmlFile.load(new URLRequest("demo.html"));
11  }
12  function onLoadHTML(evt:Event):void {
13      var txtFld:TextField = new TextField();
14      addChild(txtFld);
15      txtFld.x = 20;
16      txtFld.y = 20;
17      txtFld.width = 500;
18      txtFld.autoSize = TextFieldAutoSize.LEFT;
19      txtFld.selectable = false;
20      txtFld.multiline = true;
21      txtFld.wordWrap = true;
22      txtFld.styleSheet = css;
23      txtFld.htmlText = evt.target.data;
24  }
```

Using InDesign and XFL

This section is an enthusiastic profile of a new feature that improves interoperability between Creative Suite 4 applications. Adobe's new XFL file format promises to improve asset transfer between applications. In its current implementation, the format facilitates movement in one direction from InDesign CS4 to Flash CS4 Professional.

All you need to do to use the feature is export a layout from InDesign as XFL. You don't even have to import anything in Flash! Just opening the file in Flash will convert the layout to a movie clip containing static graphics and Static text elements to preserve visual accuracy. Figure 11-14 shows the original InDesign output, and Figure 11-15 shows the final output from Flash with no interim steps applied.

Um Veros
Nostinibh
123 Everywhere Avenue
City, ST 00000
555 555-5555
Fax 555 555-5555
www.adobe.com

Ciliquatin volorpero Tincidut
Dionsendiam nim augiam del ut at iustrud digna ad dolor at. Ut velit irit aliquis dolobore exero ero ex elendrer augait, con henim iliquam nibh et, quat vent ex exercing ea feumsan enim vel ut iure tat lore magniam, volore ver irit venit augiat. Sed tin ver iliquametum ipsustie consenibh er autetum quat lute modit dunt pratem veliquamet, quamet, veliquat in ute erat. Tat.
To od tat, consenit, cortinisi blaore tem zzrilis autpat. Ut niam, se conulla consequat. Ut inci dolobore exero blaor sit, susto core magna faciliscipit adignim nos ad tinibh et la consequi ex elit wisi.
Ud tissequat. Cummolo rtisi.
At exerci blan ver aliquis modolute duipisit, conullutpat aliquissed dolobor eriliqu ismodigna core facipsu msandre minisl iril ip endiam dignim elis atiscillum ip enis aciliquipit iuscipisit, conum del irilit num aute.
• Ut nos nit loborer in ut luptat, conullumsan ulputat amet utem augiam dolorpe rostie dio
• Odio erostin hent alit il euisim er sit volum quipsus cilit, consequ atetum vero consenibh erit at ad do odiamcommy nullaor sequat. Ut in ut wisismo loreet eugiam zzriure dolorem nonsendrem venim voluptat illamet, quipsus

Figure 11-14. Original InDesign layout

Um Veros
Nostinibh
123 Everywhere Avenue
City, ST 00000
555 555-5555
Fax 555 555-5555
www.adobe.com

Ciliquatin volorpero Tincidut
Dionsendiam nim augiam del ut at iustrud digna ad dolor at. Ut velit irit aliquis dolobore exero ero ex elendrer augait, con henim iliquam nibh et, quat vent ex exercing ea feumsan enim vel ut iure tat lore magniam, volore ver irit venit augiat. Sed tin ver iliquametum ip - sustie consenibh er autetum quat lute modit dunt pra - tem veliquamet, quamet, veliquat in ute erat. Tat.
To od tat, consenit, cortinisi blaore tem zzrilis autpat. Ut niam, se conulla consequat. Ut inci dolobore exero blaor sit, susto core magna faciliscipit adignim nos ad tinibh et la consequi ex elit wisi.
Ud tissequat. Cummolo rtisi.
At exerci blan ver aliquis modolute duipisit, conullutpat aliquissed dolobor eriliqu ismodigna core facipsu msan - dre minisl iril ip endiam dignim elis atiscillum ip enis aciliquipit iuscipisit, conum del irilit num aute.
• Ut nos nit loborer in ut luptat, conullumsan ulputat amet utem augiam dolorpe rostie dio
• Odio erostin hent alit il euisim er sit volum quipsus cilit, consequ atetum vero consenibh erit at ad do odiamcommy nullaor sequat. Ut in ut wisismo loreet eugiam zzriure dolorem nonsendrem venim voluptat illamet, quipsus

Figure 11-15. Opened XFL document in Flash

Because the fidelity of the original document is being preserved, editing the text after opening in Flash is not a trivial matter. Each line of text and, in the example used in Figures 11-14 and 11-15, even the initial capital letters in the body copy are separate Static text elements. However, the text is still editable, so simple edits are possible. If dynamic text resources are not essential to your project, this is a workflow worth exploring.

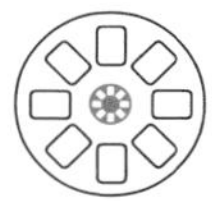

Project Progress

In this chapter, you will add text to your Help screen, as well as an ActionScript event listener that will display text when Scaly's inverse kinematics armature passes over elements on the page.

Cosmetic Assets

First create a handful of cosmetic assets that will not be controlled by ActionScript, to help convey the purpose of the screen.

Page title

1. Open your main portfolio document and scroll the Timeline panel to the *help* section of your file.
2. In the *content* layer, double-click the *HelpScreen* movie clip to edit it.
3. Use the Text tool to create a text field anywhere on the stage.
4. Create a Static text field anywhere on the stage and type `HELP` into the field.
5. Select the text within the field and, using the Properties panel, set the *Family* to `Arial Narrow`, *Style* to `Bold`, *Color* to black, and *Size* to `76`.
6. Set the *width* of the field to `200`, making the height approximately 90. Set the *x* and *y* coordinates to `160` and `65`, respectively.
7. Switch to the Selection tool and copy and paste the field to create a duplicate. Use the Modify→Transform→Flip Vertical menu command to flip the field and position it at `160`, `120` so it is immediately below the first field.
8. Change the color of the flipped field to `0x990000` (maroon) with an alpha value of `50%`.
9. Select both fields by holding down the Shift key when clicking, and group them (Modify→Group).
10. Use the Transform panel to rotate the group `-15` degrees.

Blurb

1. Create a Static text field anywhere on the Stage and type a short descriptive blurb about yourself. The provided source file uses the following text: `<your name> is a tinkerer... a mover and shaker... based in <your locale>`.

2. Switch to the Selection tool and select the field. Using the Properties panel, set its *width* to approximately **125**, and its *x* and *y* locations to approximately **450** and **150**, respectively. The text you included in the blurb will affect the size and position of the field. When you are finished, just be sure that it is within the framed area shown in the guide layer.

Flourishes

If desired, draw a few lines, words, or creative shapes in the upper- and lower-right corners of the area described by the guide layer frame. This is entirely optional, but will help the illusion that the content is under the wheel as visitors to your portfolio look through the view window.

Dynamic Text Field

The last task for the Help screen is to create a dynamic text field and populate it with ActionScript.

1. Create a Dynamic text field approximately 100 pixels tall and 100 pixels wide, and type **Drag Scaly's hand** into the field.

2. Switch to the Selection tool and select the field. Using the Properties panel, set its *x* and *y* locations to approximately **130** and **250**, respectively.

3. Set the font *Family* to **Arial Narrow**, the *Style* to **Bold**, the *Size* to **14**, and the Color to **0x990000** (maroon).

4. Click the Character Embedding button and embed uppercase and lowercase letters, numbers, and punctuation. As an alternative to embedding fonts, you can specify a font in step 3 that you expect to find on most computers, such as *Arial*, or a device font such as *_sans*.

5. Double-click any unoccupied location of the Stage to return to the main Timeline and add the following script to the script frame for the Help section (frame 227 in the *actions* layer):

```
1   this.addEventListener(Event.ENTER_FRAME, onLoop);
2   function onLoop(evt:Event):void {
3       if (arm.hand.hitTestObject(navigation)) {
4           help.helpText.text = "Navigate the portfolio using
    these buttons.";
5       } else if (arm.hand.hitTestObject(logo)) {
6           help.helpText.text = "This is version 1.0 of my
    portfolio";
7       } else if (arm.hand.hitTestObject(soundController)) {
8           help.helpText.text = "Turn the sound on and off using
    this controller.";
```

```
9          } else {
10              help.helpText.text = "Drag Scaly's hand around to
    learn more...";
11          }
12     }
```

This script will continuously populate the Dynamic text field with a string, based on where Scaly's hand is. The **hitTestObject()** method determines if one object collides with another. In lines 3, 5, and 7, the script tests to see if Scaly's hand collides with the navigation buttons, logo, or sound controller, respectively. If a collision is detected, the appropriate string is placed into the text field. If no collision is detected, line 9 allows an unconditional alternative and line 10 places a generic string into the field.

Input Text Field

The project's Lab screen will ultimately contain a text-driven animation in which characters typed into a text field will scale movie clips on screen. When the user types the letter A into the text field, the movie clip graphic for letter A will enlarge. When the next letter is typed, the A will reduce to normal size and the new letter graphic will enlarge.

To recap the screen's asset preparation from Chapter 4, all movie clips and text fields were optimized in Adobe Illustrator for Flash use. During that process, a text object was set as a Flash Input text field and given an instance name of *typeAnimInput*, using Illustrator's Flash Text panel. The text object was also placed inside an Illustrator symbol that was set as a Flash movie clip with an instance name of *typeCard*.

During the AI Import process, the relationship among all of these assets remained intact, and the necessary text field and movie clips were automatically created, complete with instance names. Because these assets were prepared in Illustrator, all you have to do to make the screen functional is add ActionScript. Scroll the timeline of your master portfolio FLA to the *Lab* section and add lines 2 through 20 of the following script to the **stop()** action in frame 187 of the *actions* layer. (The complete script appears here for clarity.)

```
1    stop();
2
3    var userInput:TextField = LabScreen.typeCard.typeAnimInput;
4    userInput.text = "";
5    userInput.addEventListener(TextEvent.TEXT_INPUT, onTextInput);
6
7    function onTextInput(evt:TextEvent):void {
8        var len:int = LabScreen.numChildren;
9        for (var i:int = 0; i < len; i++) {
10           var tempMC:MovieClip = MovieClip(LabScreen.getChildAt(i));
11           tempMC.scaleX = tempMC.scaleY = 1;
12       }
13
14       var char:String = evt.text.toLowerCase();
15
```

```
16        var letterMC:MovieClip = MovieClip(LabScreen.
    getChildByName(char + "Letter"));
17        if (letterMC) {
18            letterMC.scaleX = letterMC.scaleY = 1.5;
19        }
20    }
```

Line 3 stores a reference to the text field in a variable for simplicity. The *typeAnimInput* text field is inside the *typeCard* movie clip, which is inside the content movie clip, *LabScreen*. Line 4 empties the text field every time you enter the Lab screen.

Line 5 assigns an event listener to the text field. Each time the user types something into the field, and event will be dispatched and the corresponding listener function in lines 7 through 20 will handle the input. This is a simple way of capturing each character typed by the user.

Lines 8 through 12 set all child movie clips in the *LabScreen* movie clip to actual size. Line 8 determines how many children are in *LabScreen*. Lines 9 and 10 loop through all of those children, storing a reference to the child found at each level, and casting each to the *MovieClip* data type. Line 11 sets the *scaleX* and *scaleY* properties of each child to 1, or 100% actual size.

Because ActionScript is case sensitive, line 14 converts each incoming character to lowercase. This prevents the need to create a separate movie clip of artwork for both uppercase and lowercase letters. Later, when a movie clip is sought by name, the script will find the *aLetter* movie clip (for example) whether the user types `a` or `A`.

Finally, lines 16 through 19 enlarge the movie clip that corresponds to the most recent character typed into the field. Line 16 finds the movie clip by name, using the character and suffix "Letter" to find *aLetter*, *bLetter*, and so on. Line 17 checks to see if the movie clip exists and, if it does, line 18 sets its *scaleX* and *scaleY* properties to 1.5, or 150% actual size.

The Project Continues...

In the next chapter, you will learn about sound and write the ActionScript to enable the sound controller you created in Chapter 3.

CHAPTER 12

AUDIO

IN THIS CHAPTER

Introduction

Audio has long been an important part of the Flash world for designers and developers alike. From MP3 jukeboxes to presentation voice-overs and sound effects, audio is as much a part of some Flash files as the vectors on which they are based. Of course, used injudiciously, sound can be an annoyance. However, well-planned sound can add a lot to games, applications, and demos.

From a programming standpoint, ActionScript 3.0 introduces a whole new level of sound management. It's now possible, for example, to more easily control Flash's 32 independent channels of audio, determine the amplitude of mono and stereo sounds, and control Flash properties based on data from sounds playing in real time.

Even without ActionScript, Flash is no slouch in the audio department. You can add sounds to animation and button timelines, play them while downloading, and even edit them to a minimal degree, right within Flash. You can pan sounds back and forth between the left and right stereo channels, set them to any volume, and loop them.

In this chapter, you'll learn how to import and embed sounds for use in the Timeline and how to play external sounds with ActionScript. You'll also learn how to apply simple effects with the Flash interface and code and optimize sound compression. Finally, you'll complete the sound controller you created in Chapter 3 for the portfolio project.

Supported File Formats

For a complete picture of Flash's support for file formats, this discussion is divided into two categories: importing audio, which is also often referred to as *embedding* audio, and loading external sounds for playback at runtime.

Working with external sounds brings many benefits to the typical Flash project. To begin with, using external audio files keeps the size of your FLA files down and speeds up compiling. Furthermore, editing external files is easy and typically doesn't require any change to the SWF if an edit is required. Best of all, loading external sounds at runtime reduces download time and improves user experience.

There are also advantages to embedding sound files into your SWF. For example, you can place sounds directly in the Timeline, view their waveforms, and scrub through the sounds to try to synchronize them with other animation events. Imported audio files are also preferred for short sounds that must play without any download delay, such as button sound effects.

Importing Audio

You can import a variety of file formats into Flash, which supports uncompressed and compressed sound. The most common formats are *WAV*, *AIFF*, and *MP3*; all three are supported on both Windows and Mac. WAV and AIFF files are typically used for uncompressed sounds, although you can use compression when creating both formats. MP3 files are for compressed sounds (specific compression algorithms will be discussed in a moment).

Additional file formats are supported when QuickTime 4.0 or later is installed. *AU* (Sun Audio) and *MOV* (sound-only QuickTime movie) files can be imported on Windows and the Mac, and *SD2* (Sound Designer 2) and *SFIL* (System 7 Sounds) can be imported on the Mac only.

Importing a sound is the same as importing graphical assets, with one exception. Because there is no visual component to an audio file, the Import to Stage option functions just like the Import to Library option. When you import a sound, Flash automatically places it in the Library and adds nothing to the Stage or Timeline (you'll learn to use the sound in your files in the "Timeline Use" and "ActionScript Use" sections, later in this chapter).

NOTE

Occasionally, applications may create MP3 files that don't comply with recognized standards. If an MP3 file fails to import, try converting the audio file to WAV or AIFF format and reimporting. For more information about possible issues when importing, see this chapter's upcoming sidebar, "Sound Highs and Lows."

External Audio Playback

For widest compatibility, only *MP3* files can be loaded from external sources and played at runtime. However, as of Flash Player 9 Update 3, released in December 2007, file formats using the *AAC* compression scheme are also supported for runtime playback. Apple's iTunes and compatible hardware (iPods, iTouch, iPhone), Sony's Playstation, and Nintendo's Wii platforms use AAC. The most common file types using AAC are *M4A*, *MP4* and *AAC*.

NOTE

Although Flash CS4 Professional is the most up-to-date version of Flash as of this writing, most compatibility issues do not hinge on which version of the authoring tool you're using. While it's true that new features are introduced with each version of the authoring tool, most compatibility problems relate to which version of the player your viewers are using.

Sound Highs and Lows

Before you can import or load any files, it's important to understand the concepts of *bit depth* and *sample rate*. Sound is transmitted in a continuous sound wave, and to represent that sound at a moment in time, it must be converted into discrete *samples*, or audio snapshots. In brief, bit depth is the number of bits per sample, and sample rate is the number of samples per second.

Bit Depth

Bit depth is much like the resolution of a sound: the higher the bit depth, the higher the resolution. This translates to better clarity, often described as a reduced *signal-to-noise ratio* (the amount of noise compared to the amount of undistorted sound in a sample).

There are two common multimedia audio bit depths, 8-bit and 16-bit (high-end formats like DVD-Audio use 24-bit sound). These bit depths are frequently compared to the bit depth of graphics. A lower-quality image might be in 8-bit color, which means the image is made up of 256 individual color values. A higher-quality image might be in 16-bit color, which contains more than 65,000 individual color values in an image.

Audio bit depths work the same way: 8-bit sound contains 256 bits per sample, while 16-bit sound contains more than 65,000 bits per sample. The higher quantity of information stored in 16-bit sound means the continuous sound wave from which samples are taken can be better approximated, and there is less noise.

Sample Rate

The sample rate is responsible for the *frequency range* of a sound, or the range of low to mid to high frequencies reproduced by the sound. The higher the sample rate, the more samples are taken per second and the more accurately a sound wave can be represented. So, more highs and lows are included in the sound. As the sample rate diminishes, the sound gets muddier and the highest and lowest frequencies (cymbal and bass, respectively, for example) are dropped.

For reference, CD audio is sampled at 44.1 kHz, "multimedia audio" (sound used in average computer multimedia presentations) is typically sampled at 22.050 kHz, and speech is often sampled at 11.0125 kHz. Put in simpler terms, typical sample rates used in Flash production are based first on CD-quality audio, then halved for multimedia-quality audio, and halved again for speech.

Reducing File Size

Sound can present real challenges both for RAM use (and the resulting computer performance) and for download speeds. CD-quality audio (16-bit, 44.1 kHz, stereo) takes 10 MB of RAM and storage resources per minute. Optimizing that effectively is an ongoing challenge and is often based on job specifics and personal taste.

Perhaps the most common approach to reducing CD-quality audio is to first drop the sample rate to 22.050 kHz, then reduce stereo to mono sound and only then reduce the bit depth to 8-bit if necessary. The noise introduced by 8-bit audio makes this last step viable only for very short sound effects that must be fully loaded into RAM for playback.

WARNING

Some of these optimization decisions are optional, but one guideline is extremely important. When reducing the sample rate, always reduce by half, starting with 44.1 kHz. That is, use a sample rate of 44.1, 22.050, or 11.0125. If you fail to do so, your sound will need to be resampled on the fly due to restrictions imposed by Flash Player. Consequently, your audio will sound like squeaking mice or bubbling molasses.

Compression

Although there may be rare circumstances when presenting full CD-quality sound over the Internet is required, you are more often going to need to compress your sounds. The leaps and bounds that high-speed bandwidth has made over the past several years still aren't sufficient to guarantee that everyone has enough bandwidth to play 10 MB per minute in sound alone.

While using uncompressed files at runtime is uncommon, choosing whether or not to compress sounds prior to importing them is a recurring question. Some Flash designers feel that it's best to import uncompressed sounds (such as uncompressed WAVE or AIFF files) and then let Flash compress the audio when compiling to SWF. Others prefer to compress sounds using external tools (saving to MP3) and then apply no further compression in Flash.

Using uncompressed sounds reduces or eliminates artifacts that might otherwise have been exacerbated by compressing already compressed sound. Additionally, this approach allows you to experiment with different compression settings when publishing to SWF. On the downside, using uncompressed sound significantly increases FLA file size and SWF compile time.

Using compressed sounds means slimmer FLA files, but it reduces quality when you want to change compression settings later on. With a little practical experience, you will likely develop a workflow that you prefer.

Fortunately, Flash has a variety of *codecs* (short for compression-decompression algorithms) for use with internal sounds, and Flash supports more than one type of compression for external sounds loaded at runtime.

Compressing External Sounds

Although the current Flash Player supports AAC, the most popular external compressed file format remains MP3. There are dozens of applications that compress MP3 sounds, and they are the most common audio files used on electronic sound players such as iPods.

Apple's iTunes, the primary application used to sync with iPods, is a free, cross-platform application that can compress sounds using many Flash-compatible formats, including MP3, WAVE, AIFF, and AAC files like M4A. See this book's companion website for examples of how to create audio files using iTunes.

When compressing audio, another key term comes into play. *Bitrate* represents the number of bits transferred per second. Bitrate is not a term specific to audio, but pertains to the transfer of any digital data. It surfaces frequently in discussions about audio because it is also a measure of the quality of compressed sound. In addition to bit depth and sample rate, a high bitrate contributes to higher-quality sound. As a point of comparison, a very high-quality MP3 bitrate is 320 kbps (kilobits per second, or 1,000 bits for every second of sound), a good bitrate is 128 kbps, and a low-quality sound is 16 kbps.

NOTE

Bit depth should not be confused with bitrate. Bitrate is how many bits per second, while bit depth is how many bits per sample. Bitrate is influenced by both bit depth and sample rate.

The most important things to remember when compressing sounds are:

- Use a bit depth of 16-bit whenever possible.
- Use a sample rate of 44.1 kHz or an even division thereof (22.050, for example) to avoid resampling.

- Use stereo or monaural sound based on need and your willingness to drop stereo playback for file size optimization.
- After considering all of the above, balance bitrate with file size; through trial and error, determine the best combination of lowest file size with the highest acceptable bitrate.

Internal Asset-Specific Sound Properties

The highest degree of control you can exert over sound quality and file size comes from applying compression on a sound-by-sound basis. Once you have imported a sound, you can select it in the Library and access its properties using the Properties button (*i* icon) at the bottom of the Library panel.

The Sound Properties dialog (Figure 12-1) allows you to change the compression and, where applicable, convert stereo to mono (termed *Preprocessing* in the dialog because it occurs prior to compression), and change the *Bit rate*, *Sample rate*, and *Quality* of compression. The latter is an arbitrary measure that includes Fast, Better, and Best, and determines how long the compression takes.

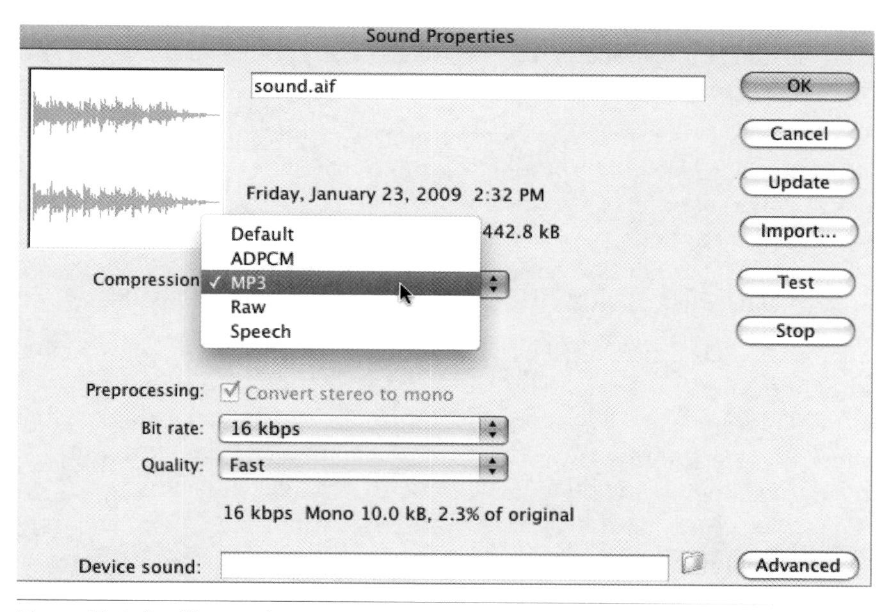

Figure 12-1. Per-file sound properties

Compression options include:

Default

Default skips asset-specific compression and applies the file-wide compression options explained in the upcoming section, "Internal File-Wide Publish Settings."

ADPCM

ADPCM applies a 4:1 compression ratio that allows you to choose not only sample rate, but also the number of bits used during compression. This is also known as IMA or IMA ADPCM in QuickTime environments.

MP3

MP3 uses the MPEG-1 Audio Layer 3 compression algorithm and allows you to customize preprocessing, bitrate, and quality settings.

Raw

Raw applies no further compression. This is useful when you want to import compressed sounds.

Speech

The *Speech* codec is optimized for speech and is not commonly suitable for music. It is also not compatible with *Flash Lite*, the Flash Platform's player for mobile devices.

NOTE

At the bottom of the Sound Properties dialog, you can specify a device sound used in Adobe's mobile Flash player, Flash Lite. The mobile devices, rather than Flash Lite, play device sounds. Because Flash can't import common device sound formats (such as MIDI), a Library sound element serves as a proxy that links to a device sound file. Flash Lite is not covered in this book.

Internal File-Wide Publish Settings

Using your file's publishing settings (File→Publish Settings), you can specify compression settings that can be applied to every sound in the entire file. This only occurs when compiling the file's SWF.

You can specify settings for *streaming* sounds (sounds that play while still being downloaded) and *event* sounds (sounds that must be fully downloaded before playback begins). The streaming option is best for larger sounds that take a long time to download, while event sounds are better suited for very short sounds that can easily be downloaded and fit into memory. Additional options will be discussed later in the "Sync Types" section of this chapter.

Because of their different uses, you can apply independent compression settings to each type of sound. This lets you choose a higher-quality setting for streaming sounds and a lower-quality setting for event sounds that must be small enough to download quickly. You can configure these settings in the Images and Sounds section of the Publish Settings dialog (Figure 12-2).

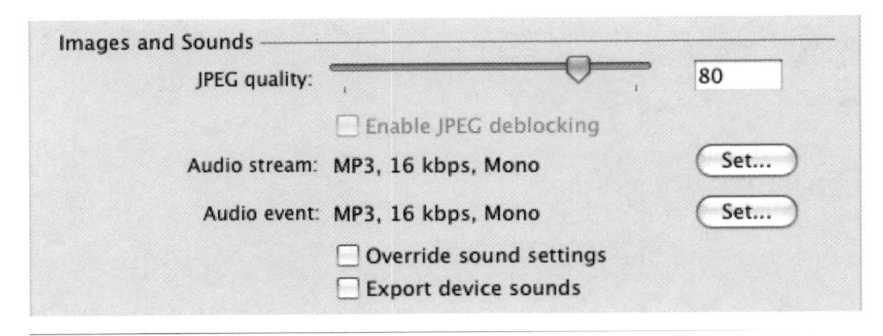

Figure 12-2. File-wide sound-related publish settings

The Sound Settings dialog (Figure 12-3) offers access to the same compression algorithms available on a per-file basis. The *Default* per-sound option, which specifies that these global settings should be used for a specific sound, has been replaced in Publish Settings with *Disable*. This option disables all sound in the SWF and is useful for testing or creating alternate versions of your file with no audio.

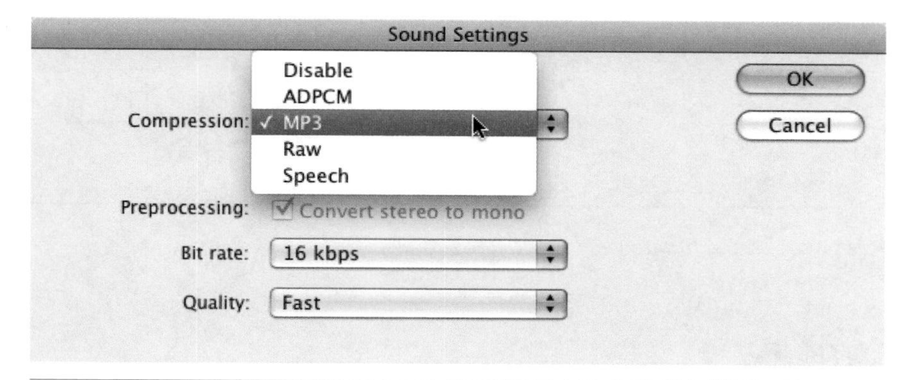

Figure 12-3. Compression options for internal sounds applied file-wide

At the bottom of the Publish Settings section (refer back to Figure 12-2), you can choose to override the sound settings. When you enable the *Override sound settings* option, Flash will override the per-asset sound settings and apply the global Publish Settings instead. This is useful for creating a low-quality alternate version of your file. You can also use this option to speed up SWF compilation during testing by switching to a Fast quality setting. When you're finished with development and ready to publish your final file, you can disable the override option, and the per-sound settings will be reapplied.

Finally, you can enable the *Export device sounds* option to support Flash Lite development, not covered in this book.

NOTE

For more information about Flash Lite, visit http://www.adobe.com/products/flashlite/.

Timeline Use

Adding sounds to the Timeline is very easy. Select the frame in the Timeline in which you want the sound to play, and choose any imported sound from the *Name* property in the Sound section of the Properties panel (Figure 12-4).

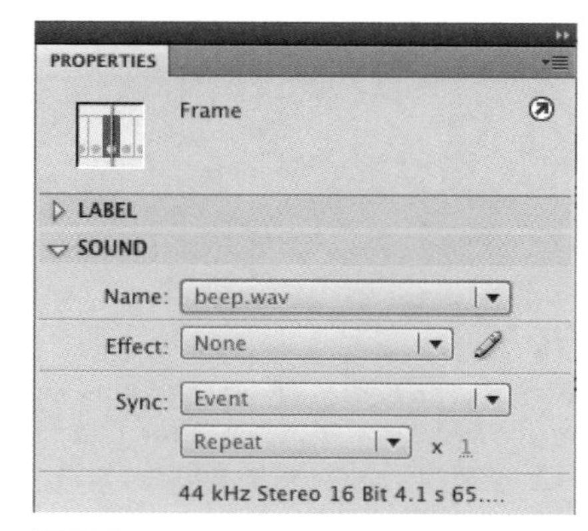

Figure 12-4. The Sound section of the Properties panel

A waveform representation of the sound will appear in that frame or span of frames. During playback, anytime the playhead is in contact with a frame with sound in it, that sound will be played. The *Sync* property in the Properties inspector (described in the next section) controls how the sound is played and when it stops.

Because the normal layer height makes this waveform difficult to see, you can increase the height of specific layers. Double-click the layer icon to open the

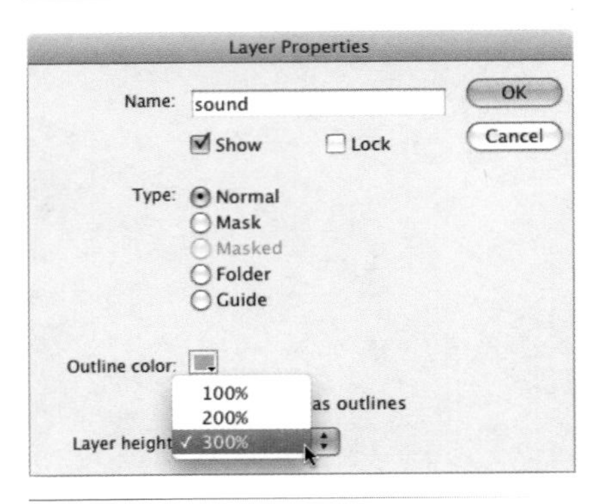

Figure 12-5. Increasing layer height to increase the scale of a displayed sound wave

Layer Properties dialog and increase the *Layer height* to 200 or 300% (Figure 12-5). The result is a taller layer and a larger visual of the sound wave, shown in Figure 12-6.

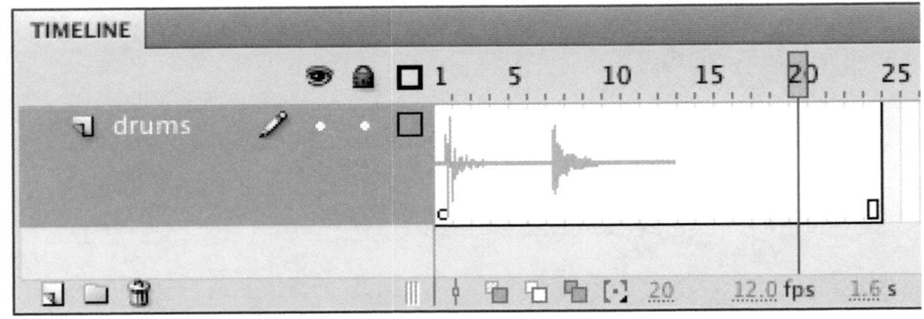

Figure 12-6. A single untreated sound in the Timeline

Sync Types

The *Sync* type you set in the Properties panel for a Timeline-based sound will have a dramatic effect on how the sound plays. With the sound frame in the Timeline panel selected, the Properties panel shows the following four sync types.

Stream

Stream sounds begin playing after just a few frames of data have been downloaded. They are best for long sounds and are well suited for animation soundtracks because Flash attempts to keep the Timeline and sound synchronized. You can also scrub through a stream sound and hear pieces of it in authoring mode.

Event

Event sounds must be fully downloaded and play through to completion no matter what. For example, if a one-second sound ordinarily requires 24 frames to play through to completion, an Event sync type will cause that sound to play until complete, even if only allocated one frame. This is important because this sync type is used to play sound effects that often occur in very short frame spans, and the sound must be allowed to finish. A sound effect in the Down state of a button, for example, would likely only occupy one frame, leaving only 1/24th of a second to play a sound.

Sound effects also usually need to play every time they are requested, making the prospect of waiting for the sound to finish problematic. For this reason, the Event sync type will play a new instance of the sound even if the prior instance is still playing. If not used carefully, this can result in a chaotic din of sound and eat up memory quickly, as in the case of placing an Event sound in a frame loop.

WARNING

The large variance in computer processor power, quantity of RAM, bus speeds, video card performance, and connection speeds (among other contributing factors) makes synchronizing audio with animation in Flash as much an art as a science. When planning synchronized sound, aim for the lowest common denominator among minimum system requirements, and test early and often.

NOTE

If you accidentally play a long sound set to the Event sync type, press the Escape key on your keyboard to stop the sound from playing.

Start

The *Start* sync type is similar to Event in every respect but one. If the playhead returns to the frame before a prior instance of the sound is finished playing, it will not play another instance of the sound.

Stop

The *Stop* sync type silences the specified sound. This will even work on Event sounds that are designed not to stop until they are finished playing.

NOTE

The companion website for this book includes sample source files that demonstrate all four sync types.

Repeating and Looping

With a sound frame selected in the Timeline panel, you can loop a sound or repeat it a finite number of times by using the menu and repeat count input below the Sync menu (Figure 12-7). *Loop* will show no visual change in the Timeline, but will loop the sound as long as the playhead is in contact with the frames in which the sound resides. *Repeat* will play the sound again as many times as you specify using the repeat count. It will also repeat the waveform in the Timeline a corresponding number of times (Figure 12-8).

Figure 12-7. Repeating a sound

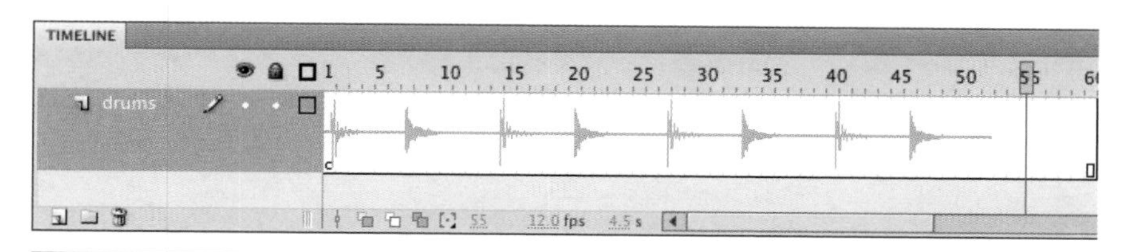

Figure 12-8. A repeating sound displayed in the Timeline

WARNING

Looping a stream sound is not recommended because the file size will increase based on the number of times the sound is looped.

Editing a Sound

Flash lets you do some basic volume and pan editing without having to use an external editor. The Effect menu in the Properties panel contains a list of preset effects, including confining the sound to the *Left channel* or *Right channel*, channel panning options (*Fade to left* and *Fade to right*), and multichannel volume fades (*Fade in* and *Fade out*). You can also edit any of these options or assign a custom option to edit an effect from scratch.

Setting Volume and Pan

When you choose *Custom* from the Effect menu or edit a preset, the Edit Envelope dialog opens. Figure 12-9 shows the dialog with the Fade Out preset. To edit the volume or pan of the sound, first click on the horizontal line that runs across the waveform to add a handle. In stereo sounds, a handle will be created in both the left (top) and right (bottom) channels. In any channel, you can then drag the handles down or up to reduce or increase volume, respectively.

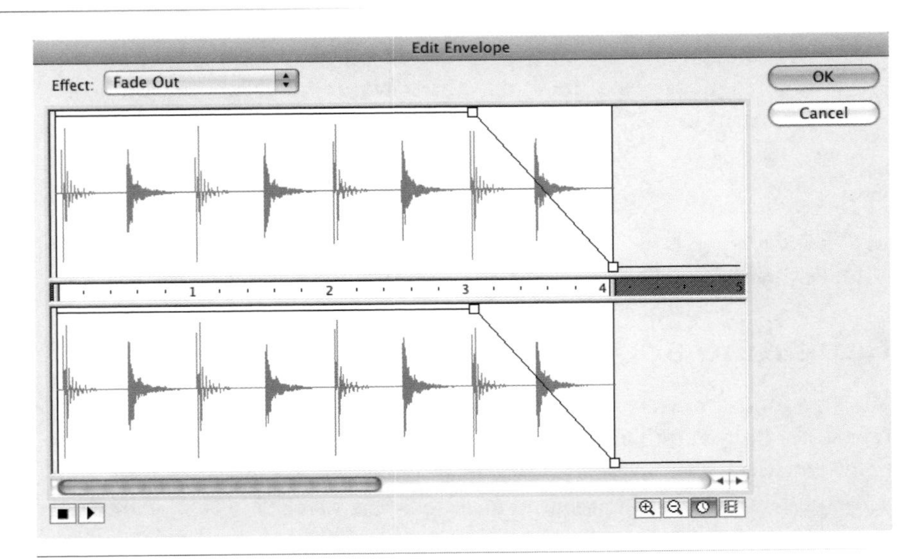

Figure 12-9. Fading a sound in the Edit Envelope dialog

Figure 12-9 shows a reduction in volume at the end of the sound, as both channels fade to 0. Figure 12-10 shows a change in pan as the sound moves from the left to right channels. The volume is decreased in the left channel and simultaneously increased in the right channel.

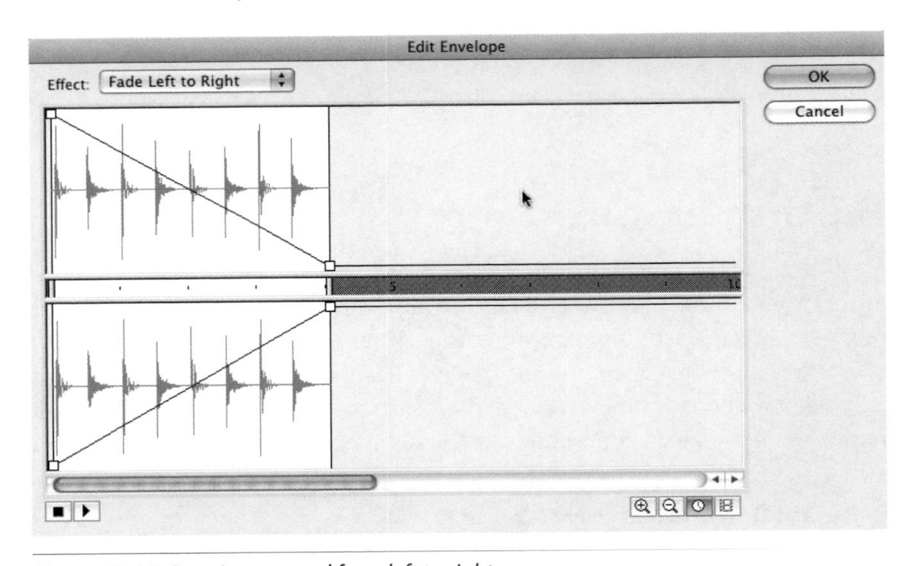

Figure 12-10. Panning a sound from left to right

During editing, you can test your work without leaving the dialog or testing your movie. Simply use the stop and play buttons in the lower-left corner of the dialog to audition your effect. No changes are made to the original file, so you can experiment to your liking.

In the lower-right corner of the dialog, you can zoom in and out of the waveform (clicking the magnifying glass icons) and switch between a time-based (watch icon) or frame-based (movie icon) display. The latter two options change the horizontal numbered strip between the channels to show either frames or seconds. In Figure 12-9, for example, the sound starts to fade at three seconds, and is muted by the fourth second.

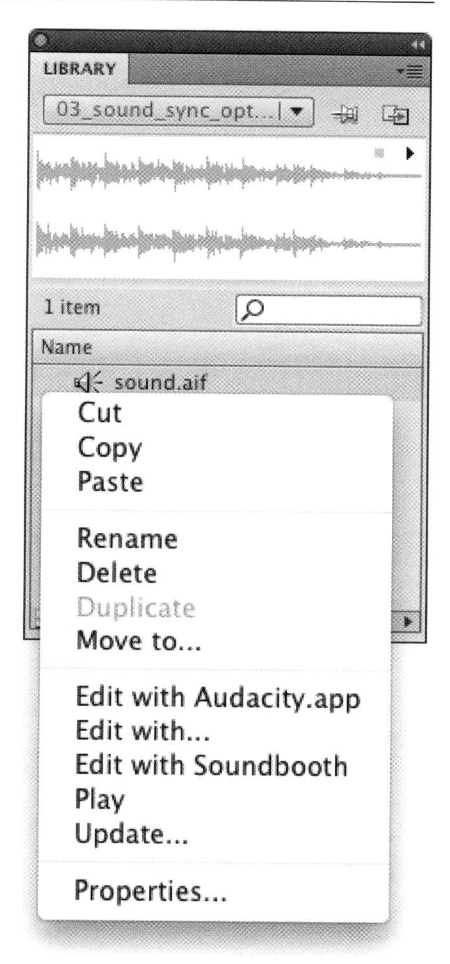

Figure 12-11. Editing a sound in an external application

Using an External Editor

Flash even provides a convenient workflow for editing your sounds in an external sound editor. Control+clicking (Mac) or right-clicking (Windows) on a sound asset in the Library displays a context-sensitive menu that reveals editing options. Figure 12-11 illustrates your editing choices: the cross-platform, open source editor Audacity; an editor you select on the fly; or Adobe Soundbooth, an entry-level sound editor targeted at Flash users.

When you select one of these options, the sound opens in the selected external editor so you can then edit the file. When you save the file, it will automatically be updated in Flash.

ActionScript Use

Using ActionScript, you can exert much greater control over sound than you could if it were a part of the Timeline. For example, you can mix 32 discrete audio channels, set the volume and pan of each sound, and query the amplitude of the left and right stereo channel of every sound.

Playing a Library Sound

Playing an embedded sound from the Library is very similar to adding a symbol instance to the display list. The process begins by assigning the sound a Linkage class (Figure 12-12).

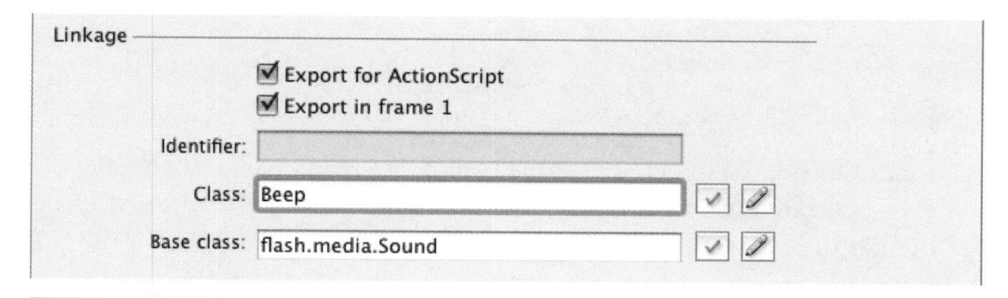

Figure 12-12. Assigning a sound's Linkage class

1. Create a new file using File→New. You won't be using this file in the project, so the book template is not needed.
2. Import *beep.wav* from the companion source files.

3. Select the *beep.wav* sound in the Library and click the Properties button (the *i* icon at the bottom of the Library panel) to open the Sound Properties dialog.
4. If the Linkage section of the dialog (Figure 12-12) is not already visible, click the Advanced button to view it.
5. Enable the *Export for ActionScript* option and enter **Beep** in the *Class* field. Note that the class you are creating extends the **Sound** class, so the **Beep** class will behave like a sound in many respects.
6. Add the following script to the first frame of the Timeline and test your movie:

```
var beepAlert:Beep = new Beep();
beepAlert.play();
```

Your file should play a short beep sound from the embedded *beep.wav* file. If needed, compare your file with the furnished *play_embedded_sound.fla* file.

Playing an External Sound

In addition to playing embedded sounds, you can load external audio files to play at runtime. This is advantageous for many reasons, including reducing SWF file size and making it easier to edit, update, or replace the external file.

The following exercise will span the next three sections of this chapter, adding functionality to your script as the chapter progresses:

1. Create a new file using File→New. You won't be using this file in the project, so the book template is not needed.
2. Save the file as *sound_scripting.fla* in a directory of your choice, and copy the companion source file *song.mp3* into the same directory.
3. Type the following script into frame 1 of the FLA:

```
1   var snd:Sound = new Sound();
2   var sndURL:URLRequest = new URLRequest("song.mp3");
3   snd.load(sndURL);
4
5   var channel:SoundChannel = new SoundChannel();
6   channel = snd.play();
```

4. Lines 1 through 3 load the external sound. First, an instance of the Sound class is created in line 1. In line 2, a **URLRequest** instance is created using the path to the sound file. Lines 5 and 6 play the sound into a unique sound channel.
5. Save and test your movie. It will play the external audio file.
6. Optionally, you can play the sound only after it fully loads. This is helpful for slow connections where the sound playback speed may surpass the download speed. If desired, replace line 6 of your script with the following four lines of code:

```
snd.addEventListener(Event.COMPLETE, onSoundLoaded);
function onSoundLoaded(evt:Event):void {
    channel = snd.play();
}
```

Now when the sound finishes loading, the *complete* event will be dispatched. When the event is received, the sound will play again from the beginning, into the same channel.

With regard to line numbering, the next two sections of this exercise assume this option was not added. If you chose to play the sound upon loading, adjust your file's line numbers accordingly.

Setting Volume and Pan

You set the volume of a sound using a range of 0 to 1. A value of 0 mutes a sound and a value of 1 sets the sound to full volume. The pan of a sound is set using a slightly different scale, from –1 to 1. The sound pan can be centered (0), entirely in the left channel (–1), or entirely in the right channel (1).

1. Continue with the previous example and add the following code to the script in frame 1:

```
7   var trans:SoundTransform = new SoundTransform();
8   trans.volume = 0.5;
9   trans.pan = -1;
10  channel.soundTransform = trans;
```

2. Both **volume** and **pan** are properties of the **SoundTransform** class. To change either value, you must first create an instance of the class. One way to accomplish this is by using the **new** keyword, shown in line 7. This creates a neutral instance with default values of full volume and center pan. After altering these values (to half volume in line 8 and full-left pan in line 9) you must then apply the transformation to the sound channel's **soundTransform** property (line 10). Without this final step, you are only adjusting the **SoundTransform** instance, not the sound channel itself.

3. Save and test your movie. This time, the sound will play at half volume, in the left stereo speaker only.

NOTE

An alternative approach to creating a SoundTransform *instance is to query the* soundTransform *property of a sound, as in this example:*

```
var trans:SoundTransform =
channel.soundTransform;
```

This starts the transform values with the existing volume and pan of the sound. This is useful for relative transformations, such as reducing the current volume (no matter what that value is) by 25%.

Visualizing Volume

The final topic covered in this chapter lets you visually represent the volume of an audio file during playback. Two properties of the **SoundChannel** instance, **leftPeak** and **rightPeak**, contain the volume of the left stereo channel and right stereo channel, respectively (when analyzing monaural sounds, both values are equal).

NOTE

The term "channel" is necessarily used in two contexts in the discussions in this chapter. First, stereo playback is achieved by dividing the sound data into left and right channels, while monaural sounds have no such separation. Second, every sound, whether stereo or mono, is played into its own discrete sound channel for mixing purposes. Each file can have 32 individual sound channels, and one master mixing channel, similar to a mixing desk that you might see in a recording studio.

These two uses of the word "channel" are not the same. For example, your file can have 32 channels of sound, all of which are in stereo. This does not create 64 channels of sound because you can't manipulate the left and right stereo divisions of an audio file independently.

In this example, two circles that represent left and right stereo speakers will expand and contract with the sound channel's volume during sound playback:

1. Continue with the previous example. Using the Oval tool, draw a circle on the Stage.
2. Select the circle with the Selection tool and convert it to a movie clip (Modify→Convert to Symbol). Name the movie clip **speaker** and choose a center registration point.
3. Copy and paste the on-Stage instance of the *speaker* symbol and give one speaker an instance name of **leftSpeaker** and the other an instance name of **rightSpeaker**.
4. Add the following code to the script in frame 1, save your work, and test your movie:

```
11  this.addEventListener(Event.ENTER_FRAME, onLoop);
12
13  function onLoop(evt:Event):void {
14      leftSpeaker.scaleX = 1 + channel.leftPeak;
15      leftSpeaker.scaleY = 1 + channel.leftPeak;
16      rightSpeaker.scaleX = 1 + channel.rightPeak;
17      rightSpeaker.scaleY = 1 + channel.rightPeak;
18  }
```

This script creates an event listener that listens for an enter frame event (lines 11 and 13) and adjusts the **scaleX** and **scaleY** of each speaker based on the volume of the sound channel's left and right stereo channels, respectively.

Instead of setting these properties to the **leftPeak** and **rightPeak** values directly, they are added to a value of 1. This prevents the speakers from shrinking, possibly to a scale of 0, during quiet portions of the audio. That is, rather than displaying scales that correspond with volumes of 0, 0.5, and 1, the resulting scales are 1, 1.5, and 2. When the volume is muted, the speaker size is 100%, and at full volume the speaker size is 200%, or double the original dimensions.

NOTE

This book's companion volume, Learning ActionScript 3.0: A Beginner's Guide *(O'Reilly), includes an example that visualizes the entire frequency spectrum of a sound, drawing waveforms during playback.*

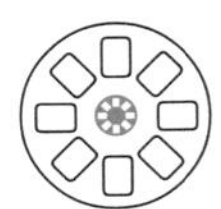

Project Progress

In this chapter you will add the ActionScript required to activate the sound controller widget of the portfolio project.

Scripting the Sound Controller

In prior chapters, you created the sound controller and integrated it into the master project file. The following script will load an external MP3 file, toggle between playing and disabled states, and visualize the amplitude of the left and right stereo channels.

Open your master portfolio FLA and add the following script to the frame 99 of the *actions* layer. Frame 99 is the last intro frame (prior to the *Home* screen) and initializes both the navigation and sound controls:

```
1   var snd:Sound = new Sound();
2   snd.load(new URLRequest("assets/song.mp3"));
3
4   var channel:SoundChannel = new SoundChannel();
5   channel = snd.play();
6
7   this.addEventListener(Event.ENTER_FRAME, onSoundPlayback);
8   function onSoundPlayback(evt:Event):void {
9       soundControl.lPeak.barMask.scaleX = channel.leftPeak * 4;
10      soundControl.rPeak.barMask.scaleX = channel.rightPeak * 4;
11  }
12
13  soundControl.addEventListener(MouseEvent.CLICK, onSoundToggle);
14  function onSoundToggle(evt:MouseEvent):void {
15      if (channel.leftPeak > 0) {
16          channel.stop();
17      } else {
18          channel = snd.play();
19      }
20  }
```

Lines 1 through 5 load and play a sound, as described in the "Playing an External Sound" section, earlier in this chapter. The only difference here is that, for the portfolio project, you are placing the sound file in your external *assets* directory, with the other external assets used throughout the book.

Lines 7 through 11 visualize the volume of the sound's left and right stereo channels. To understand how this script works, it helps to recall the structure of the sound controller movie clip from Chapter 3. It has an instance name of *soundControl* and contains two child instances of a volume meter, one for each channel, named *lPeak* and *rPeak*. Within the meter is another movie clip named *barMask* that sits in a mask layer. By manipulating the **scaleX** property of the mask movie clip (lines 9 and 10), you expose different widths of the meter, and therefore different temperatures of color, during sound playback. Every enter frame (lines 7 and 8), the **leftPeak** and **rightPeak** properties return a percentage value, between 0 and 1, of the full volume of each channel.

When playing relatively quiet sounds, your visualization meter may appear a bit listless. If you want to liven things up a bit, you can increase the effect the actual peak values have on your artwork. Simply multiply the values of the **leftPeak** and **rightPeak** properties by an adjustment factor, as seen in lines 9 and 10. For example, these lines multiply low-level settings of 0.1 by 4 to become 0.4. The larger values increase the widths of the masks, showing larger meters.

Lines 13 through 20 toggle the sound playback states. With every mouse click (lines 13 and 14), a simple test is performed. If there is any volume in the left channel, the sound is stopped. If the left channel is silent, the sound is played again from the beginning. This basic script lets you turn the sound on and off.

NOTE

The conditional test in the function **onSoundToggle()** *relies on the volume in the left channel for simplicity, so it's not well suited to sounds with lots of quiet passages. The companion website demonstrates other techniques for turning sound on and off.*

The Project Continues...

In the next chapter, you will encode video, review its use in the project's Gallery screen, and learn ActionScript alternatives for displaying and controlling video in your own files.

CHAPTER 13

VIDEO

Introduction

Whether you realize it or not, Flash has become the most widely used video playback platform on the Web, and it is becoming more and more common in the distributed desktop video playback market. According to Nielson's online video ratings, YouTube alone, which uses Flash Player, accounted for over 5 billion streams, and more than 77 million unique visitors just in the month of July 2008. Add in just a handful of other top video sites that use Flash, such as Nickelodeon, Disney, Hulu, and CNN, and it's easy to see how ubiquitous Flash video has become.

The number of streams, or even unique visitors, isn't really the most telling statistic, however. According to August 2008 data compiled by independent research firm comScore (commissioned by Adobe), Flash Player is used to view 86% of online videos in the United States, and 80% of online videos worldwide, making Flash the number one technology for viewing video online.

Why is Flash video so omnipresent? The main reason is that it uses Flash Player and doesn't require any additional installation just for viewing video. Depending on which version of Flash Player you're considering, up to 99% of all online users have Flash Player already installed on their systems. It is also operating system–neutral, making development for the platform much simpler and more economical.

A particularly good reason, however, for the popularity of Flash video is that it's easy for everyone—not just experienced video professionals—to create and use. This chapter will show you how to prepare video for Flash delivery and also how to add it to your files. Adding a video to your Flash projects can be as simple as dropping a component onto the Stage, or as dynamic as necessary through ActionScript control.

IN THIS CHAPTER

Video Formats

Before you can use video in Flash, a little preparation is required. In addition to any editing that you might have to do to make your videos suitable for distribution, you will also have to *encode*, and most likely *compress*, your files into a Flash-compatible format.

Although the terms *encoding* and *compression* are often used interchangeably, there is a difference between the two. In simple terms, *encoding* is translating information from one format to another. In the video world, this might refer to converting from analog to digital, or translating from one digital format into a more appropriate specification needed for the job.

Compression, on the other hand, is the process of reducing the size of an asset. Compression and decompression algorithms (*codecs*) are used to reduce file size during authoring, and decompress at runtime for optimal performance and visual quality. The reason encoding and compression are sometimes used interchangeably is because most video is compressed during encoding. However, it is not uncommon to work with uncompressed video, particularly during editing.

When used in Flash, uncompressed video is not a practical reality, so video is both encoded and compressed into two primary formats:

FLV

FLV is a proprietary format that can be used by Flash Player version 6 and later. The audio in FLV files is usually compressed using the MP3 codec, but use of uncompressed audio is also possible. Video can be compressed using one of two codecs: Sorenson Spark (H.263), available for Flash Player versions 6 and later, and On2 VP6, available for Flash Player versions 8 and later.

Which video codec you choose is typically based on need. Within Flash, On2 VP6 is considered by most to deliver better quality than Sorenson Spark at the same data rate. It can also encode alpha data for runtime compositing effects like chroma key (green screen), in which videos have a transparent background and can be blended with dynamic backgrounds. However, On2 VP6 is more processor-intensive, making Sorenson Spark a favored codec for older or less powerful computers.

F4V

The latest versions of Flash Player, beginning with version 9 update 3 and later, can also use the H.264 codec for video and the AAC codec for audio. These are the codecs used for MPEG4 video—commonly used by later generation iPods and most other recent hardware and software vendors for audio and video playback.

Although these files are typically referred to as F4V files, Flash does not rely on the extension of the file for playback. As such, other files that have been properly encoded using H.264 and AAC, such as QuickTime movies, can also be handled by Flash Player 9 Update 3 and later.

Encoding Software

There are numerous software packages available to encode video into Flash-compatible formats. This chapter will focus on three. Developed by the inventors of FLV codecs, Sorenson Squeeze and On2 Flix are both commercial products that offer commercial-grade versions of their codecs. Adobe Media Encoder ships with Flash CS4 Professional and will be discussed in greater detail.

On2 Flix

Flix from On2 (*http://www.on2.com/*) comes in Standard, Pro, and Exporter configurations. The Standard model is a great economical encoder that encodes FLV for Flash video and FXM for JavaFX video. It supports constant bitrate encoding, one-pass variable bitrate encoding, input for a wide variety of original sources, and output using both FLV and SWF formats.

The Exporter is a QuickTime extension that supports exporting from QuickTime-savvy applications such as Apple FinalCut Pro, Adobe AfterEffects and Premiere, and Apple iMovie and QuickTime Movie Player. It supports most of the Flix Standard features, as well as two-pass VBR encoding and alpha channel support.

The Pro version offers all the features of the Standard version plus two-pass encoding, high-definition (HD) output, support for *cue points* (markers that can be used for seek points and to trigger events using ActionScript), alpha channels, and batch encoding, and an HTML/player configuration that outputs final player files at the click of a mouse. It also supports a really cool feature that vectorizes your video during encoding. This creates vector shapes from every pixel-based video frame, a technique used in recent commercials and feature films such as *A Scanner Darkly*.

Using Flix Pro (Figure 13-1), you can specify video and audio compression settings, save and load setting presets, trim the video by setting in and out points, add overlays (for logos and similar needs), deinterlace interlaced sources like MPEG and DV, apply image filters (like brightness, contrast, hue, and saturation), apply noise reduction, and more.

NOTE

As the name implies, constant bitrate (CBR) transfers a constant number of bits of data every second. Variable bitrate (VBR) encoding, on the other hand, adjusts the data transfer as the encoding progresses in an attempt to juggle maximal quality and minimal file size. The process encodes larger amounts of data when the video changes substantially from frame to frame, and encodes less data when the changes between frames are subtler.

Using VBR, encoding can be completed in one pass or two. In the latter case, the first pass is used to analyze the file and the second to encode based on the data gathered in the prior pass. The two-pass method usually yields better results, but can also take much longer.

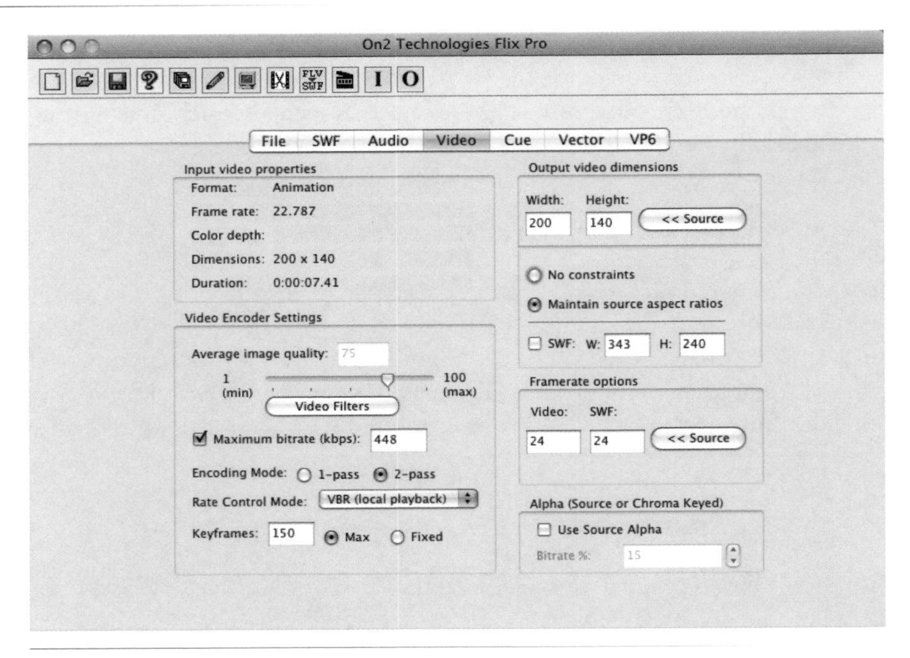

Figure 13-1. Applying video settings in On2 Flix

Sorenson Squeeze

Sorenson Squeeze (*http://www.sorensonmedia.com/*) is a full-featured video compression tool that comes in a variety of configurations. It comes in standard, Pro, and Multi-User versions that export a large number of video formats, as well as an economical version specially optimized for Flash video.

The standard version is very feature-rich, including Sorenson Video 3 Pro, Sorenson Spark Pro, Sorenson MPEG4 Pro, and Sorenson H.264 Pro video codecs. It also supports On2 VP6 Pro video encoder and BIAS SoundSoap noise reduction plug-in, both sold separately. The former supports all the Flash features enabled by the VP6 codec, and the latter is a powerful and amazingly friendly sound-reduction plug-in that washes the noise right out of your sound. Both of these features are bundled in the Pro version of Squeeze.

Squeeze has a customizable interface (Figure 13-2) and also includes additional cool features like batch encoding; drag and drop; a watch folder feature (which will watch for and automatically compress any files dropped into a specific folder); the ability to digitize video; watermarking, hue, saturation, sharpen, and other filters; CBR and one- and two-pass VBR encoding; high-definition (HD) output (including Blu-Ray); and more.

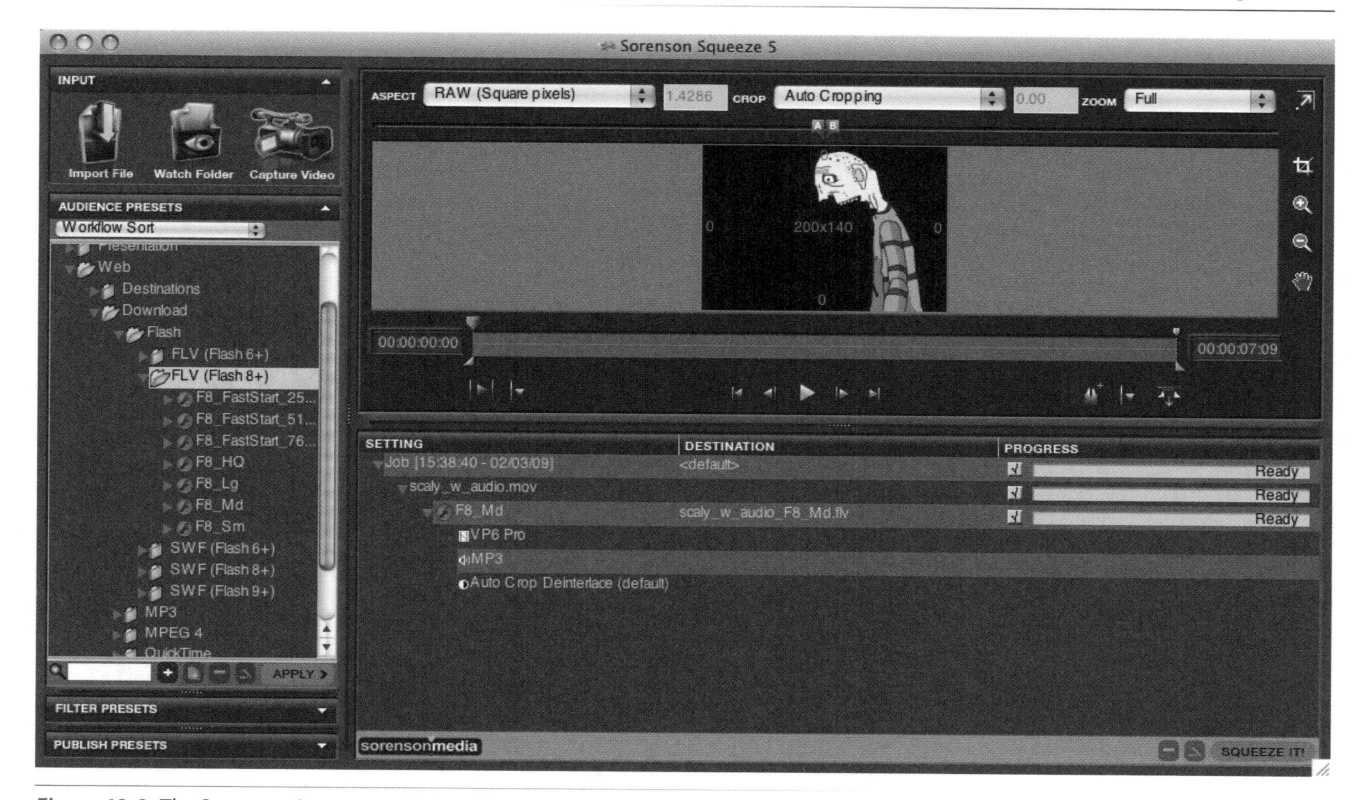

Figure 13-2. The Sorenson Squeeze main interface

Adobe Media Encoder

The Adobe Media Encoder ships with Flash CS4 Professional and is both feature-rich and simple to use. It supports encoding to a wide variety of video formats including QuickTime, MPEG2, MPEG4, and H.264, as well as audio-only formats (AIFF, WAV, MP3) and Blu-Ray high-definition versions of MPEG2 and MPEG4. For use with Flash, the application can encode to FLV with both Sorenson Spark and On2 VP6, as well as to F4V using H.264 and a wide variety of other codecs for use in non-Flash-related video work.

It supports audio and video settings, filters, cue points, trimming with in and out points, cropping, batch encoding, and even FTP features that let you upload your finished files to a server.

This section will go through the basics of using Apple Media Encoder, so you may want to open the application and locate the *scaly_vid.mov* furnished source file to encode your own video in the process. Where applicable, encoding instructions like the first few following this paragraph will run throughout this section. If you choose not to encode your own file, feel free to skip them as you work through the material.

The first task in encoding a video is adding a file to the encoding queue. Launching the application will display a basic screen (Figure 13-3) that allows you to add one or more files for encoding. If you wish to try Adobe Media Encoder for yourself, start by adding a file.

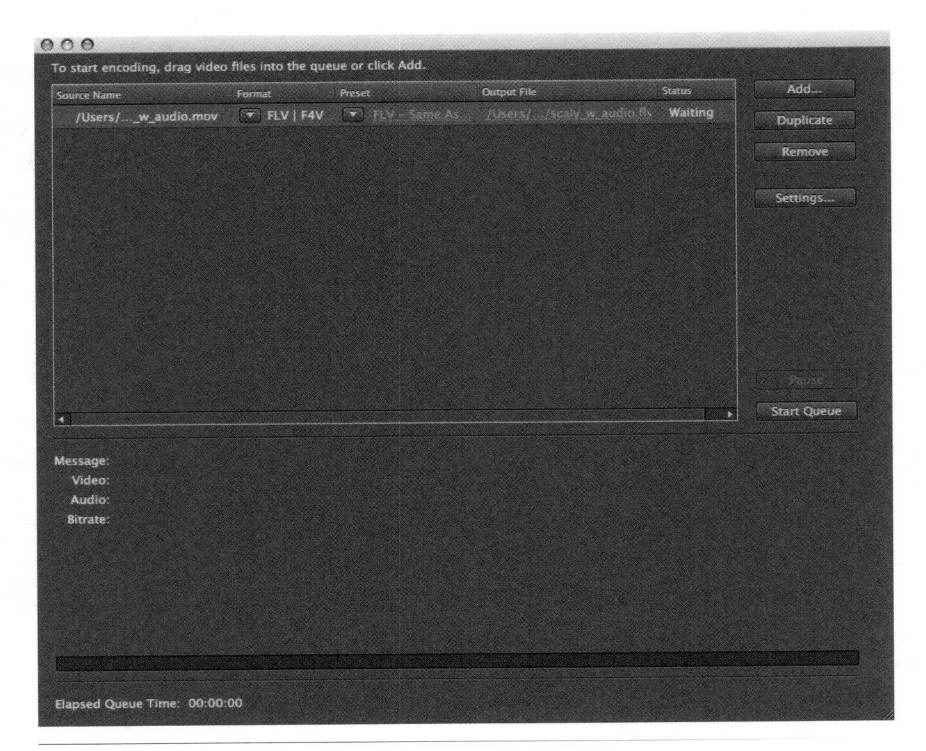

Figure 13-3. Adding a video in Adobe Media Encoder

1. Click Add in the upper-right corner of the application.
2. Locate the *scaly_vid.mov* file from the furnished source files.
3. Click Open in the file browser dialog to add the video to the encoding queue.
4. You can choose an encoding format, apply encoding presets, and set the destination path for each video right from the encoding queue. However, we will take a more detailed look at each available option, so with the video selected, click Settings to open the Export Settings window (Figure 13-4).

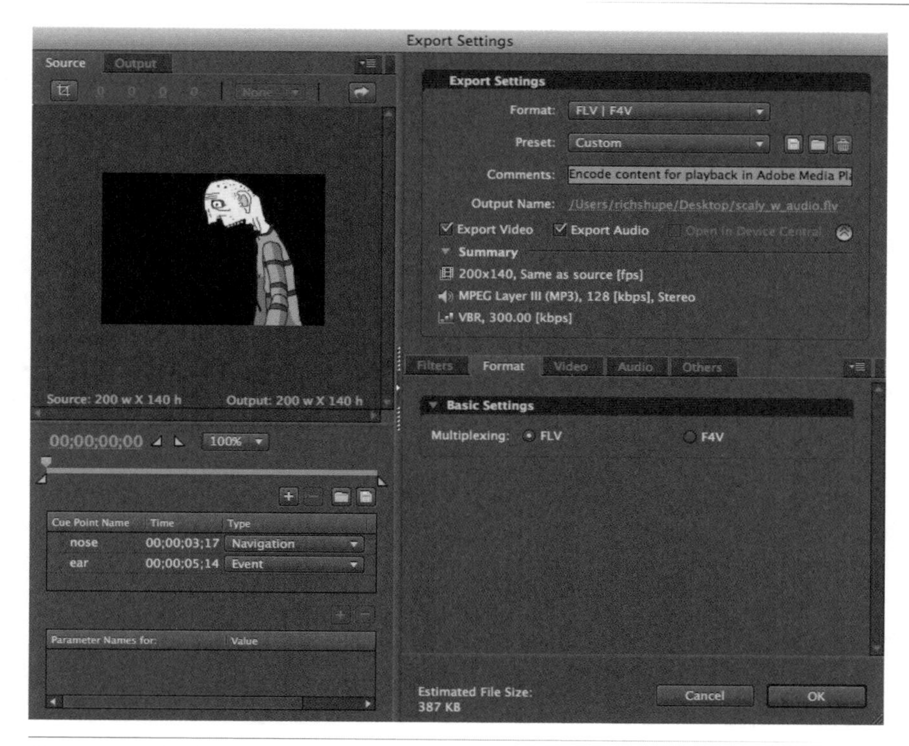

Figure 13-4. Configuring export settings in Adobe Media Encoder

The Export Settings window is the primary Adobe Media Encoder interface, and is divided into quadrants. The upper-left corner is the source and output preview area. This is where you crop the video and preview a sample frame using the application's current settings. The lower-left corner is where you trim the video to a smaller running time and add cue points.

The lower-right corner is where you apply all other settings, and the upper-right corner is where you select presets from menus and enable or disable video or audio encoding. It also summarizes the major settings for at-a-glance updates.

5. Before moving on, choose the *FLV | F4V* option from the *Format* menu.
6. As a starting point, choose the second option from the encoding *Preset* menu, *FLV – Same As Source (Flash 8 and Higher)*.
7. By default, the output file will be saved in the same directory as the source file. If you want to change this directory, click on the *Output Name* link and specify a new location.
8. Make sure *Export Video* and *Export Audio* are enabled.

The tabs in the lower-right corner of the interface allow you to customize your encoding settings. For Flash video, the Format tab lets you switch

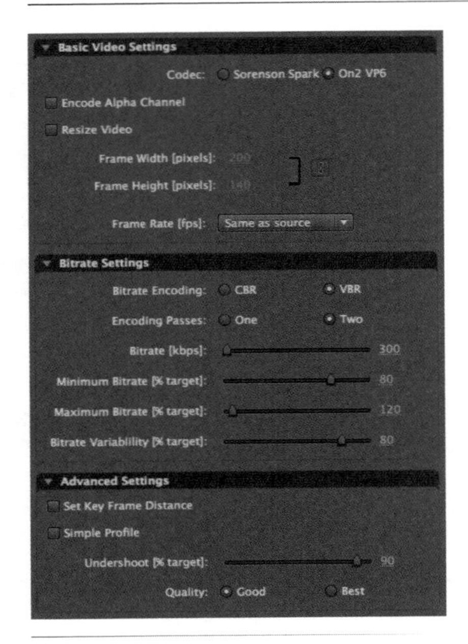

Figure 13-5. Configuring video settings in Adobe Media Encoder

NOTE

For more information on bitrate, see Chapter 12.

WARNING

When choosing how often to insert a key frame, remember that high-quality frames increase the size of the final output file. Furthermore, if you've set a maximum data rate for the video, adding key frames can reduce the overall quality of the entire video because the high-quality frame reduces the amount of data available for the surrounding frames. Try to use key frames only when needed, as evidenced by trial and error.

between FLV and F4V formats. Because you selected the preset in step 6 of this sample exercise, FLV output has already been chosen.

The Filters tab lets you to preprocess your video with graphics filters. Gaussian Blur ships with the application and can be applied in horizontal, vertical, or both directions. Although not needed in this case, applying a very slight blur can sometimes soften the video and reduce compression artifacts.

The Video tab is where you pick a codec and customize its settings. These are the first settings you're likely to customize.

Video settings

The video settings are divided into three categories and arranged in a scrollable list (Figure 13-5).

The Basic Video Settings let you choose the *Codec*, whether or not to *Encode Alpha Channel* data included in the video (when the On2 VP6 codec is selected), specify pixel width and height values to *Resize Video*, and set the *Frame Rate*.

The Bitrate Settings let you set the *Bitrate Encoding* to constant bitrate encoding and one- and two-pass variable bitrate encoding. You can also set a custom *Bitrate*. When you choose two-pass VBR encoding, you can set a target bitrate, minimum and maximum bitrate limits, and the degree to which the bitrate can vary from the target setting.

Finally, the Advanced Settings provide optimization options beginning with how often a key frame is created (*Set Key Frame Distance*). Unrelated to animation key frames, a video key frame is a forced, high-quality frame that is used to refresh the video display. It is usually best to let the codec work its magic here, but if you are having problems with visual artifacts, particularly when compressing your source very aggressively, you can manually insert a key frame every *n*th frame to try to improve the appearance of the frame periodically.

Simple Profile, available only when encoding with On2 VP6, will optimize encoding for playback on less powerful machines. *Undershoot*, also available only when encoding with On2 VP6, lets you set a percentage of the target data rate to strive for, so that you can leave a little room for extra data to be allocated to sections of the video that are difficult to compress.

Finally, *Quality* allows you to strike a balance between how good the file will look and how long it will take to encode. *Best* yields the best results using the longest encoding time, *Speed* sacrifices a bit of quality for a faster encoding time, and *Good* falls somewhere in the middle.

Continuing the example exercise, change the following video settings:

9. For Basic Video Settings, choose the `On2 VP6` *Codec*, do not enable *Encode Alpha Channel* or *Resize Video*, and select `Same as source` for the *Frame Rate*.

10. For Bitrate Settings, select VBR for *Bitrate Encoding*, Two *Encoding Passes*, and 300 kbps for *Bitrate*, and accept the default minimum, maximum, and variability bitrate encoding percent target values of 80, 120, and 80, respectively.
11. For Advanced Settings, disable *Set Key Frame Distance* and *Simple Profile*, set the *Undershoot (% target)* to 90, and select Good *Quality*.

Audio settings

Using FLV, you don't have many choices for the audio settings (Figure 13-6) because FLV uses the MP3 codec when compressing audio.

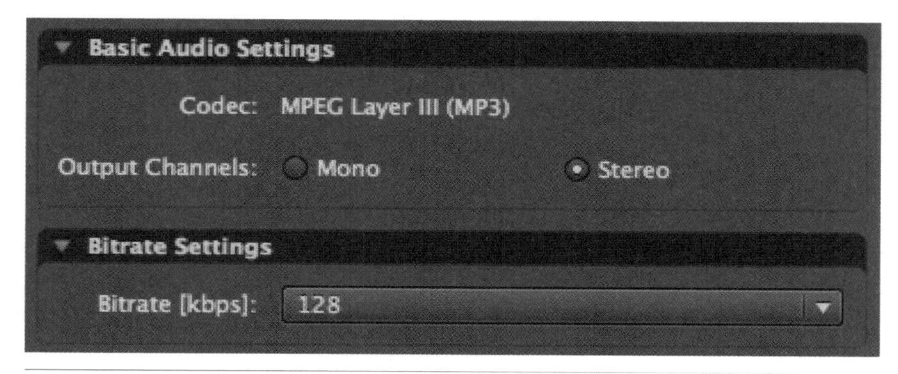

Figure 13-6. Configuring audio settings in Adobe Media Encoder

12. In the Basic Audio Settings section, choose Stereo for *Output Channels*.
13. For Bitrate Settings, select a 96 kbps *Bitrate*.

Trimming

Although you don't need to trim the sample video, you can quickly learn how to reduce the playing time of a video at encoding time without having to return to your video editor.

At the top of the lower-left quadrant of the interface are the trimming controls (Figure 13-7). A horizontal bar representing the length of the video anchors these controls.

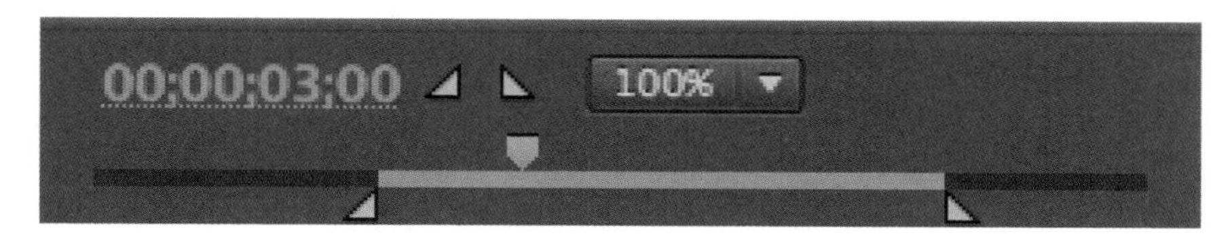

Figure 13-7. Setting In and Out points in Adobe Media Encoder

By default, the bar is orange all the way across. However, you can drag the right-angle triangles at left and right below the bar to set the in and out points of the video, respectively. These controls will make a new video with a shorter running time that starts at the in point and ends at the out point. You can preview frames before setting the in or out points by dragging the playhead wedge above the bar.

You can also click the orange time link to set a time-accurate position in the video. The time units are separated by semicolons and are, from left to right, hours, minutes, seconds, and frames. The latter will change based on the frame rate you specified in the Basic Video Settings.

Once you're displaying the time you like, you can click the right-angle triangle buttons to the right of the time control; these correspond to the in- and out-point draggable markers, and clicking these buttons will automatically set the points for you.

The drop-down menu to the right of the trimming time and in/out buttons is not related to trimming the video length. Instead, it is situated below the video preview frame and controls the size of the preview.

Cue points

Using the same playhead wedge and time controls described in the preceding section, you can set cue points. *Cue points* are markers that you can embed in a video. At playback, they can serve as seek positions to which you can navigate, or they can trigger events using ActionScript.

There are three types of cue points. You can embed *event* and *navigation* cue points directly into the video so that they will travel with the video file forever. This is a convenient one-time authoring process that provides for multiple future uses, but it also means that you must reencode the video to change the cue points. *ActionScript* cue points are applied with ActionScript and you can freely edit them any time after encoding. This is convenient but can also be slightly less accurate.

Event cue points are used to trigger events in ActionScript, allowing you to synchronize surrounding assets with the video playback. Navigation cue points can trigger events in ActionScript as well but also insert key frames into the video so you can jump to a specific time during playback. Because key frames can increase file size or reduce the overall quality of the video, use navigation cue points only when required for seeking at runtime.

To practice, you'll add two cue points, which appear in the lower-left corner of Figure 13-4. If you prefer to use a furnished data file, click the folder button just above the Cue Points area and find the *scaly_vid_cue_points.xml* file supplied with the companion source files. This will import the file and prepopulate the cue point data.

14. Use the playhead marker to approximately locate 3 seconds and 17 frames. Alternatively, click the time control to enter 3 seconds and 17 frames precisely.
15. Click the plus-sign button to add a cue point.
16. Click the default name under *Cue Point Name* and rename the cue point `nose`, then choose Navigation from the cue point's Type menu.
17. Repeat steps 14 through 16 to add an Event cue point at 5 seconds and 14 frames.
18. When you're finished, always click the document button above the cue point area to save your file. If you ever need to reencode your video after it's been removed from the encoding queue, or even to apply the same cue point settings to an alternate file, loading this file will save you a lot of time reentering cue point data. In this case, name the file *scaly_vid_cue_points_01.xml*.

NOTE

Later on, use a text editor to look at the cue point text that you exported. It is a relatively straightforward XML file. If you are comfortable with XML, editing this file before using Adobe Media Encoder can be a fast way to create many cue points.

When you become more comfortable with ActionScript, you can also add parameters and corresponding values to each cue point for use in ActionScript. This book's companion volume, *Learning ActionScript 3.0: A Beginner's Guide* (O'Reilly) includes a section in its video chapter that shows you how to use parameters when captioning a video.

Cropping

Although not necessary for the sample video, you can crop edges off videos during encoding. Figure 13-8 shows using the cropping tools in the upper-left quadrant of the Adobe Media Encoder interface to crop a few pixels off the left and right edges of the video. You can set this automatically by using the aspect ratio menu (showing 4:3 in Figure 13-8), by manipulating the numbers for each edge to the left of the menu, or by dragging the corner handles of the crop rectangle on the video preview.

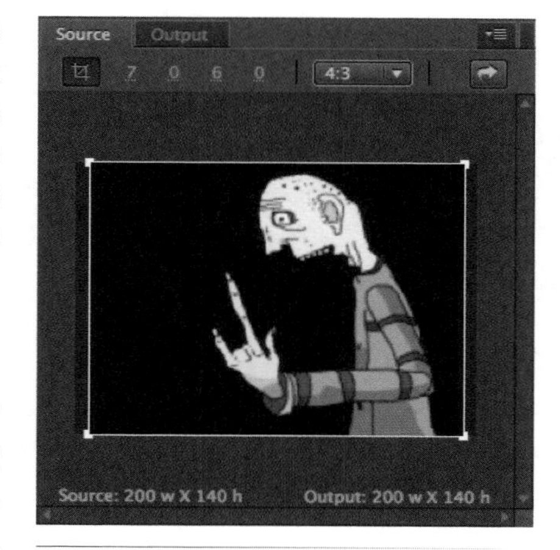

Figure 13-8. Cropping a video in Adobe Media Encoder

Other

As an added nicety, you can use Adobe Media Encoder to automatically FTP finished videos to an online server (Figure 13-9), if you are familiar with using FTP and have access to the server in question. You can configure the settings like any other FTP client in the Others tab of the Encoder interface. You can even test your connection so you're not disappointed by a failure after a long encoding process, and optionally send the local files to the trash after uploading.

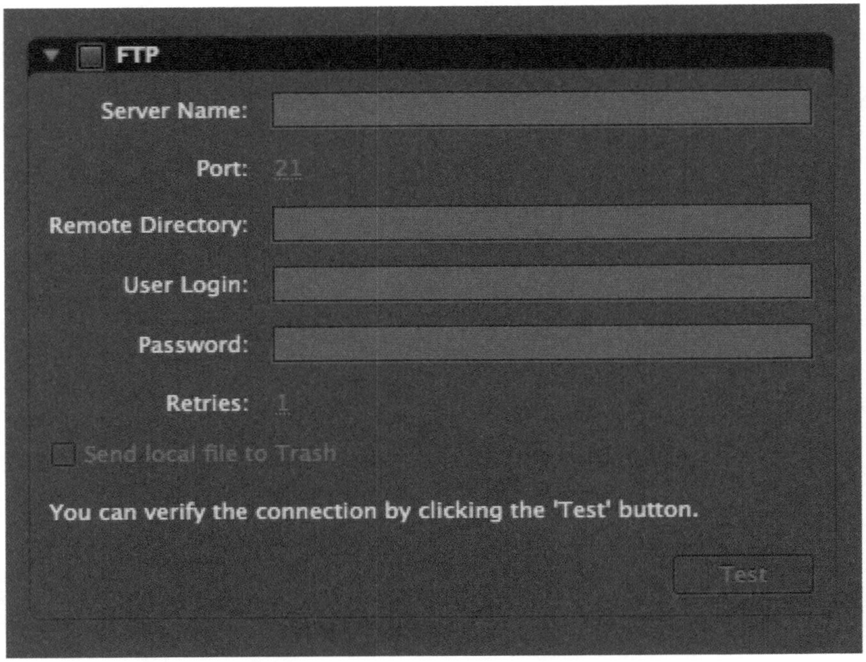

Figure 13-9. Configuring FTP settings in Adobe Media Encoder

Encoding

After you configure all the settings, you can click OK in the lower-right corner of the Adobe Media Encoder Export Settings window and begin encoding. Back in the main interface (Figure 13-10), you can click Start Queue to encode any videos you have in the encoding queue. A preview will appear as the progress bar advances through the encoding process. As each video finishes, its status will be updated to Finished, and the files will be ready for use.

NOTE

If you ever need to reencode, you can take advantage of a really handy feature: Adobe Media Encoder will keep your settings alive until the video is deleted from the encoding queue. To reencode, select the video in the queue and use the Edit→Reset Status menu command. The video will then be ready for encoding using all its prior settings.

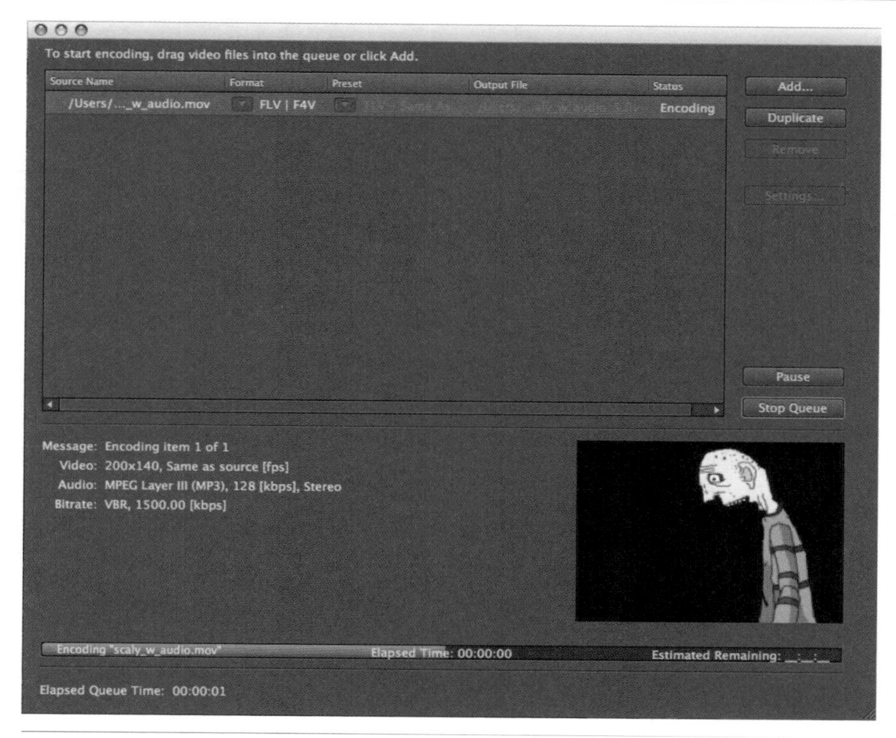

Figure 13-10. Encoding a video in Adobe Media Encoder

Playing Video with the FLVPlayback Component

The simplest way to add video to your file is by using the *FLVPlayback* component (Figure 13-11). The component will load external FLV or F4V files at runtime, and even includes a customizable player control skin. The skinning system lets you pick which controller features to offer the user, the skin's color and degree of transparency, and whether to position the skin under or on top of the video itself.

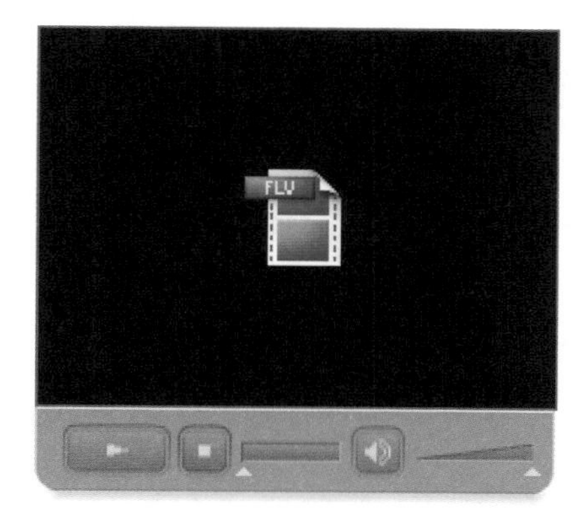

Figure 13-11. The FLVPlayback component

As you learned in Chapter 9, components make it easy to expand the functionality of your projects without having to use a lot of ActionScript. Adding video to your files using the component that ships with Flash CS4 Professional is a snap: all you need to do is drop the component onto the Stage, use the Component Inspector to set a few values, such as skin selection and source path, and you're on your way.

This is the approach outlined in the step-by-step "Project Progress" section at the end of this chapter, so here you'll increase your Flash chops by using ActionScript to configure the component. The end results are not very different, and this method still relies on the component to work. However, using ActionScript to manipulate the component offers much more control, as well as the ability to change the way the component behaves at runtime. For example, if you use ActionScript, you are not locked into values you enter in the Component Inspector and can change the video source at any time.

To complete this exercise, you need a little bit of setup:

1. First, have the *scaly_vid.flv* source file handy and save a new Flash file into the same directory in which the video resides. Sharing the same directory is not required, but it will simplify discussions about pathnames.

2. Next, the *FLVPlayback* component must be in your Library. Drag the component from the Components panel to the Stage. After the next step, you'll delete the component from the Stage just like a movie clip or button instance, and the component will stay in your Library, ready for use.

Now you need a skin for the video player. A quick way to get the skin you need is to specify the skin in the Component Inspector and then test your movie. When compiling the SWF, Flash will automatically copy the skin to the same directory as your SWF.

3. Open the Component Inspector, select the component on the Stage, and double-click the *skin* parameter value.

4. From the *Skin* menu, select the SkinUnderPlaySeekMute skin. This will add a play/pause toggle, a seek bar, and a mute button to your skin.

5. Click OK to return to the Stage, then test your movie. Having specified no source for your video, only the skin will appear in your test movie. However, if you look in the directory you are using for your test file, you will see that Flash has copied the specified skin to this location for you.

6. Finally, delete the component from the stage and your setup is complete. You can now instantiate the component at runtime using ActionScript.

7. Add the following script to frame 1 and test your movie:

```
1   import fl.video.FLVPlayback;
2
3   var vid:FLVPlayback = new FLVPlayback();
4   addChild(vid);
5   vid.width = 200;
6   vid.height = 140;
7
8   vid.source = "scaly_vid.flv";
9   vid.skin = "SkinUnderPlaySeekMute.swf";
10  vid.skinBackgroundColor = 0x996600;
11  vid.skinBackgroundAlpha = 0.5;
12  vid.autoPlay = false;
```

Line 1 of this script tells the compiler where to find the FLVPlayback classes required when compiling your SWF. Unlike classes that are part of the `flash` packages, component classes in the `fl` packages must be imported, even in the Timeline.

Lines 3 through 6 create a new instance of the *FLVPlayback* component, add it to the display list at position (0, 0), and match the width and height of the component to the source video dimensions.

Lines 8 through 12 configure the component. Line 8 sets the source using the *scaly_vid.flv* file that is in the same directory as your SWF; lines 9 through 11 specify the skin you selected and give it a color and alpha value. Line 12 sets the autoplay feature to `false`, so you must use the skin's play button to start the video.

NOTE

If you are not satisfied with the feature set offered by the FLVPlayback component, or if you want to save the approximately 50–75k the component adds to your file size, you can also create ActionScript-only video playback solutions. The book's companion website and the companion volume, Learning ActionScript 3.0: A Beginner's Guide, *have more information on the topic.*

Full-Screen Video

The FLVPlayback component provides support for true full-screen video playback. That is, rather than relying on prior techniques of scaling the SWF, browser window, or both to make the video as large as possible, Flash Player versions 9 and later will play the video directly to the monitor, consuming all available display space. During full-screen playback, you can then press the Escape key to return to actual-size playback.

Figure 13-12. FLVPlayback Full Screen button

To enable this feature using the FLVPlayback component, you must do three things. First, you must choose a skin that includes the full-screen option, such as SkinUnderPlayFullscreen or SkinUnderAll. These skins will include the button shown in Figure 13-12.

Next, you must enable full-screen support. Full-screen playback is a user preference and a possible security risk. For example, the user may not want to switch to full-screen mode. In a worst-case scenario, an unethical developer may use the feature to take over the user's screen and simulate an interface into which the user might type sensitive information, like a password.

For these reasons, full-screen mode can only be initiated by the user, and must be enabled by the developer to work at all. Adobe has also taken additional measures, such as disabling the majority of keyboard input while in full-screen mode, to reduce security risks.

The simplest way to enable full-screen support is to change the HTML template used in the file's Publish Settings. Using the Publish Settings dialog (File→Publish Settings), make sure HTML is enabled in the Formats section. Change the *Template* option in the HTML section to *Flash Only – Allow Full Screen* (Figure 13-13). This template will set the *allowFullScreen* Flash object and embed attributes to true without requiring you to edit the HTML files directly. Any file published with this setting will support full-screen mode if the *FLVPlayback* skin contains a full-screen button.

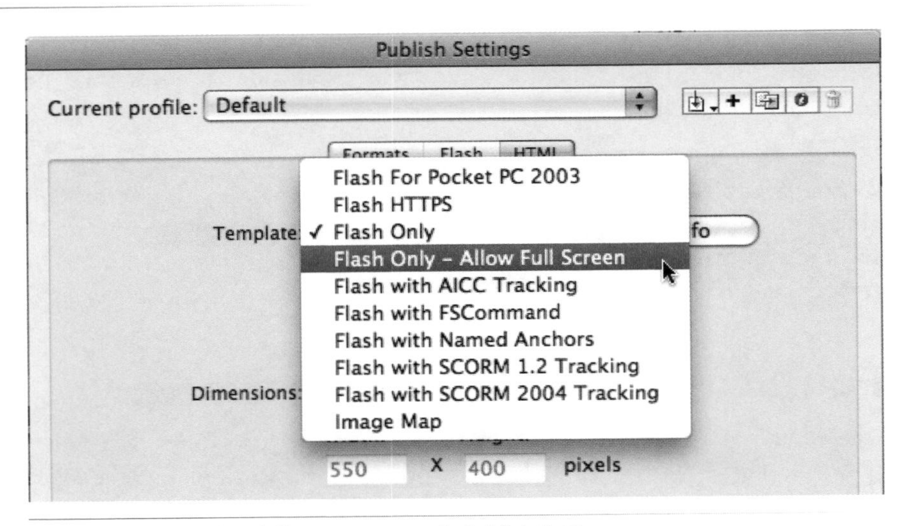

Figure 13-13. Enabling full-screen support in Publish Settings

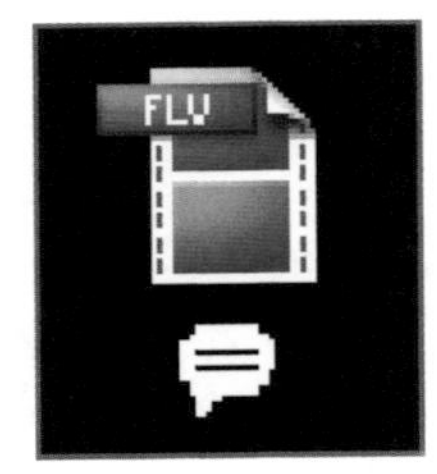

Figure 13-14. FLVPlayback Captioning component

Figure 13-15. FLVPlayback caption button

Third, you must preview your file in a browser. Full-screen playback does not work when testing your file within Flash because the aforementioned HTML setting must exist for Flash Player to allow the change. Therefore, instead of using Control→Test Movie to view your file, you must use the File→Publish Preview→HTML menu command (often set as the default preview option). This will switch to your default browser and show you the file in an HTML setting so you can test the full-screen mode.

Captioning Video

Flash also ships with a component called *FLVPlayback Captioning* (Figure 13-14) that will add captions to your video. The captions are typically precreated in an external XML file, which is loaded at runtime. Like the full-screen video option, the skin you select for the *FLVPlayback* component must include the caption button (Figure 13-15) to support this feature.

It is almost effortless to use the captioning feature within Flash. If there is only one instance of the *FLVPlayback* component on the Stage, the *FLVPlayback Captioning* component will automatically link to it. All you have to do is use the Component Inspector (Figure 13-16) to specify the path to the external caption XML file, and you're in business.

Creating the caption file requires nothing more than a text editor, although a dedicated captioning application can help you determine the duration and times of occurrence for each caption. The text file is saved in the TimedText standard format. You can find more information about TimedText from a number of sources, including the Flash help system, the World Wide Web Consortium (*http://www.w3.org/AudioVideo/TT/*), and this book's companion website.

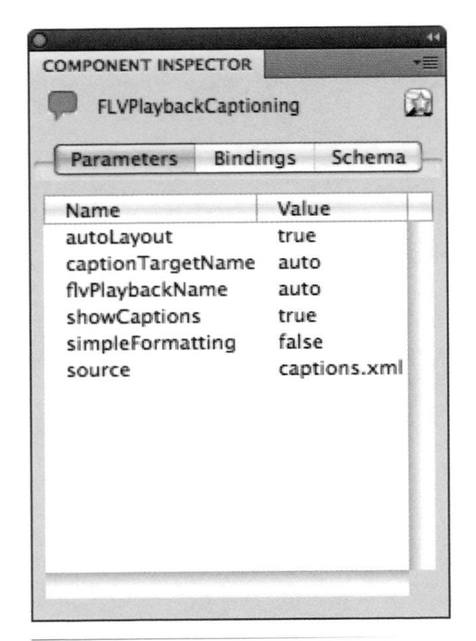

Figure 13-16. Configuring the FLVPlayback Captioning component

In short, a TimedText file is an XML document that looks much like HTML. The following is a simple example that can be used with the Scaly video:

```
1   <?xml version="1.0" encoding="UTF-8"?>
2   <tt xmlns="http://www.w3.org/2006/04/ttaf1" xmlns:tts="http://www.
    w3.org/2006/04/ttaf1#styling">
3       <head>
4           <styling>
5               <style id="1"
6               tts:textAlign="center"
7               tts:fontSize="18"
8               tts:fontWeight="bold" />
9           </styling>
10      </head>
11      <body>
12          <div>
13              <p begin="00:00:03.00" dur="00:00:02.00"
    style="1">Nose</p>
14              <p begin="00:00:05.00" dur="00:00:02.00"
    style="1">Ear</p>
15          </div>
16      </body>
17  </tt>
```

Lines 1 and 2 serve administrative functions. Lines 3 through 10 include a head section that contains styling information. The single style added in lines 5 through 8 is called *1* and uses basic properties similar to cascading style sheet (CSS) properties to center and bold 18-point type.

Lines 11 through 16 specify the captions. Lines 13 and 14 show the beginning time, duration, style, and text of each caption. Finally, lines 15, 16, and 17 balance the opening XML tags for each respective section.

NOTE

This book's companion volume, Learning ActionScript 3.0: A Beginner's Guide, *includes additional information about video captioning, including how to make the background behind the caption text transparent, how to apply more than one style to a single caption, and even how to swap caption files at runtime.*

Streaming Versus Progressive Download

Throughout this chapter, you've worked with videos in their most common configuration—a delivery method called *progressive download*. Without assistance from additional server-side software, Flash videos are downloaded in progressive stages, making the material that has already been delivered available for playback *while the remainder of the file continues to download*. The file is usually available for playback after a sufficient buffer has been filled to cushion against bandwidth interruptions or slowdowns.

Even though progressive delivery reduces playback delays, this is still a file download process. It requires no special software and can occur using a standard remote web server, or even in a local environment on your own computer—just like loading an HTML file from a remote server or your local machine.

There are, however, a few limitations imposed by this technique. For example, the user is limited to interacting with only the portion of the video that has already been downloaded. Additionally, the file typically remains on the user's machine until his or her web cache is cleared, making it difficult to prevent appropriation of content.

A *streaming* file, however, is literally delivered in a continuous stream, rather than in progressive stages, from a special server designed specifically for streaming video. Because the video is delivered in tiny pieces in a continuous stream, the user can jump around anywhere in the video, including later portions that she has not already seen.

Further, the streaming video does not typically remain on the user's machine, and you can add security measures, such as digital rights management, user registration, and more. Additional interactive features may also be available, including interaction among multiple users, and video and audio recording.

Streaming software, like Adobe's Flash Media Server and the open source servers red5 (from Infrared5) and Wowza Media Server (from Wowza Media), are examples of streaming Flash video servers. You can also work with vendor partners such as Influxis (*http://www.influxis.com/*) that can run Flash video streaming servers for you, eliminating or reducing the need to learn how to use the servers yourself.

NOTE

For more information on streaming video, check out Flash Video for Professionals: Expert Techniques for Integrating Video on the Web *by Lisa Larson and Renee Costantini (Sybex).*

The term "streaming" is bandied about inaccurately with surprising frequency when it comes to Flash video. In many cases, when Flash users refer to streaming video they are really talking about progressive download delivery. It will help you plan your projects, create more accurate budgets, and avoid possible communication problems if you, your colleagues, and your clients understand the difference between these two terms.

In this book, and for the portfolio project, you will be using progressive download video, so streaming options are something you can explore at a later time.

Embedding Videos in a SWF

While loading external videos at runtime is the vastly preferred method for adding Flash video to your projects, there is an alternate approach. You can also embed a video directly into a SWF so that even the video asset is internal.

When to Embed

NOTE

Although not commonplace, using small, embedded videos for special effects is analogous to creating a movie clip from an imported image sequence, as discussed in Chapter 3.

Embedding is useful for special cases when you're dealing with very small videos. For example, if the videos are very short (under 10 seconds is generally recommended), they can be used for simple animations and for special effects, such as animating small sources with alpha channels (like fire).

To embed videos in Flash CS4 Professional, the video must already be encoded in FLV format. Once you have an FLV asset, you can import it just like any other asset (File→Import to Stage), starting with a file-browsing process and instructions to Flash regarding how to handle the import (Figure 13-17).

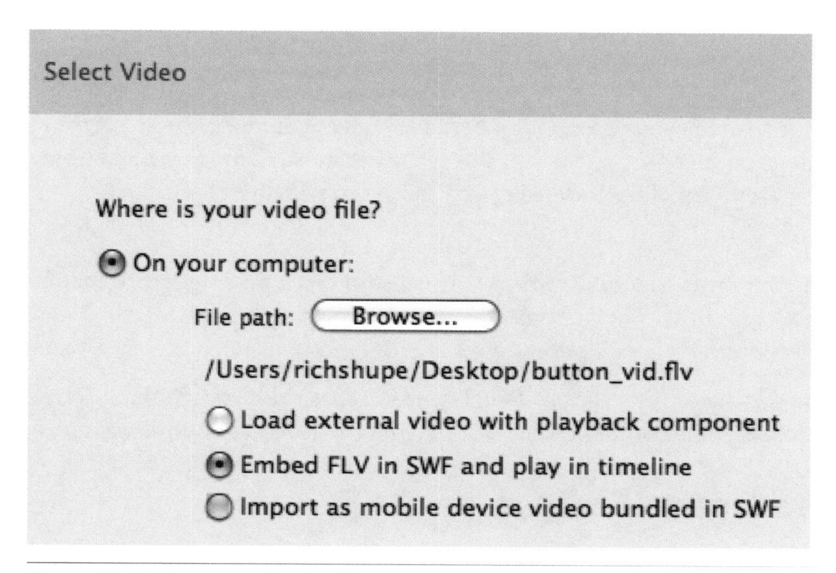

Figure 13-17. Step one of embedding a video: importing an FLV file

Importing to the Stage or Library has the same effect. The import process presents you with options (Figure 13-18), including which type of symbol to create upon import, whether or not to add the video to the Stage, and whether or not to include audio. Using the *Expand timeline if needed* option, you can even automatically extend the timeline to the required number of frames to contain the full length of the video.

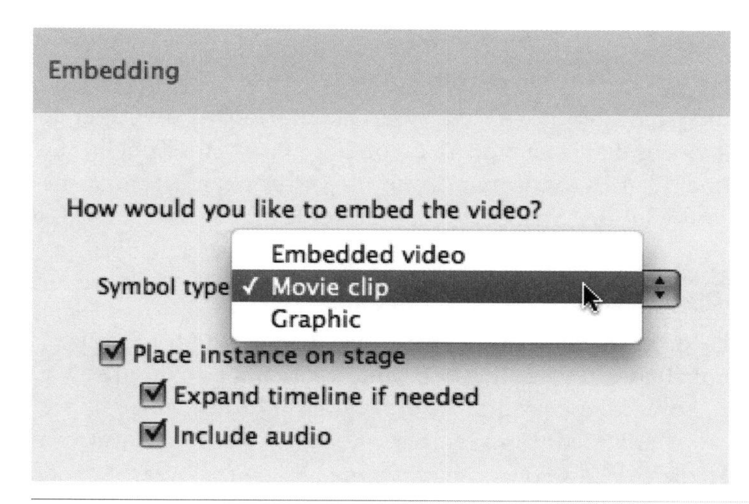

Figure 13-18. Step two of embedding a video: choosing the symbol type and placement of the video

When choosing which symbol type to create, you have three options:

Embedded video

This option places the video into the Timeline that is active at the moment of importing the asset. This is preferred if you must try to scrub or synchronize the video with other assets in the same Timeline.

Movie clip

This is usually your best option because it gives you the greatest degree of ActionScript control. Also, you can place the movie clip in a single frame of a parent Timeline, if desired, and the video will still play.

Graphic

This option is the least attractive because the graphic's parent Timeline must contain as many frames as the graphic to play the whole video. However, it's a viable option if you do not plan to use ActionScript.

After importing to your chosen Timeline, you can scrub through the video as if it were any other Timeline layer, previewing the video in any frame.

When Not to Embed

It's very important to remember that embedding video is not a recommended alternative to loading external FLV or F4V files. Embedding videos longer than 10 seconds is discouraged for a number of reasons. Here are some highlights:

Large videos increase download time and memory requirements

When a video is embedded, the entire SWF must be downloaded prior to playback, so large videos can contribute to excessively long download times. Additionally, embedded videos must fit entirely into memory, which causes problems when large videos are played on computers with limited RAM.

Problems aside, large videos may not even fit into a SWF

Although there have been reported workarounds, Adobe's official statement is that videos that span more than 16,000 frames in the Timeline are not supported. That's a lot of frames, but long videos can eat up frames quickly. If you consider a video frame rate of 30 frames per second, a video longer than approximately nine minutes cannot be embedded.

Embedding videos increases development time

Embedded videos dramatically increase the time required to compile SWFs. Long videos can cause publish times to exceed one hour, so you can imagine how tedious testing video integration can be. More importantly, you can't easily edit and replace embedded videos. Instead, you must discard the embedded video, reimport a replacement, and recompile the SWF. Frequent changes can be torturous.

Significant synchronization issues exist

The video and audio in longer videos can fall out of sync and can end up drifting apart by several seconds by the time the video has completed playing. This discrepancy can begin after seconds, not minutes, of playback and can quickly render the video unusable. Another significant limitation is that when a video is embedded, the frame rates of the SWF and video must match. For example, when using lower-quality video assets that were compressed at 5–12 frames per second, the SWF must also run at that frame rate, limiting the speed of other animations, transitions, and ActionScript performance linked to frame rate.

To emphasize the fact that embedding video is not a preferred practice, the video-importing dialog contains the following message:

> *WARNING: Embedded video is likely to cause audio synchronization issues. This method of importing video is ONLY recommended for short video clips with no audio track.*

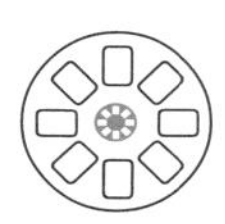

Project Progress

In Chapter 9, you learned how to add functionality to your files with little to no programming using components. In this chapter, you'll finish your Gallery screen, and thus the last piece of content in your portfolio, by adding a video using the *FLVPlayback* video component (Figure 13-19).

Figure 13-19. The project FLVPlayback in use in the Gallery screen

Adding Video

You are strongly encouraged to encode the provided *scaly_vid.mov* source file to learn more about creating Flash-compatible video files. If you have Adobe Media Encoder (the video-encoding application that ships with Flash CS4 Professional) installed, you can follow the steps described previously in the "Encoding Software" section of this chapter. If you prefer not to compress the video yourself, feel free to use the provided source file, *scaly_vid.flv*.

1. Place the *scaly_vid.flv* video from the furnished source files into the *assets* folder of your main portfolio project directory.
2. Open your main portfolio file or pick up from the provided source file, *portfolio_12_final.fla*, and scroll the Timeline panel until you can see the Gallery section.
3. Double-click the Gallery movie clip to edit it, and then double-click the *foreground* movie clip to revisit the components.
4. Scroll to the far right of the movie clip so you can see the frame in which the *FLVPlayback* component will reside. Select the *components* layer and drag the *FLVPlayback* component from the Components panel (under Video this time, not User Interface) to the frame on the Stage.

5. Using the Properties panel, set the size of the component to a width and height of `200` and `140`, respectively, and position the component neatly within the frame, *nearly touching the top edge of the frame*. Playback controls will appear below the video; you must leave room for them.

6. With the component selected, open the Component Inspector (Figure 13-20). Click the *source* value field and type `assets/scaly_vid.flv`. This selects the video asset, but you still need to choose your controls.

7. Click the *skin* value field and, from the skins menu, choose `SkinUnderPlaySeekMute.swf`. This will select one of the skins that sits under the video (instead of atop the video) and includes a play/pause button, seek bar, and mute button. This is a small video, so you don't need (or even have room for) all the features available.

8. To test your movie, go to the Gallery screen, advance to the last component, and make sure your video plays. You should see a skin and a video. If you see a video but no skin, check step 6. If you see a skin that seeks forever or fails to perform in any way, check step 5.

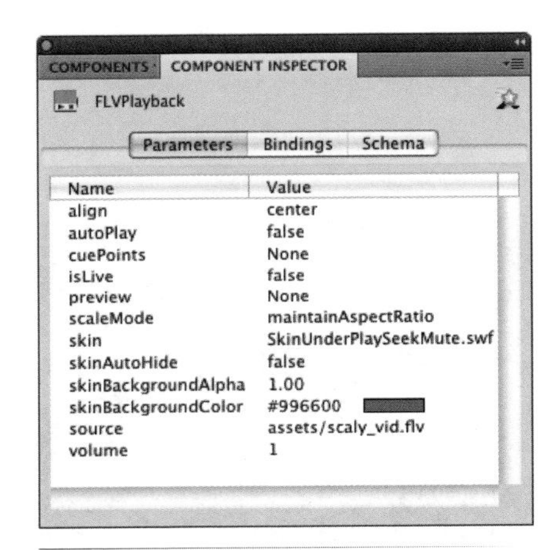

Figure 13-20. Configuring a FLVPlayback component using the Component Inspector panel

9. Theoretically, you are now done, but you want to add one small enhancement. The skin selection process in step 6 automatically copied the skin SWF to the same directory as your main portfolio FLA. Look in that directory and move the skin into the assets folder. Then go back to Flash to change the location of the skin.

10. Click the Component Inspector's *skin* value field (Figure 13-20) and, from the skins menu that appears, choose `Custom Skin URL`. Then type `assets/SkinUnderPlaySeekMute.swf` in the URL field.

11. Follow the instructions in step 7 again, but this time, if there is a missing skin, check step 9 for any errors.

12. Save your file and compare it against the chapter project file, *portfolio_13_final.fla*.

The Project Continues...

Your project is now almost complete! It's time to ready all your hard work for deployment. In the next chapter, you'll create a preloader to keep the initial interest of viewers with slower Internet connections. You'll then wrap everything up for distribution via the Web. Finally, you'll package your portfolio for desktop distribution using the Adobe Integrated Runtime (AIR) engine.

CHAPTER 14

PUBLISHING AND DEPLOYING

IN THIS CHAPTER

Introduction

Welcome! It's the final chapter of the book, and you're almost home. You worked hard, you know your way around Flash, and you created a portfolio that ties together everything you've learned. All that remains now is getting it out into the world.

In this chapter, you'll publish a SWF and a host HTML file that will run in your friendly neighborhood web browser. As an added bonus, you'll also create an AIR (Adobe Integrated Runtime) desktop application that will wrap the portfolio into one neat little package. You can download the AIR application and run it locally or even leave a copy behind at job interviews.

First, however, you'll create a preloader to keep your viewers aware of any loading delays caused by slow Internet connections. When the preloader is finished and you've deployed to both HTML and AIR, your mission will be complete. In this chapter, you'll work on the project in three separate segments so you can focus on each topic individually.

Using a Preloader

When a viewer to your site is struggling with a slow Internet connection, waiting a long time for assets to load can be disconcerting. At best, the experience is frustrating. At worst, your visitor may think there is a problem with your site and move on.

In time, you will be able to optimize your online applications to keep SWF sizes small, load more assets from external sources, and reduce embedded assets by accomplishing more with code. Until you're comfortable and experienced enough to do these things, you'll have to work around large assets.

The portfolio project that wound its way through this book was designed to show you how. The user interface's large viewing wheel contributes significantly to the size of your SWF. By displaying a progress bar during the initial load time, you can indicate to visitors that your content is coming and also indicate how long they will have to wait.

To do this, you will create a preloader that sits as the only asset in the first frame of any FLA. When the file begins to load, the preloader will stop in frame 1 and begin monitoring the percentage of your SWF that has loaded. When that value reaches 100%, the SWF will be fully loaded and the code will continue playing the file.

The Assets

The only asset essential to this task is a movie clip that will act as a progress bar. To provide the viewer with some context as to how much of the loading process remains, however, you need to create the progress bar in a specific manner:

1. Create a new file (without using the book template) and give it a black Stage color.
2. Select the Rectangle tool in the Tools panel.
3. Make sure Object Drawing mode is off.
4. Using the Tool panel's color chips, select a gray stroke color (#999999) and white fill (#FFFFFF).
5. Draw a rectangle anywhere on the Stage.
6. Switch to the Selection tool and double-click the fill of the rectangle, or click outside the rectangle and drag over the entire rectangle with the Selection tool to select both the fill and stroke.
7. Using the Properties panel, give the rectangle a width of `150` pixels and height of `10` pixels. To control these values independently, click the adjacent link icon to unlock the aspect-ratio lock.
8. Convert the shape to a movie clip, name it `preloader`, and give it an upper-left registration point. Click OK to create the symbol.
9. This movie clip is a symbol you can drop into any project to use as a preloader. You will create your progress bar and add your code inside this movie clip.
10. Double-click the movie clip to edit its contents. In the Timeline panel, name the existing layer `progBar`.
11. Using the Selection tool, carefully double-click the stroke of the shape you created. Be sure the fill remains unselected. Zoom in if selecting only the stroke is difficult.
12. Cut the stroke (Edit→Cut), create a new layer in the Timeline panel called `outline`, and paste the stroke in place (Edit→Paste in Place) in that layer. Be sure the stroke ends up at the same x and y position so it appears to

surround the fill. Lock this layer. By separating the stroke and fill, you will create not only a progress bar, but also a rectangular frame that the progress bar will fill up, providing context to the viewer.

13. Select the fill and convert it to a movie clip, name it `bar`, and choose a *center-left registration point*. This is important because you will incrementally scale this bar to the right with ActionScript to show loading progress. If the registration point is not on the left side, it may not appear to grow correctly. For example, if you use a center registration point, it will expand in both the left and right directions. Click OK to create the symbol.

14. Select the instance of the `bar` movie clip on the Stage and give it an instance name of `progBar`. Double-click anywhere on the black background of the Stage to return to the main timeline. You're now done with the assets and are ready to move on to coding. Compare your file to *preloader_01.fla* of the companion source files.

The ActionScript

When writing ActionScript within your FLA, you'll usually write code in the main timeline. This consolidates code and reduces the need to hunt for scripts elsewhere. In this case, however, you're designing a generic preloader that you can reuse often simply by dropping it into the first frame of each file. For this reason, you're going to place the ActionScript inside the preloader, making it self-contained like a component.

When writing code inside a movie clip that refers to the main timeline, you need to adjust your point of reference. The main timeline is the *parent* of a movie clip inside it. If you use the identifier **`this`**, you will be referring to the movie clip in which you wrote the script. If, instead, you use **`this.parent`**, you can refer to the main timeline.

1. Double-click the *preloader* movie clip you created to edit its contents.

2. At the top of the Timeline panel, create a new layer called `actions`, type the following script into this layer, and save your work:

```
1   var main:MovieClip = MovieClip(this.parent);
2   var mainInfo:LoaderInfo = main.loaderInfo;
3   var loadPercent:Number = 0;
4
5   main.stop();
6   progBar.scaleX = 0;
7
8   this.addEventListener(Event.ENTER_FRAME, onCheckLoaded);
9   function onCheckLoaded(evt:Event):void {
10      loadPercent = mainInfo.bytesLoaded / mainInfo.bytesTotal;
11      progBar.scaleX = loadPercent;
12      if (progBar.scaleX == 1) {
13          removeEventListener(Event.ENTER_FRAME, onCheckLoaded);
14          main.play();
15      }
16  }
```

So, what does this new ActionScript do? Line 1 stores a reference to the main timeline—the parent of the preloader movie clip you are creating. To prevent errors from the compiler, this line casts the main timeline as a **`MovieClip`** data type. This lets the compiler know that the main timeline is, in fact, a movie clip, so using movie clip properties and methods is legal.

Line 2 stores a reference to the main timeline's **`loaderInfo`** property. As its name implies, this property contains information pertaining to the loading of a display object. In most cases, you will query this property when checking on an instance of the **`Loader`** class that you create to load an external asset. However, this also works when checking on the loading progress of the main timeline.

The **`loadPercent`** variable in line 3 is initialized to a value of 0 and will contain the percentage of the main timeline that has loaded throughout the preloading process.

Line 5 stops the main timeline from playing so the preloader can do its work. The **`progBar`** instance name in line 6 refers to the progress bar movie clip that you created earlier. This line initializes the horizontal scale of the progress bar to 0 before the loading begins.

Lines 8 and 9 create an event listener that listens for the enter frame event, thereby updating the progress bar many times per second. Line 10 divides the current number of bytes loaded by the total available bytes to calculate the percentage loaded each time the listener function is called. Line 11 sets the horizontal scale of the progress bar to this percentage. Thus, when 50% of the SWF is loaded, the progress bar will be scaled to 50% of its total width.

Finally, the conditional statement in lines 12 through 15 checks to see if the amount loaded has reached 100%. If so, the listener is removed and the main timeline is played. This makes the preloader disappear and the portfolio's user interface begins to animate in.

Testing Your Preloader

With the assets and code in place, the preloader is complete and you can now test it using Flash's download simulator. The *preloader_02.fla* companion source file has been provided for this purpose. It contains a second frame with a large asset in it. You can copy and paste the preloader from your file into frame 1 of *preloader_02.fla* for testing, or simply use *preloader_03.fla*, which has a finished preloader included.

1. Test your movie. The movie will load instantly because everything is local.

2. When you test a movie in Flash, you run a copy of Flash Player launched from within the Flash application interface. The application menus have changed so that you can use player features designed to simulate online

connections. Access the View→Download Settings menu command to see a variety of connection speeds and select 56k (modem speed).

3. Test the movie again, while still in player mode, by using the View→Simulate Download menu command (do not close the SWF and return to Flash to retest). This will test the SWF again, but this time, throttle the speed to simulate a 56k Internet connection. You should see the progress bar increase in horizontal scale until it reaches full width and then advance to the next frame to show the content.

Project Progress Preloader

Because you added the necessary ActionScript to the preloader movie clip itself, all you need to do for the prelaoder to work is add it to frame 1 of your main portfolio FLA:

1. Open your main portfolio FLA and add a layer to the timeline beneath the labels layer. The placement of the layer isn't critical; it just helps keep things tidy by keeping frame labels and ActionScript at the top of your timeline. Name it `preloader`.
2. Create a keyframe (F6) in frame 2, isolating frame 1 from the rest of the layer's frames.
3. If your preloader FLA is still open, make sure you're in the main timeline. Otherwise, open your preloader FLA, or *preloader_03.fla* if you prefer to use the furnished source file.
4. Copy the *preloader* movie clip from your preloader FLA and paste it into the first frame of the preloader layer in your portfolio FLA.
5. Save your work and test the preloader using the same process outlined previously in this chapter, in the "Testing Your Preloader" section.

Your project is now complete! Now it's time to deploy it to the world.

Distributing External Assets

When a SWF is compiled and readied for deployment, your distribution doesn't have to include any of the assets that were imported into the FLA. For example, you don't need to include your portfolio's interface PSD when you upload your files to a server. However, you must include all external assets that will be loaded at runtime.

Whenever any of the Project Progress exercises throughout the book have needed to load external assets, you've consolidated the external files in the *assets* folder in your main project directory. When uploading your HTML deployment package, and when preparing your AIR application, you will need to include this directory and the assets within.

Because the preparation for deployment via web browsers and AIR is different, the external assets will be mentioned in both exercises.

Deploying for Web Browsers

Most of the time, Flash designers and developers create Flash projects that are uploaded to a server and accessible via the World Wide Web. This typically requires not only the creation of the SWF that is compiled every time you *test* your movie (Control→Test Movie), but also a host HTML file. For your convenience, both of these files are created when you *publish* your FLA (File→Publish). To configure properties of both files, use the Publish Settings dialog (File→Publish Settings).

Publish Settings: Format

The Publish Settings dialog is a file-specific set of properties divided into segments based on which assets you want to publish. By default, the Flash and HTML options are enabled (Figure 14-1), and as you select additional publish options, additional sections become active.

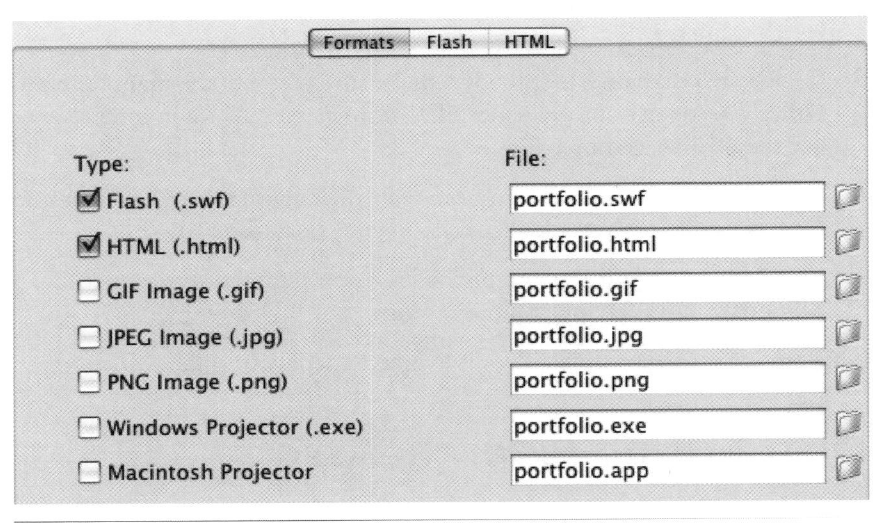

Figure 14-1. Determining which file types to publish in the Formats section of Publish Settings

NOTE

Although you can publish both Windows and Macintosh executables on either platform, the resulting application will only run on the targeted operating system.

The GIF, JPG, and PNG options publish the first frame of the movie in their respective formats, with GIF supporting the added option of publishing animations as animated GIFs. The Windows and Macintosh Projector options publish an operating system–specific executable, which adds the player code to your SWF so the application is self-sufficient.

To deliver your project across the Web, however, you typically need to enable the Flash and HTML options. Exceptions to this rule include situations where only the SWF file is required, such as when integrating your SWF into an existing HTML page or uploading to an online content management system (CMS) that is already responsible for generating the HTML host files. The next two sections cover both SWF and HTML generation, and you can apply what is needed to your work on a project-by-project basis.

Regardless of which formats you choose to publish, you can use default names when creating the files (which use the name of the FLA as base file names) or create custom names for any format. In both cases, you can also specify a path for the asset by clicking the folder icon to the right of the name field.

Publish Settings: Flash

The Flash section of the Publish Settings dialog (Figure 14-2) provides instructions to the Flash compiler that are used when creating SWFs. It's in this section that you target the minimum version of Flash Player required to run your SWF, set file-wide image and sound compression values, manipulate security settings, and more.

In general, you can use the default settings for SWF export. When changes are needed, the most common areas of interest include image and sound compression and local playback security. These and other options are explained in the following list.

General

Player

The *Player* setting specifies the minimum player version required to display your content. If no CS4-specific features are used, for example, you may be able to target Flash Player 9. If ActionScript 3.0 is not required, you may be able to target Flash Player 8, and so on. This is also where you target AIR as a platform to create AIR files.

Script

The *Script* setting specifies the version of ActionScript used in your FLA. Flash Player versions prior to 6 can parse ActionScript 1.0 only. Player versions 6 and later can parse ActionScript 2.0, and versions 9 and later can process ActionScript 3.0. AIR can use only ActionScript 3.0.

The Settings button for this option lets you configure ActionScript settings, which can include how strictly the compiler validates your code, whether or not to show warnings in addition to errors when testing, whether or not instance names added in the Properties inspector are automatically declared in your scripts, and more.

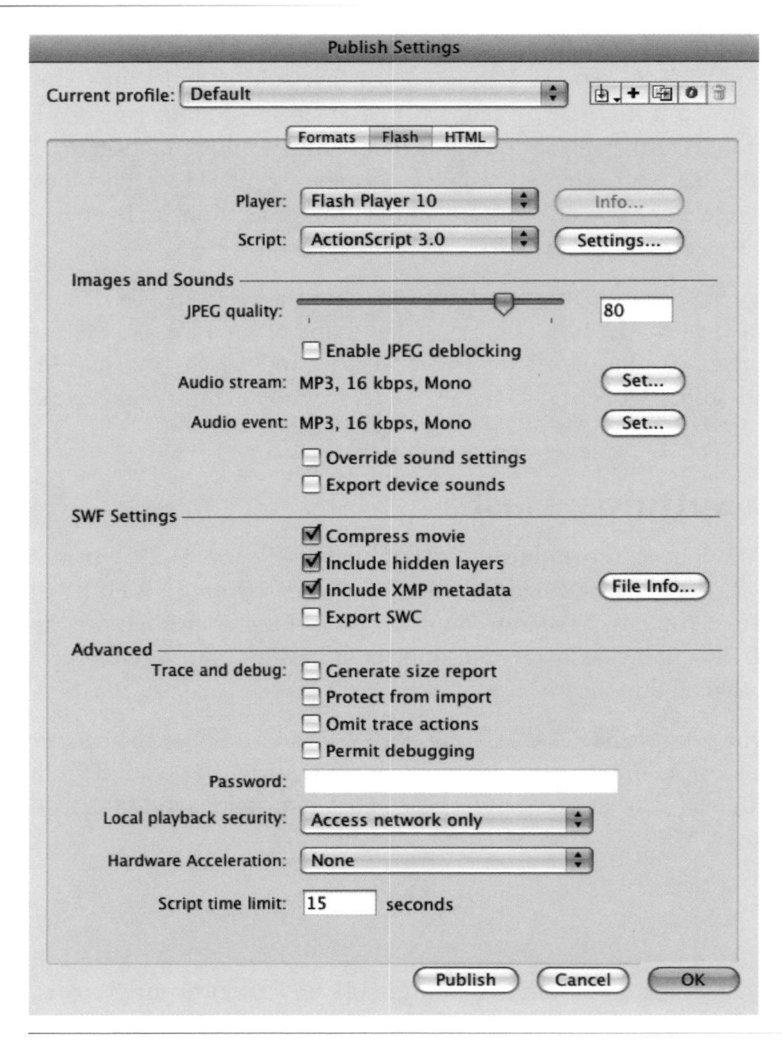

Figure 14-2. Flash publish settings

Images and Sounds

JPEG quality

The *JPEG quality* setting applies a global compression setting applicable to all bitmaps that do not have a custom compression defined on a per-bitmap basis in the Library Bitmap Properties options. See Chapter 4 for more information.

Audio stream/event

These settings allow you to dictate the compression settings for *stream* and *event* sounds. If you use the *Override sound settings* option, you can override the per-sound compression values applied in the Library's sound properties options. See Chapter 11 for more information.

SWF Settings

Compress movie

The *Compress movie* option applies additional compression to your SWF file during compiling and should be used for all player versions 6 and later. It has the greatest impact on text- and script-heavy files.

Include hidden layers

The *Include hidden layers* feature, by default, includes layers hidden in the Timeline panel when compiling a SWF. However, with this feature disabled, you can selectively, even temporarily, prevent layers from being included in the SWF. For example, you could temporarily prevent sound layers from being compiled, saving time during development.

Include XMP metadata

The *Include XMP metadata* option can include an extensive amount of metadata in the SWF, readable by Adobe Bridge and other XMP-aware applications. The File Info button lets you add descriptive information about the file as a whole (title, author, rating, and so on), video and sound information, mobile playback data, and lots, lots more.

Export SWC

The *Export SWC* option lets you compile a protected file that is typically used for distributing components or script libraries.

Advanced

Generate size report

The *Generate size report* option exports a comprehensive text file detailing the size of the data contained in your SWF. It is broken down by frame number, scene, symbol, asset type, external file, data type, and more. Reading this text file can help you identify unnecessarily large assets ripe for optimization.

Protect from import

Enabling the *Protect from import* option prevents unauthorized users from importing graphic assets from your SWF. Use this feature carefully because it will also prevent *you* from importing your SWF. See "Password," later in this list.

Omit trace actions

A helpful ActionScript debugging technique is to trace text into the Output panel during authoring. Common uses include tracing the values of variables and references to objects so you can determine if your scripts are functioning properly. As a last step before distribution, you can use the *Omit trace actions* option to prevent others from discovering your traces in the wild.

Permit debugging

Enabling the *Permit debugging* feature allows the Flash Debugger to debug SWF files from a remote location like a server. This is an intermediate to advanced skill that is reserved for ActionScript debugging and requires use of the debug version of the Flash Player. See the next entry, "Password," for more information.

Password

To prevent unauthorized use of the debugger or unauthorized SWF import, you can add a *password* to your file. In this case, a valid password must be entered when attempting to debug or import a protected SWF; otherwise the process will fail.

Local playback security

As a security safeguard, your Flash file cannot access both local files from your hard drive and files or locations on the Internet. The *Local playback security* option lets you pick which of these realms will contain your file. In general, *Access network only* is the option of choice when uploading your files to a server.

Hardware acceleration

For processor-intensive projects you can take advantage of *Hardware acceleration*. Two options are available. *Direct* lets Flash Player draw directly to the screen instead of letting the browser handle the display. *GPU* uses the graphics card to handle video playback and compositing.

Script time limit

As a preventive measure, Flash Player will allow viewers of your SWF to abort scripts that take too long to run. This prevents crashes due to programming problems like circular logic and endless loops.

Publish Settings: HTML

The HTML section of the Publish Settings dialog (Figure 14-3) creates an HTML host file that will contain your SWF and display it when users visit your site with a web browser. This is where you add Flash Player version detection, set the size of the SWF display, and more.

In general, you can use the default settings for HTML export. When changes are needed, the most common adjustments are to scale your SWF when resizing the browser and enable full-screen mode. These and other options are explained in the following list.

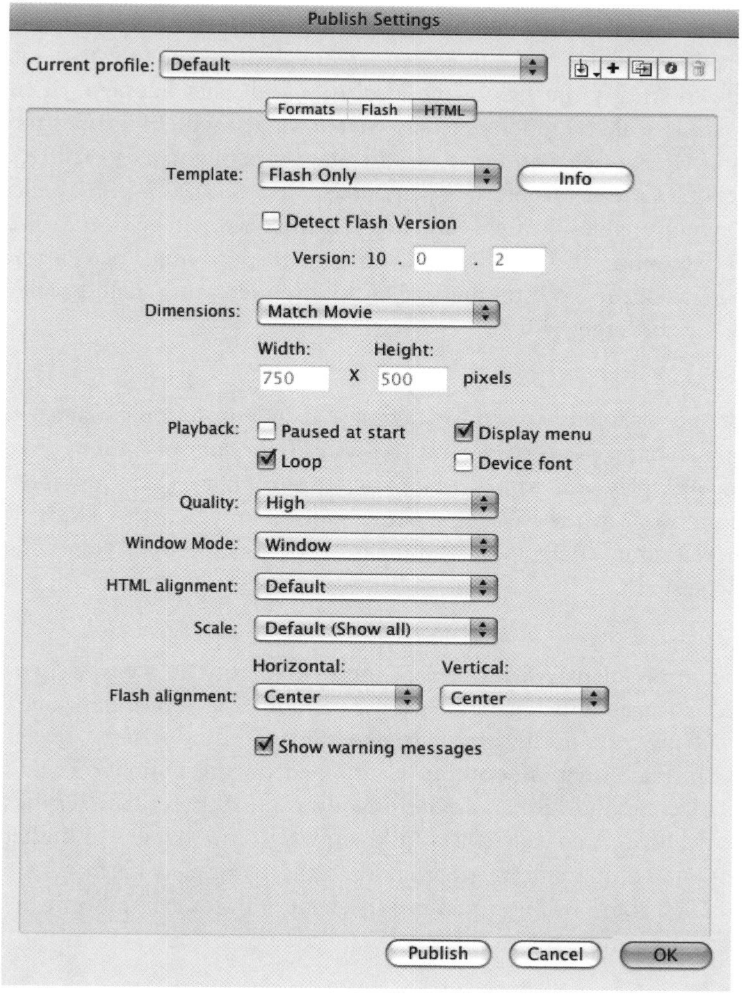

Figure 14-3. HTML publish settings

Template

Template lets you choose from precreated HTML templates that enable or support Flash Player options. Some templates include JavaScript support features like player version detection and communication between the SWF and learning management systems (LMSs), among other things. By and large the Flash Only option will serve you well. Another notable option is Flash Only – Allow Full Screen, which enables full-screen mode for SWF and video content.

When player version detection is needed, the *major version*, or the leading digit (such as Flash Player 10, or Flash Player 9) is dictated by the version you specified in the Flash segment of the dialog. The *minor version*, or the dot release that follows the main player version (such as Flash Player 10.0.2), can then be typed into the field.

Dimensions

The *Dimensions* menu and the accompanying *Width* and *Height* fields allow you to specify the size of the Flash file and what happens when the browser is scaled. Options include *Match Movie* (which disables the Width and Height fields and matches the Stage size of your SWF), *Pixels* (which enables the fields and lets you enter a desired size in pixels) and *Percent* (which enables the fields and lets you enter a size in percent form). The first two options will not scale the image, but specifying a percentage bases the size of the SWF on the size of the browser window and scales your SWF accordingly.

Playback

Playback options include pausing the movie at startup, looping playback, displaying a context-sensitive menu of Flash Player control options (like zoom, rewind, play, and so on) when you Control+click (Mac) or right-click (Windows) on the SWF, and the Windows-only feature of substituting device fonts (default serif and sans-serif fonts installed on your operating system).

Quality

Quality controls display features that improve quality at a cost to performance (such as antialiasing). Settings include *Low* (no antialiasing), *Medium* (some antialiasing, but no bitmap smoothing), *High* (always antialiased, but bitmap smoothing is dropped during animation), and *Best* (antialiasing and bitmap smoothing always on). Two other settings are also included. *Auto Low* starts in low quality, but changes to higher quality if the computer can accommodate the associated performance hit. *Auto High* starts in high quality, but changes to lower quality if the computer can't handle the features.

WARNING

Window Mode has long been one of the most error- or conflict-prone settings in Flash because it must coexist with a flurry of browser versions and features and the rendering quirks that come with them. If you are experiencing display problems within a browser, first set the Window Mode to Window and see if that improves or solves your problem.

Window Mode

Window Mode controls how the Flash file can visually interact with the surrounding HTML. The default value, *Window*, renders an opaque background in the SWF and sets the HTML background color to that of the Stage. HTML content can't flow over or beneath the Flash content. *Opaque Windowless* sets the background of the SWF to opaque, but lets HTML content stack on top of or be eclipsed by the SWF. *Transparent Windowless* renders the SWF background as transparent, letting HTML appear in front of and behind the SWF.

HTML alignment

HTML alignment controls how the SWF display window is positioned within the HTML page, relative to other HTML elements on the page, such as text, images, and so on. The values include *Default*, *Left*, *Right*, *Top*, or *Bottom*. Default centers the content in the browser window and crops all four sides of content if the browser window is smaller than the SWF dimensions. The remaining options align the SWF along the specified edge of the browser window and crop the other three sides if the browser window becomes smaller than the SWF dimensions. See "Flash alignment," later in this list.

Scale

Scale controls how the SWF is scaled, if percentage is specified in the Dimensions setting. When the user changes the browser window size, the following settings apply. *Default (Show All)* shows the entire stage while maintaining aspect ratio. Borders may appear above and below, or at left and right, of your stage boundaries if the dimensions of the browser window do not match the aspect ratio of your stage. *No Border* also scales your SWF while maintaining aspect ratio, but doesn't allow borders to appear. As a result, the SWF display area in the HTML page may crop the SWF. *Exact Fit* matches the exact size of the SWF display area without preserving aspect ratio. Distortion will result if the browser window size doesn't match the aspect ratio of the SWF stage. Finally, *No Scale* prevents the SWF from scaling.

Flash alignment

In contrast to HTML alignment, which aligns the SWF display area within the HTML page, *Flash alignment* aligns the content within the SWF display area, cropping as needed. Horizontal options include *Left*, *Center*, and *Right*, and vertical options include *Top*, *Center*, and *Bottom*.

Show warning messages

This setting, which should remain enabled whenever possible, will turn on a message system that will warn you if there is a conflict in any settings you choose. For example, if you specify an HTML template that displays an alternate image upon failure to detect Flash, but you don't specify the creation of that file in the Formats section of the settings dialog, a warning will be displayed.

NOTE

The HTML alignment setting does not change the position of a SWF if it is the sole element on the HTML page. Like the `align` *attribute of the HTML* `img` *tag, for example, it controls the relative positioning of the SWF in conjunction with other HTML assets. Consequently, this setting will not appear to have any effect until additional content is added to the HTML page or the generated SWF tags are integrated into another HTML document.*

Deployment

Once you configure your Flash and HTML settings, you can publish your file using the button in the Publish Settings dialog or the File→Publish menu option. Flash CS4 Professional will compile your SWF and create a corresponding host HTML page. These two files, along with any local external assets designed to load at runtime, must be collected for upload. After uploading the HTML, SWF, and external files to a server, you can point your browser to the address of the HTML file and view your finished work.

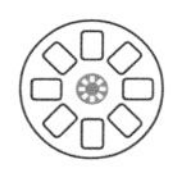

Project Progress HTML

To prepare your portfolio project for deployment, you must configure the Flash and HTML publish settings and collect the *assets* folder containing your external files for runtime loading:

1. Check to make sure your main portfolio FLA and the *assets* folder are in your main project directory.
2. Open your main portfolio FLA.
3. In the Flash publish settings, set the following options (omitted options are inconsequential):
    a. Player: `Flash Player 10`
    b. Script: `ActionScript 3.0`
    c. JPEG quality: `80`
    d. Compress movie: `on`
    e. Include hidden layers: `on` (unless you specifically used this feature to your advantage to enable/disable features during testing—this was not a planned part of the project progress, so if you are unsure of your actions, enable this feature).
    f. Protect from import: `on`
    g. Omit trace actions: `on`
    h. Local playback security: `Access network only`
    i. Hardware Acceleration: `None` (feel free to experiment with this feature, but a setting of None is most compatible with all possible systems that may view your portfolio).
4. In the HTML publish settings, set the following options:
    a. Template: `Flash Only`
    b. Detect Flash Version: `on, testing for version 10.0.2`
    c. Dimensions: `Match Movie`
    d. Playback: `Loop and Display menu`
    e. Quality: `High`
    f. Window Mode: `Window`
    g. HTML alignment: `Default`
    h. Scale: `Default (Show All)`
    i. Flash alignment: `Horizontal Center` and `Vertical Center`
5. Click Publish to create the SWF and HTML files.
6. Inside your main project directory, you should now find the following three items: the HTML and SWF files created in the publish process (most likely titled *portfolio.html* and *portfolio.swf*), and the *assets* directory, which contains the external files you plan to load at runtime.

7. Upload these three items to your server. Do not upload the FLA file or any external files that were used during authoring (such as files that were imported into the FLA or compiled into the SWF).
8. Compare your site with the online version found at the companion website (Figure 14-4).

Figure 14-4. The finished Portfolio project viewed in a browser

Deploying for AIR

AIR is an application designed to expand the realm of rich Internet applications to include the desktop. For example, AIR is capable of delivering Flash SWFs, HTML, and JavaScript—technologies typically used for web development—in a desktop environment.

AIR applications consist primarily of two separate parts: a standalone player and a data file. AIR, itself, is a player that is installed on a user's computer just like any other program. It contains all the runtime code needed to play your project, but includes no file-specific data. Your project file is the opposite side of the equation. It contains all the file-specific information, but no runtime code. Your AIR application, as it is most often called, is essentially a mini-installer that installs the data portion of your project onto a user's hard drive. The installation process makes it appear, for simplicity and ease of use, as though your project is a standalone application. In reality, when

NOTE

Users who don't have AIR installed, or who want the latest and greatest version, can find it at http://get.adobe.com/air/.

NOTE

Another advantage of using AIR to deliver your project is that it is not subject to as many security restrictions as a SWF running in a browser. For example, AIR grants limited access to the local filesystem that is forbidden to browser-bound SWFs.

NOTE

Flash CS4 Professional ships with a publishing profile for Adobe AIR 1.1. However, at the time of this writing, an update that supports publishing to Adobe AIR 1.5 was available. To update your AIR publishing capabilities, visit http://www.adobe.com/support/flash/downloads.html and look for "Adobe AIR 1.5 Update for Flash CS4 Professional." Alternatively, consult the update options in Flash's Help menu or use Adobe Updater to check for any available updates.

you double-click your project file, the AIR player launches automatically and loads the data file.

It's important to note that Adobe did not conceive AIR as a replacement for Flash projectors or as competition for third-party projector enhancers. As described, AIR doesn't create self-contained executable files complete with runtime code. Therefore, it can't practically serve as the primary executable on a disc-based project (CD-ROM or DVD-ROM) because users must have AIR installed on their computers for your application to operate.

AIR doesn't have as broad a feature set as some projector enhancers, such as Screentime's cross-platform mProjector. However, AIR does make it possible to deliver your Flash projects outside the confines of a web browser and with a level of professionalism previously unavailable directly from Flash. Best of all, it's free and integrated right into Flash.

Publish Settings

The first step in preparing a file for desktop delivery is to set the file's targeted player to AIR rather than Flash Player. As discussed previously in the "Deploying for Web Browsers" section, the player version is set in the Flash section of the file's Publish Settings dialog (File→Publish Settings), shown in Figure 14-5. In this case, however, instead of choosing a version of Flash Player (such as Flash Player 10, the default for a new FLA created in Flash CS4 Professional), choose Adobe AIR as your target player.

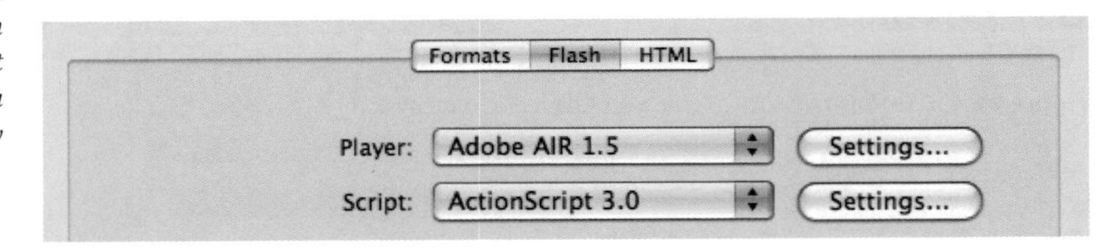

Figure 14-5. The FLA's publish settings, setting Adobe AIR 1.5 as the target player

When you set the player to Adobe AIR, a Settings button will become visible adjacent to the Player drop-down list. Click this button to compile your SWF and open the AIR Application and Installer Settings dialog shown in Figure 14-6. As its name implies, the dialog is divided into two main sections featuring settings pertinent to your application and to the AIR installation process.

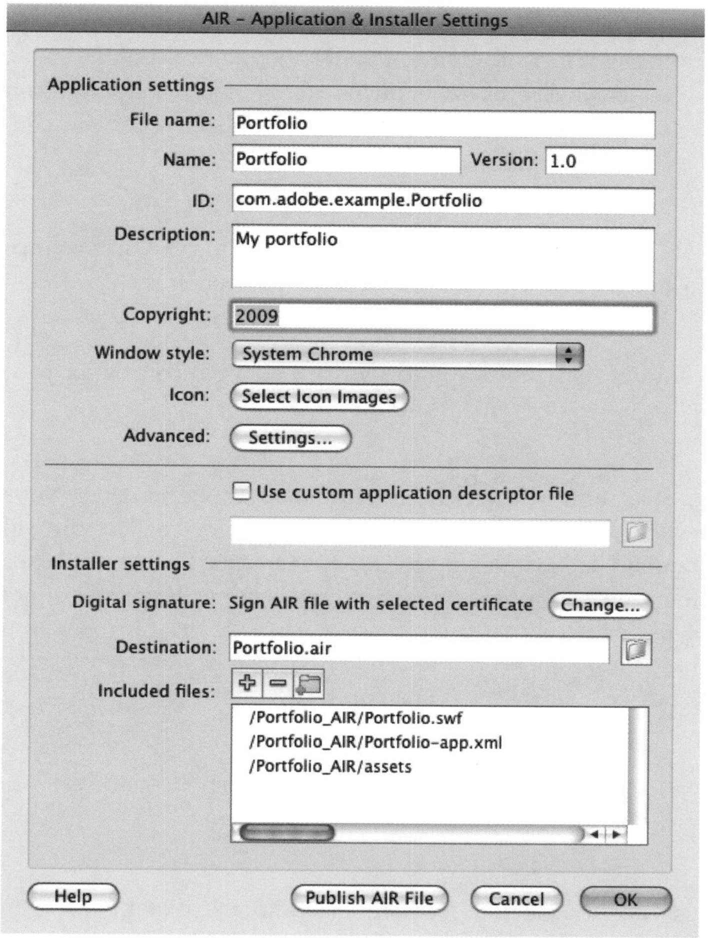

Figure 14-6. AIR application and installer settings

Application settings

The application settings contain information about your project, as well as about how your file will be displayed.

File name

The *File name* is, literally, the name of the AIR file that users will see.

Name

The *Name* of your project will be displayed in the application menu, system Dock (Mac) or Start menu (Windows), and window title bar.

Version

You can optionally assign a *Version* number to an application to keep track of updates.

ID

The *ID* identifies your application to the AIR engine by a unique value. The default is `com.adobe.example.<application name>`, but you can change it if desired. The value must be 212 characters or fewer and contain only a–z, A–Z, 0–9, dot (.), and dash (-).

Description

The optional *Description* is a string describing your project. Users will see this in an installer window during the installation process.

Copyright

The optional *Copyright* allows you to specify a copyright string for your project.

Window style

The *Window style* setting dictates how the window will be rendered. None renders the window with no interface elements at all. System Chrome will render a standard rectangular operating system window with controls such as name bar, close, minimize, and so on. Custom Chrome (opaque) allows you to create your own window interface in the FLA. Custom Chrome (transparent) allows you to create your own window interface in the FLA, but makes the Stage transparent.

NOTE

If you choose not to use the System Chrome setting for the AIR window style, you will need to create your own custom controls for features such as dragging, minimizing, and similar functionality provided by the operating system.

Icon

The optional *Icon* feature lets you specify custom application icons as external files in four standard sizes (Figure 14-7).

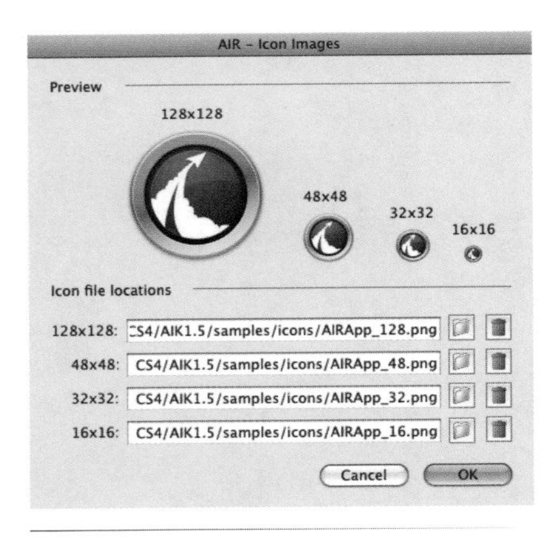

Figure 14-7. AIR application icon images

Advanced

When you click the *Advanced* button, an additional dialog will open (Figure 14-8). This dialog lets you specify advanced features, such as the initial size and position of the application window and where the application is installed.

Associated file types

The *Associated file types* feature allows you to specify which file types your AIR application handles. For example, FLA files are associated with the Flash application. You must specify the name and file extension of the file type, but you can also add a MIME type, description, and even file icons.

Initial window settings

You can use *Initial window settings* to specify the *width*, *height*, *x* location and *y* location of the window. You can also specify whether the window is initially *visible*, and if it's *minimizable*, *maximizable*, or *resizeable* along with its *minimum* and *maximum width* and *height* values.

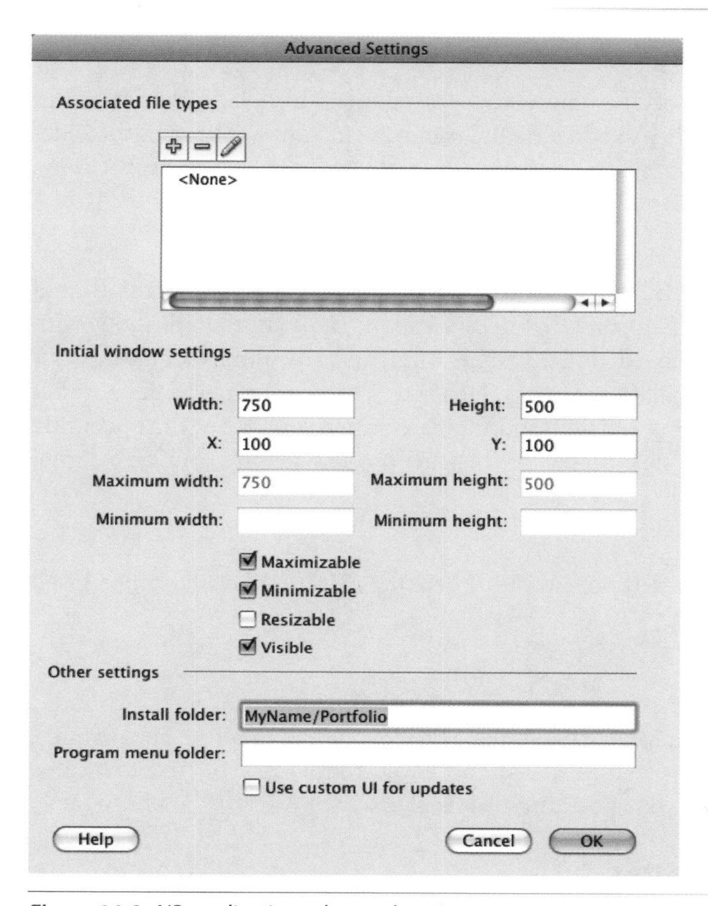

Figure 14-8. AIR application advanced settings

Other settings

Other settings include the folder or directory into which your application should be installed (the folder will be created if it doesn't already exist), and into which Program menu (Windows only) your file should be added.

You can also dictate the use of a custom interface for handling updates. By default, AIR will display a standardized dialog that asks the user what to do when a new version of the application is launched. You can prevent this from happening and show your own custom interface to retain control over how an application handles updates.

NOTE

Creating custom update user interfaces is an intermediate to advanced skill. The companion website can point you to additional Help resources if this becomes a point of interest.

Use custom application descriptor file

In the middle of the Application and Installer Settings dialog is the *Use custom application descriptor file* option. This allows you to replace the manual completion of user interface elements in the AIR publish settings process with a preconfigured XML file. For more information about the structure of this file, see this book's companion website.

Installer settings

Installer settings specify how your application is bundled. These settings include the capability to add a digital signature to your application that, ideally, instills confidence in users during the installation process, and specifying which external assets are bundled with your application.

Digital signature

A *Digital signature* is a security certificate that you can purchase from a digital signing company that allows you to "sign" your applications with your identity. This identifies you or your firm as the publisher of the work during the installation process. This may give your users confidence that they are installing an application from a trusted source. See the upcoming "Digital certificates" section for instructions for creating your own digital signature.

NOTE

The companion website has more information about digital signing options, including links to VeriSign, Thawte, GlobalSign, and ChosenSecurity, all of which sell digital signature certificates for Adobe AIR.

Destination

The *Destination* setting specifies where the AIR installer file will be saved when it is built.

Included files

The *Included files* feature lets you add external assets to an AIR bundle. This allows you to distribute one AIR file to your audience. For example, you will use this feature later in the "Project Progress AIR" section of this chapter to add your portfolio's external assets to your AIR application.

Digital certificates

To sign your AIR application, you need to use or create a digital certificate. You must specify a digital certificate to build an AIR application.

In the *Installer settings* section of the Application and Installer settings dialog, click the Set button (which is named Change if the certificate has already been chosen) to the right of *Digital signature*. Doing so will open the dialog shown in Figure 14-9. You have the option of either using or creating a certificate, or creating an interim file to which you will later apply a certificate.

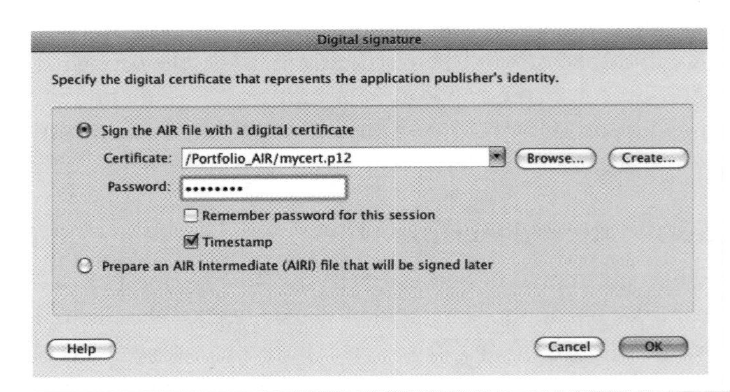

Figure 14-9. AIR digital signature settings

Sign the AIR file with a digital certificate

If you've already acquired a digital signature or want to create your own, you can choose to *Sign the AIR file with a digital certificate*. Click Browse to locate an existing certificate or click Create to use your own digital signature. If you want to do the latter, see the next section, "Creating a Self-Signed Digital Certificate."

Either way, you must enter the password you used when acquiring or creating the signing certificate to authenticate its use. You can optionally remember the password for the current authoring session to prevent the need to enter the password each time you make a change.

The *Timestamp* option is also very important, and you should enable it whenever you have Internet access available. When you create an AIR application, the packaging tool checks to see if the signing certificate is valid when the installer is built. That timestamp is then embedded into the installer.

When a user attempts to install the application, the installer looks for the timestamp. If found, as long as the certificate was valid at that time—even if the certificate has since expired—the installation can continue. On the other hand, if no timestamp is found, the installer will only work as long as the certificate is valid.

While it sounds like you would never proceed without a timestamp, the feature was made accessible to developers because it relies on an authentication server. Because you can disable the feature, you can still create an installer if the server is inaccessible.

NOTE

While creating your own certificate is quick, free, and easy, be aware that the publisher of the work will be identified as unknown during the installation process. Though this is more common than you think, it still may give your users pause when installing your application.

Prepare an AIR Intermediate (AIRI) file that will be signed later

If you haven't yet acquired or don't wish to create a signing certificate during the development process, you can set up your publish settings to create an interim file. This will allow you to test your application in the authortime AIR preview application (called *adl*, for AIR Debug Launcher), but you will not be able to build an installer that users can run to install your application.

At any point, you can return to this setting and either load a signing certificate or create your own, after which you can build an installer.

Creating a self-signed digital certificate

Creating your own digital certificate is as easy as filling out a few simple form elements, shown in Figure 14-10. Fill out the *Publisher name*, *Organization unit*, *Organiztion name*, and *Country*, then enter and confirm a password. You can choose from four types and strengths of encryption when creating the certificate and specify where it will be saved.

Figure 14-10. Creating a self-signed digital certificate

When you create the certificate, the *Publisher name* appears in the installer as the developer of the application, and the password is used when selecting the certificate, as described earlier in the "Digital certificates" section.

Deployment

After setting up the Adobe AIR publish settings in your document, each time you test your file, it will launch and run in the AIR Debug Launcher (adl). If you completed the digital signature process rather than choosing to create an AIR intermediate file, it will also build an AIR application.

The application is the lone file that you need to distribute to your audience. When users download and double-click this file, it will look for the AIR engine on the user's computer. If the AIR engine is found, it will run the application's built-in installer and prompt the user to install the file.

The first step in this process is to show the user the digital signature so he can determine if the file is trustworthy (Figure 14-11). The user can choose to abort or continue with the installation process.

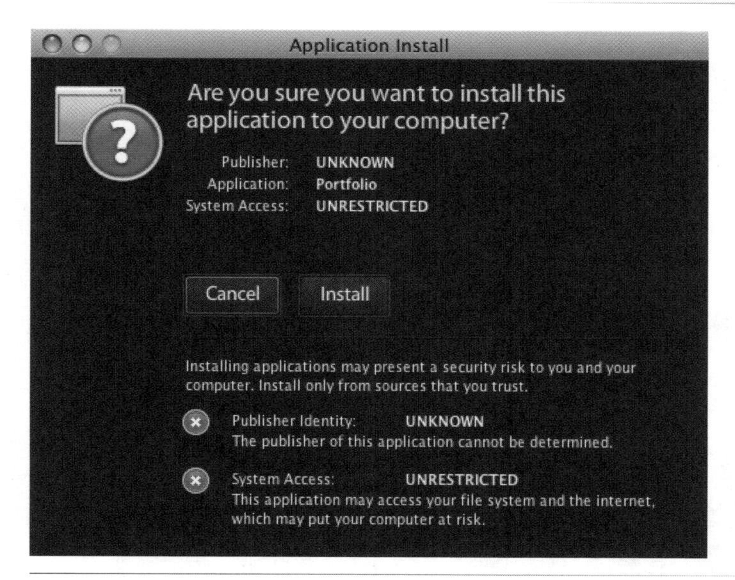

Figure 14-11. Seeking permission to install an AIR application

If the user decides to continue, the installer will display a final dialog (Figure 14-12) that shows the name and description of the application and the preset installation location (which the user can change, if desired), and then offers to start the application after installation.

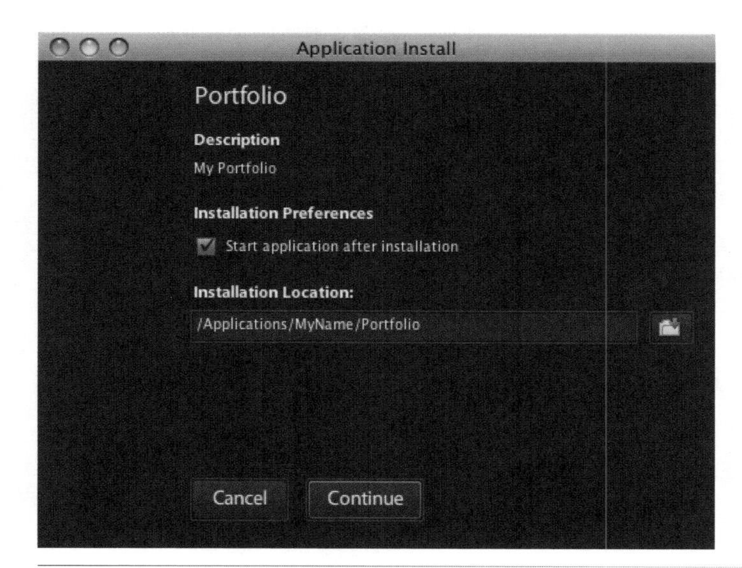

Figure 14-12. AIR installation complete

After the installation, the user can launch the application any time with the installed file, just like most desktop programs. Unless your project requires online access, the AIR application can function without relying on an Internet connection.

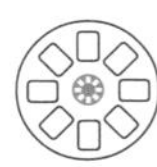

Project Progress AIR

Now it's time to apply what you've learned to the final Project Progress session of this book. To whet your appetite, you can see what you'll be creating in Figure 14-13. Additionally, all the screenshots in the AIR section of this chapter were taken from the project files, so you can reference them any time you like.

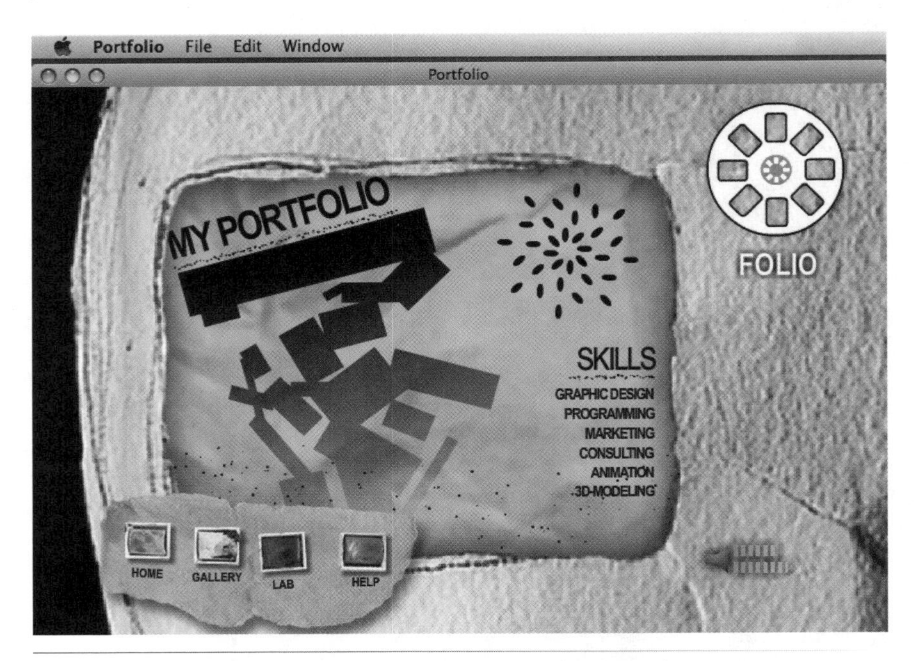

Figure 14-13. The finished Portfolio Project running as an AIR application

NOTE

If you are using Flash CS4 Professional without any updates, your player version will likely be Adobe AIR 1.1. If you installed the AIR update mentioned earlier in this chapter, the version will probably be Adobe AIR 1.5 or later.

1. Open your final portfolio FLA and access the publish settings (File→Publish Settings). In the Flash section, choose Adobe AIR from the Player menu and click the Settings button. Your file will compile and the Application and Installer Settings dialog will appear.

2. Under Application settings, enter **`Porfolio`** for *File name* and *Name*, enter **`1.0`** for *Version*, and enter **`com.adobe.example.Portfolio`** for *ID*. Continue by entering **`My Portfolio`** for *Description*, and enter the current year, company name, or other copyright notice for *Copyright*. Finally, select *System Chrome* for *Window style*. Verify your settings using Figure 14-6.

3. Click the Settings button next to the Advanced option. In the Advanced Settings dialog, under Initial window settings, enter **`750`** for *Width*, **`500`** for *Height*, and **`100`** for *X* and *Y*. Enable *Maximizable*, *Minimizable*, and *Visible*, but not *Resizable* (this would allow the user to change the size and shape of the stage, revealing more of the viewing wheel and possibly other behind-the-scenes material).

4. Continue by entering **`MyName/Portfolio`** for the *Install folder* and, if using Windows, you may optionally enter a *Program menu folder*. Feel free to substitute your own name in the *Install folder* path, but know that you'll need to allow for this change when comparing to screenshots and searching for your installed application. Compare your results with Figure 14-8 and close the Advanced Settings dialog.
5. Back in the Application and Installer Settings dialog, under Installer settings, click the Set (or Change) button found next to the *Digital signature* option.
6. In the Digital Signature dialog that opens, select the *Sign the AIR file with a digital certificate* option and click the Create button.
7. In the Create Self-Signed Digital Certificate dialog that opens, match the chapter figures by entering **`My Name`** for *Publisher name*, **`Design`** for Organizational unit, **`My Company`** for *Organization name*, and **`US`** for *Country*. Feel free to substitute your own values, as long as you remember that the screenshots will differ.
8. Continue by entering and confirming a password, and choose **`1024-RSA`** for the encryption *Type*.
9. In the *Save as* option, choose a location to save the certificate. Compare your settings with Figure 14-10. Finish with the Create Self-Signed Digital Certificate dialog by clicking OK.
10. Returning to the Digital Signature dialog, enter the password you chose in step 8, and enable Timestamp. Compare your settings with Figure 14-9 and click OK to close the dialog.
11. Finally, back in the Application and Installer Settings dialog, choose a *Destination* for your AIR file.
12. Continue by clicking the Add Folder icon in the *Included files* section. Add the *assets* folder in which you've been storing all your external assets. This will add the directory in the same relative location, allowing your portfolio project to find the external graphics, sound, video, and SWF used in the project. Compare your settings with Figure 14-6.
13. You are now finished with the AIR settings. Click the Publish AIR File to create your AIR application and then dismiss the Application and Installation Settings dialog by clicking OK. Click OK again to close the main Publish Settings dialog.
14. In the directory you specified in step 11, double-click the *Portfolio.air* file. Click through and compare the Application Install screens to Figures 14-11 and 14-12. After clicking Continue on the final screen, your application will launch.

NOTE

After configuring a FLA to publish as an AIR application, a shortcut to the AIR-specific settings will be visible in the File menu. Thereafter, you can use the File→AIR Settings menu command to alter any AIR settings.

15. As the very last steps in the development of this project, check the install location you specified in step 4 (which is, by default, found in the Applications folder on the Mac and Program Files folder on Windows) to find the application launcher your AIR installer created. Hereafter, you can bypass the installer and simply double-click the launcher like any other application. The only time you'll likely need to reinstall is if you update your file and create a new version of the application.

NOTE

AIR applications can also be installed directly from within a browser if you don't want to require your audience to be responsible for the installation. For more information, see the companion website.

All that remains now is to distribute your files. Conveniently, you bundled everything up into your single *Portfolio.air* file, so all you need to do is zip it up and put it online for your legions of fans, clients, and supporters to download!

What's Next?

Congratulations! Your project is complete. You've worked your way through a detailed, powerful application with great breadth and depth. You've touched an all the major features Flash CS4 Professional has to offer and developed a feature-rich portfolio application. What's next?

If you want to expand your knowledge and skill set beyond using Flash as a linear, timeline-based animation tool, your next step is to master ActionScript. ActionScript will add an entirely new dimension to your projects. Even some of the simplest uses of ActionScript—like runtime changes based on date and time, processing user input, and randomization—can add a degree of expressiveness and life to your projects that will never be matched by use of the Timeline.

With these features, however, comes a steeper learning curve. ActionScript 3.0 is a powerful, object-oriented programming (OOP) language that takes practice to master. Fortunately, Flash doesn't force you to use object-oriented techniques to benefit from ActionScript. Unlike Flex, or even many ActionScript editors that can be used in conjunction with Flash, Flash CS4 Professional lets you program with simple procedural techniques right in the timeline.

This book's companion text, *Learning ActionScript 3.0: A Beginner's Guide* (O'Reilly), has been written with this in mind. It introduces syntax and programming concepts in the timeline and slowly introduces OOP practices over the course of the book. Chapter 6 of this book (which was excerpted in part from *Learning ActionScript 3.0*) introduced you to the basics of events and the display list. Now it's time to jump in and learn the rest of the language.

Even if you decide not to pursue learning ActionScript at this time, the skills you learned in timeline animation, component use, sound and video, 3D, inverse kinematics, text, and more will get you started down a road to Flash design and development. Just keep working and improving, and experiment every day!

INDEX

B

C

D

E

F

G

H

I

J

K

L

M

N

O

P

Q

R

S

T

U

V

W

X

Z

O'REILLY®

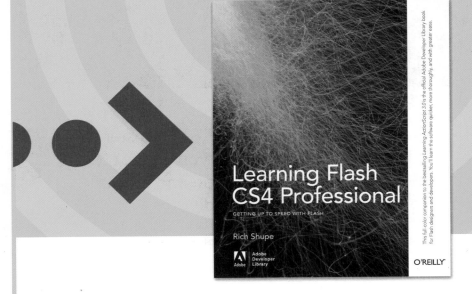